Advanced Principles of Optics

Advanced Principles of Optics

Editor: Roderick Swayne

NY RESEARCH PRESS
New York

Published by NY Research Press
118-35 Queens Blvd., Suite 400,
Forest Hills, NY 11375, USA
www.nyresearchpress.com

Advanced Principles of Optics
Edited by Roderick Swayne

International Standard Book Number: 978-1-63238-735-6 (Hardback)

Cataloging-in-Publication Data

Advanced principles of optics / edited by Roderick Swayne.
p. cm.
Includes bibliographical references and index.
ISBN 978-1-63238-735-6
1. Optics. 2. Light. 3. Physics. I. Swayne, Roderick.
QC355.3 .A38 2020
535--dc23

Printed in China.

Contents

Preface

Optics is a discipline of physics that deals with the study of behavior and properties of light. It examines the light's interactions with matter and also facilitates the construction of instruments that use and detect it. Optics explains the behavior of ultraviolet, visible and infrared light. It treats light as an electromagnetic wave and considers that other forms of electromagnetic radiations like microwaves, X-rays, and radio waves contain similar properties as light. Practical optics is practiced by using simplified models such as geometric optics that treats light as a collection of rays which travels in a straight line. These rays may bend when they pass through reflective surfaces. Another model of light is physical optics, which focuses on wave effects like diffraction and interference. Optical science is significant in many disciplines such as astronomy and various fields of engineering, medicine and photography. This book aims to shed light on some of the unexplored aspects of optics and the recent researches in this field. Some of the diverse topics covered in this book address the varied branches that fall under this category. Those in search of information to further their knowledge will be greatly assisted by this book.

The researches compiled throughout the book are authentic and of high quality, combining several disciplines and from very diverse regions from around the world. Drawing on the contributions of many researchers from diverse countries, the book's objective is to provide the readers with the latest achievements in the area of research. This book will surely be a source of knowledge to all interested and researching the field.

In the end, I would like to express my deep sense of gratitude to all the authors for meeting the set deadlines in completing and submitting their research chapters. I would also like to thank the publisher for the support offered to us throughout the course of the book. Finally, I extend my sincere thanks to my family for being a constant source of inspiration and encouragement.

Editor

1

Optical Image Encryption using Devil's Vortex Toroidal Lens in the Fresnel Transform Domain

Hukum Singh, A. K. Yadav, Sunanda Vashisth, and Kehar Singh

Department of Applied Sciences, The Northcap University, Sector 23-A, Gurgaon 122 017, India

Correspondence should be addressed to Hukum Singh; hukumsingh@ncuindia.edu

Academic Editor: Augusto Beléndez

We have carried out a study of optical image encryption in the Fresnel transform (FrT) domain, using a random phase mask (RPM) in the input plane and a phase mask based on devil's vortex toroidal lens (DVTL) in the frequency plane. The original images are recovered from their corresponding encrypted images by using the correct parameters of the FrT and the parameters of DVTL. The use of a DVTL-based structured mask enhances security by increasing the key space for encryption and also aids in overcoming the problem of axis alignment associated with an optical setup. The proposed encryption scheme is a lensless optical system and its digital implementation has been performed using MATLAB 7.6.0 (R2008a). The scheme has been validated for a grayscale and a binary image. The efficacy of the proposed scheme is verified by computing mean-squared-error (MSE) between the recovered and the original images. We have also investigated the scheme's sensitivity to the encryption parameters and examined its robustness against occlusion and noise attacks.

1. Introduction

In today's interconnected world, the security of information exchanged over a network is a critical issue. The safe storage and transmission of data continues to be a challenge because of increasing threats to the confidentiality, integrity, and availability of information. In the recent decades, optical processing techniques have provided effective solutions to some of these problems and therefore have become an active research field [1–8]. Optical security systems based on optoelectronics can perform highly accurate encryption and decryption in almost real time. The most widely known technique, first proposed by Refregier and Javidi [1], for optical image encryption is based on double random phase encoding (DRPE). The DRPE is an optically symmetric-key technique that encrypts a given image using two RPMs: one in the spatial plane and the other in the frequency plane. It may be implemented digitally or optically and has potential applications in many areas such as security verification systems, watermarking, information hiding, and multiple-image encryption.

In an attempt to strengthen security, the DRPE was further extended to several other transform domains such as the fractional Fourier [9, 10], Fresnel [11–23], gyrator [24, 25], and fractional Mellin [26, 27]. However, majority of these techniques use random phase keys making them vulnerable to a variety of plaintext and ciphertext attacks. It is also known that the conventional DRPE technique suffers from the problem of optical axis alignment. Several studies have attempted to overcome these problems by using structured phase mask (SPM), instead of RPM. The use of SPM offers additional advantage of having more encryption keys for enhanced security [28–31]. The SPM is generally made from a Fresnel zone plate (FZP) and a spiral phase plate (SPP). Barrera et al. [28, 29] introduced an SPM called toroidal zone plate (TZP). The TZPs are easier to position in the decoding and provide their own centering mask. They are diffractive optical element (DOE) and are very difficult to replicate. They have the properties of multiple keys in a single mask which provide extra security parameters. Rajput and Nishchal [30] used wavelength-dependent SPM instead of RPM in fractional Fourier transform domain to study a single-channel asymmetric color image encryption scheme. Their scheme alleviates the alignment problem of interference and does not need to iterate encoding and offers multiple levels of security. Abuturab [31] introduced a new method for encoding color information based on Arnold transformation and double structured phase mask in gyrator transform domain.

In the present work, we propose for the first time a new scheme for image encryption using a DVTL-based phase mask in the FrT domain. The use of FrT possesses an advantage over Fourier domain by providing additional parameters such as propagation distances (z_1, z_2) and propagation wavelength (λ_p) which constitutes the keys to the encryption process. The paper is organized as follows: in Section 2, we present a brief mathematical description of FrT, formation of DVTL, and the encryption-decryption scheme. Section 3 contains the results based on computer simulations for validation and evaluation of the scheme's performance. Finally, the conclusions of the study are summarized in Section 4.

2. Principle

2.1. The Fresnel Transform. The Fresnel transform (FrT) of an input image $f(x, y)$ at a propagation distance z, when it is illuminated by a plane wave of wavelength λ can be written [32, 33] as

$$F_z(u, v) = \mathrm{FrT}_{\lambda,z}\{f(x, y)\} = \iint_{-\infty}^{+\infty} f(x, y)\, h_{\lambda,z}(u, v, x, y)\, dx\, dy, \tag{1}$$

where the operator $\mathrm{FrT}_{\lambda,z}$ denotes the Fresnel transform with parameters λ and z and $h_{\lambda,z}$ is the kernel of the transform given by

$$h_{\lambda,z}(u, v, x, y) = \frac{1}{\sqrt{i\lambda z}} \exp\left(i\frac{2\pi z}{\lambda}\right) \exp\left\{\frac{i\pi}{\lambda z}(u - x)^2 + (v - y)^2\right\}. \tag{2}$$

A useful property of the FrT is

$$\mathrm{FrT}_{\lambda,z_1}\left\{\mathrm{FrT}_{\lambda,z_2} f(x)\right\} = \mathrm{FrT}_{\lambda,z_1+z_2}\{f(x)\}. \tag{3}$$

The distance parameters z_1 and z_2 are selected according to the size of the aperture to satisfy the Fresnel approximation. The distributions of complex amplitude in the adjacent planes are determined by a Fresnel transform with respect to z_1, z_2, and λ.

2.2. DVTL-Based Phase Mask. A phase mask based on devil's lens can be described by one-dimensional Cantor function [34], a particular case of devil's staircase. A triadic Cantor set in the interval [0, 1] can be defined as [34, 35]

$$F_S(x) = \begin{cases} \dfrac{l}{2^S} & \text{if } p_{S,l} \le x \le q_{S,l} \\[2ex] \dfrac{1}{2^S}\dfrac{x - q_{S,l}}{p_{S,l+1} - q_{S,l}} + \dfrac{l}{2^S} & \text{if } q_{S,l} \le x \le p_{S,l+1}, \end{cases} \tag{4}$$

where $F_S(0) = 0$, $F_S(1) = 1$, S is the order of Cantor function, and l defines the number of horizontal sections of the function, having a value from 0 to $2^S - 1$. Here, q and p are the start and end points of each segment of the Cantor set. For some basic values of S, the values of $q_{S,l}$ and $p_{S,l+1}$ are provided in [34, 36–40]. A DVTL can be constructed by combining a devil's lens, a Fresnel toroidal lens, and a vortex lens. A devil's lens is a circularly symmetric pure phase diffractive optical element and is defined as

$$D_S(\zeta) = e^{(-i2^{S+1}\pi F_S)}, \tag{5}$$

where $\zeta = (r/a)^2$ is the normalised quadratic radial coordinate and a is the lens radius. Thus, the phase variation along the radial coordinate is quadratic in each zone of the lens. The phase shift at the gap regions defined by the Cantor set is $-l2\pi$, with $l = 1 \cdots 2^S - 1$.

The radial Hilbert transform mask (RHM) is another SPM which can serve to make an image edge-enhanced relative to the input image in addition to increasing the key space. The radial Hilbert transform [41, 42] is expressed in terms of a vortex function as

$$V_m(\emptyset) = e^{(im\emptyset)}, \tag{6}$$

where $\emptyset$ is the azimuth angle and m is an integer denoting the order of transformation, also called topological charge. It is apparent that the opposite halves of any radial line of the mask have a relative phase difference of $m\pi$ radian. Therefore, for each radial line, we have the equivalent of a one-dimensional Hilbert transform of order m. The radial Hilbert transform can be helpful in aligning the axis of the optical setup.

Just as a Fresnel lens, the toroidal lens is based on quadratic phase change and is given by

$$L_{\lambda,f_0}(\zeta) = e^{\{-i\pi(r-r_0)^2/\lambda f_0\}}, \tag{7}$$

where f_0 is the focal length and λ is the wavelength of incident light. Now, a DVTL-based phase mask is obtained by taking the product of the three functions $D_S(\zeta)$, $V_m(\emptyset)$, and $L_{\lambda,f_0}(\zeta)$ as follows:

$$(\mathrm{DVTL})_{S,l,\lambda,f_0} = e^{i\{-2^{S+1}\pi F_S + m\emptyset - \pi(r-r_0)^2/\lambda f_0\}}. \tag{8}$$

A plot of DVTL, which is a combination of Devil's lens (DL), a vortex lens (VL), and Fresnel toroidal lens (TL), is shown in Figure 1.

2.3. The Encryption and Decryption Scheme. A flowchart of encryption and decryption process of the proposed scheme is presented in Figure 2. In this scheme, the input image $I(x, y)$ to be encrypted is first bonded with an RPM defined as $e^{2i\pi n(x,y)}$ in the input plane where $n(x, y)$ is uniformly distributed in [0, 1]. The resulting complex image is subjected to a Fresnel transform $\mathrm{FrT}_{\lambda,z_1}$. Then, in the frequency domain, it is bonded with an SPM based on $(\mathrm{DVTL})_{S=3}$ defined as $e^{i\{-2^{S+1}\pi F_S + m\emptyset - \pi(r-r_0)^2/\lambda f_0\}}$. Thereafter, it is subjected to another Fresnel transform $\mathrm{FrT}_{\lambda,z_2}$. Mathematically, the encryption process can be written as (see Figure 2(a))

$$E(x, y) = \mathrm{FrT}_{\lambda,z_2}\left\{\left[\mathrm{FrT}_{\lambda,z_1}\{I(x, y) \times \mathrm{RPM}\}\right] \times (\mathrm{DVTL})_{S=3}\right\}. \tag{9}$$

The decryption process (Figure 2(b)) is the reverse of encryption with the following steps: the conjugated encrypted image E^* is first subjected to the Fresnel transform $\mathrm{FrT}_{\lambda,z_2}$. The resulting complex image is multiplied by

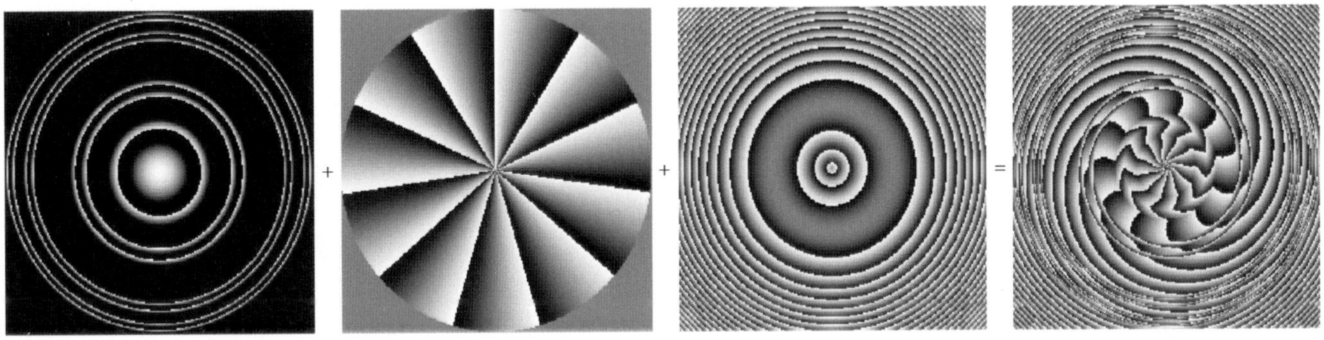

(a) (b) (c) (d)

Figure 1: Phase plots of each component which when combined result in Devil's vortex toroidal lens (DVTL); (a) DL for $S = 3$; (b) VL for $m = 11$; (c) TL; (d) DVTL.

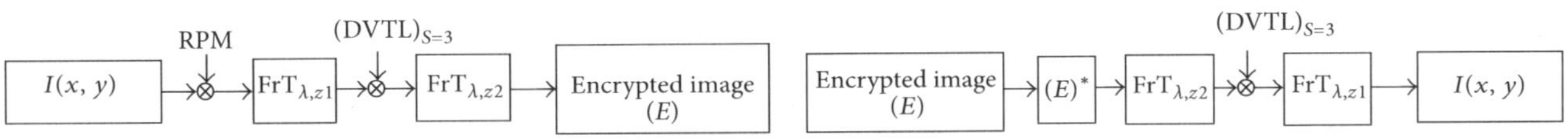

(a) (b)

Figure 2: Flow chart of the proposed scheme: (a) encryption process and (b) decryption process.

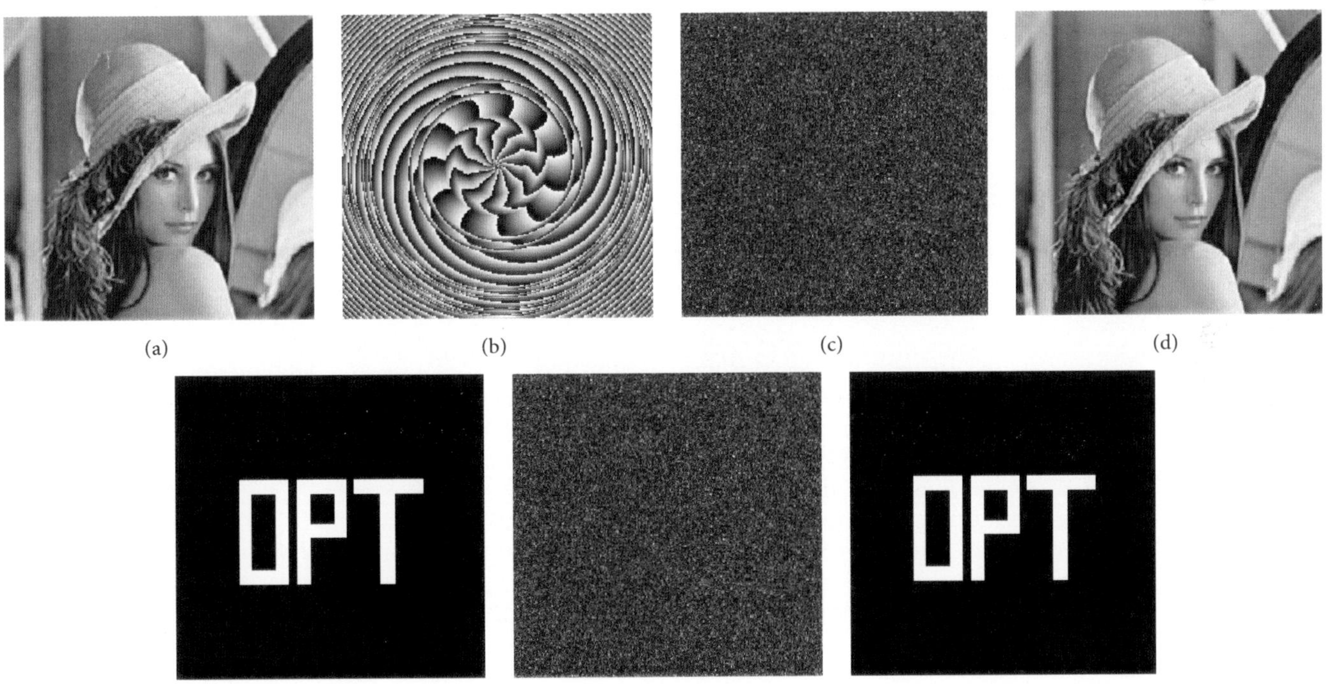

(e) (f) (g)

Figure 3: Results of validation of the proposed scheme for grayscale and binary images: (a) input image Lena of size 256 × 256 pixels; (b) $\mathrm{DVTL}_{S=3}$ phase key at pixel spacing = 0.023, $\lambda = 632.8$ nm, $f = 350$ mm, $m = 11$, and $S = 3$; (c) encrypted image; (d) decrypted image; (e) input binary image; (f) encrypted binary image; (g) decrypted binary image.

$(\mathrm{DVTL})_{S=3}$ and then the $\mathrm{FrT}_{\lambda,z_1}$ is performed. Mathematical expression for decryption is given by

$$I(x, y) = \mathrm{FrT}_{\lambda,z_1}\left[\mathrm{FrT}_{\lambda,z_2}\{E^*(x, y)\} \times (\mathrm{DVTL})_{S=3}\right], \quad (10)$$

where * denotes the complex conjugate.

3. Simulation Results and Discussion

The proposed scheme has been verified by performing numerical simulation on a MATLAB 7.6.0 (R2008a) platform. We have considered two test images of size 256 × 256 pixels as input, one grayscale image of Lena (Figure 3(a)) and the other binary image of OPT (Figure 3(e)). The structured

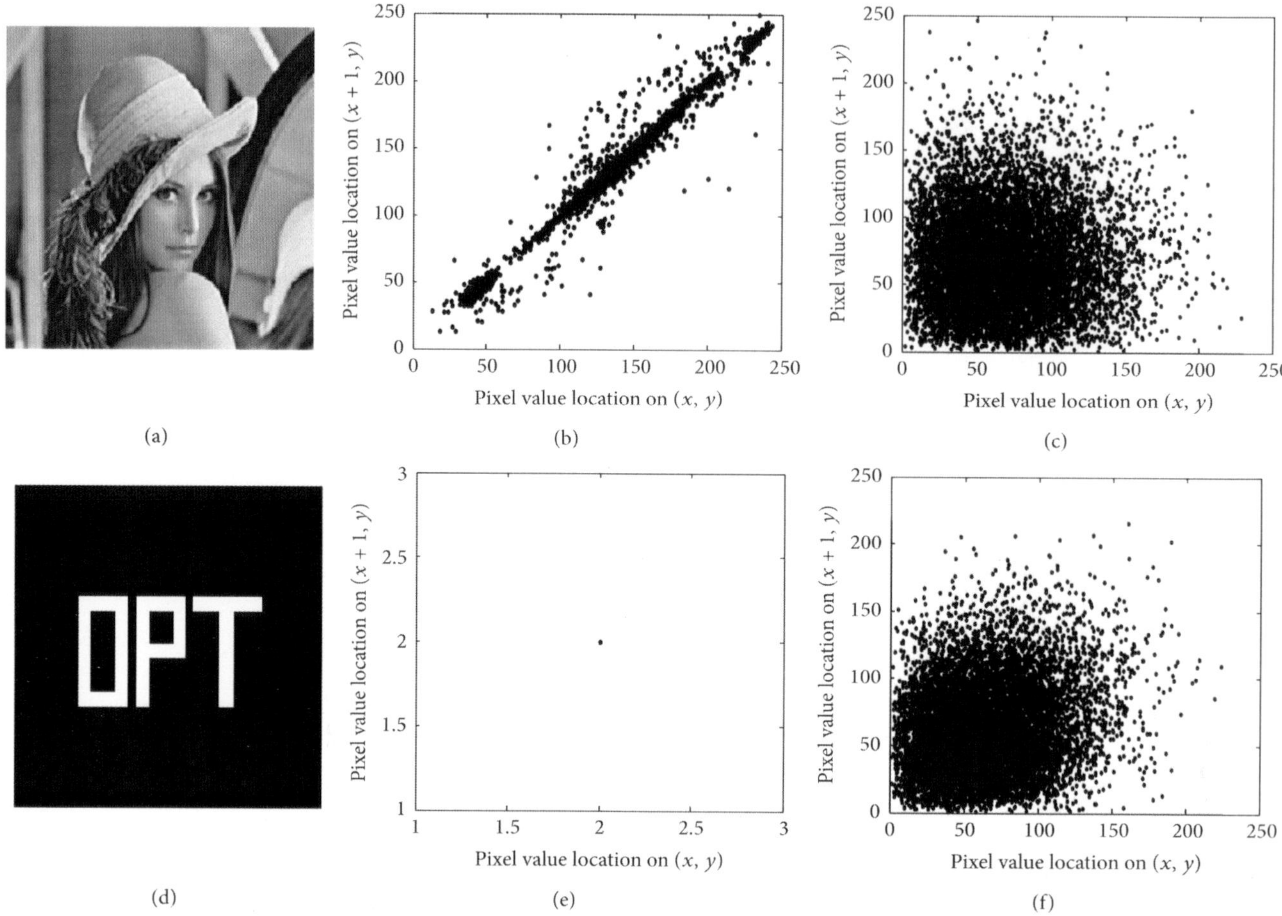

FIGURE 4: Plots of correlation distribution for randomly chosen 10,000 pixel pairs: (a) input image Lena; (b, c) correlation distribution respectively of input and encrypted image of Lena; (d) input image OPT; (e, f) correlation distribution, respectively, of input and encrypted image of OPT.

phase mask (DVTL) used in the frequency plane during encryption is shown in Figure 3(b). It is generated with parameter values as wavelength (λ) = 632.8 nm, focal length (f) = 350 mm, m = 11, r_0 = 1.15 mm, and S = 3. Figures 3(c) and 3(f) show the encrypted images of the grayscale and binary inputs, respectively, and are pure stationary white noise. The corresponding decrypted images are presented in Figures 3(d) and 3(g). The FrT parameters used in the present simulations are z_1 = 25 mm, z_2 = 35 mm, and propagation wavelength λ_p = 632.8 nm. These values have been taken arbitrarily and considered for the sake of simplicity. It can be seen from Figures 3(d) and 3(g) that the input images are faithfully recovered using all the correct parameters.

3.1. Correlation Coefficient (CC) Analysis. Correlation coefficient is a criterion used in the literature [26, 43, 44] to measure the similarity of two images quantitatively. It is obvious that arbitrarily chosen pixels of original images are generally highly correlated in horizontal, vertical, and diagonal directions. We know that CC of the encrypted images is much weaker than that of original images. However, a secure image encryption algorithm must produce an encrypted image having low CC between adjacent pixels. We have randomly selected 10,000 pairs of adjacent pixels (horizontal, vertical, or diagonal) for computation of CC from the input and the encrypted images separately. Then, the CC of each pair is calculated by the following relation [25]:

$$\mathrm{CC} = \frac{\sum_{i=1}^{N}(x_i - \overline{x})(y_i - \overline{y})}{\sqrt{\left(\sum_{i=1}^{N}(x_i - \overline{x})^2\right)\left(\sum_{i=1}^{N}(y_i - \overline{y})^2\right)}}, \tag{11}$$

where $\overline{x} = \sum_{i=1}^{N} x_i$ and $\overline{y} = \sum_{i=1}^{N} y_i$ are, respectively, the mean values of x_i and y_i.

The CC values of adjacent pixels in the horizontal, vertical, and diagonal directions of original images and their encrypted versions are given in Table 1. It is clear that, for the original images, the CC values are very high as compared to those of encrypted images. This clearly indicates that the adjacent pixels in the original images are strongly correlated. However, for the encrypted images, CC values are nearly zero, which means that the adjacent pixels in the horizontal, vertical, or diagonal directions are weakly correlated. Figure 4(a) shows the original image of Lena, whereas Figures 4(b) and 4(c) show the scatter plots of correlation distribution of horizontally adjacent pixels, respectively, in the original-, and the encrypted image. Similar information about the binary image OPT is shown in Figures 4(d)–4(f).

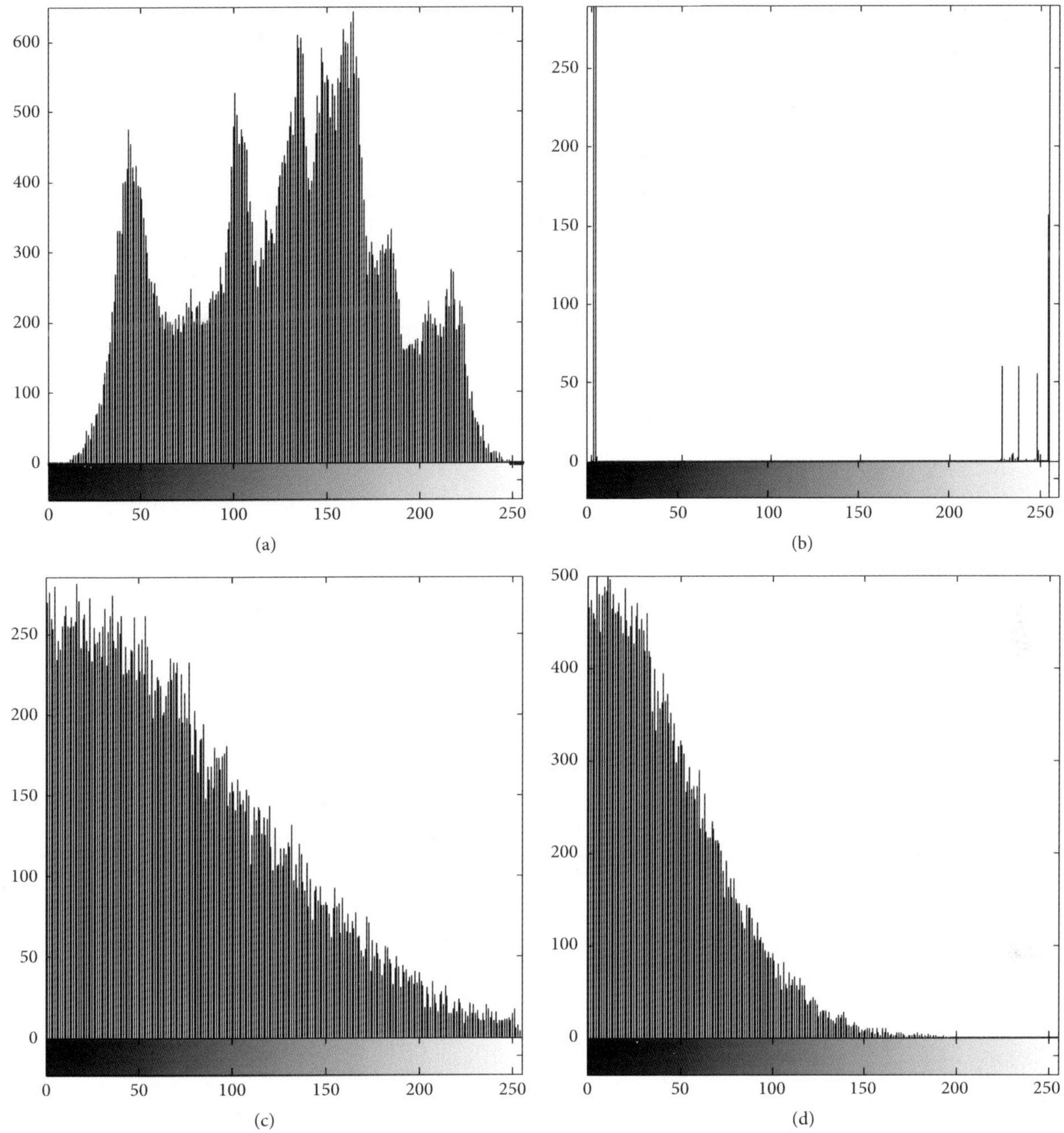

FIGURE 5: Histogram of (a) input image Lena, (b) input image OPT, (c) encrypted image Lena, and (d) encrypted image OPT.

TABLE 1

Image	Original image			Cipher image		
	Horizontal	Diagonal	Vertical	Horizontal	Diagonal	Vertical
Lena	0.9562	0.9509	0.9817	0.0093	0.0121	0.0086
OPT	0.9497	0.9354	0.9695	0.0106	0.0020	0.0092

3.2. Statistical Analysis

3.2.1. Histogram Analysis. Statistical analysis has been performed on the proposed image encryption algorithm. Image histogram is another very important feature in image analysis. Figures 5(a) and 5(b) show the histograms of Lena and OPT which are obviously quite different from each other. On the other hand, we observe that the histograms (Figures 5(c) and 5(d)) of their encrypted images are quite similar. After a number of simulations, we can conclude that the ciphertext of different original images have similar histograms. So,

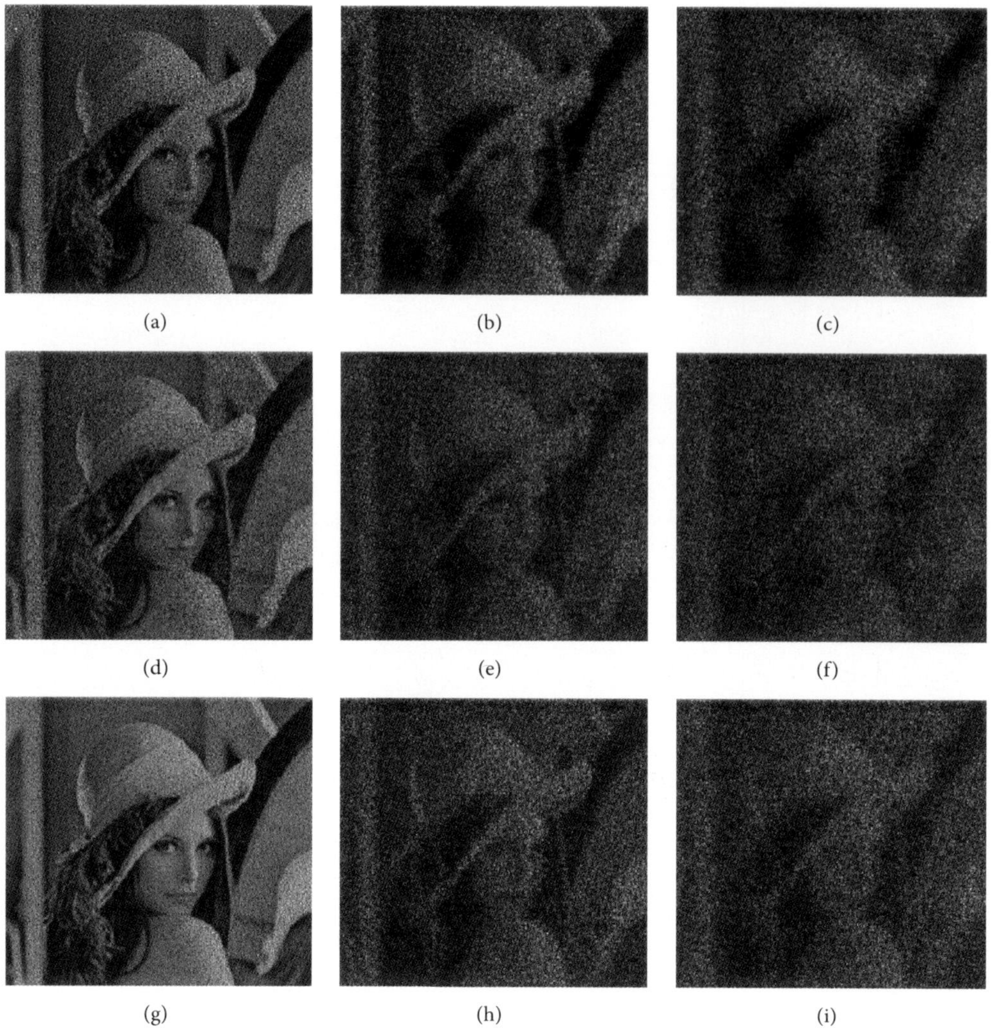

FIGURE 6: Results with incorrect parameters of FrT: (a–c) decrypted images with incorrect values of $z_1 = 25.3, 26$, and 27 mm, (d–f) decrypted images with incorrect values of $z_2 = 35.3, 36$, and 37 mm, and (g–i) decrypted images with variation in λ_p by 10, 20, and 40 nm.

attackers cannot obtain useful information according to the statistical properties.

3.2.2. Image Entropy Analysis. Entropy is a statistical measure of randomness that can be used to characterize the texture of the input image. A secure encryption should provide a situation in which the encrypted image does not provide any information about the original image. Image information entropy measures the distribution of image gray values. The more uniform the gray value distribution is, the bigger the information entropy is.

The entropy $H(s)$ of message source s can be calculated [26, 43, 44] as

$$H(s) = -\sum_{i=0}^{255} p(s_i) \log_2 p(s_i), \qquad (12)$$

where $p(s_i)$ represents the probability of symbol s_i. The ideal value for the cipher information entropy is 8. The information entropy of the cipher-image of Lena and OPT generated by the proposed algorithm is 7.7484 and 7.1122.

3.3. Sensitivity Analysis. We have also examined the scheme's sensitivity to the individual encryption parameters by considering wrong parameter values. The recovered grayscale images for wrong parameters are shown, respectively, in Figures 6 and 7 for various parameters of FrT and DVTL. Figures 6(a)–6(c) correspond to minor-to-small deviation from the correct value of the propagation distance z_1 (=25.3 mm, 26 mm, and 27 mm, resp.). Similarly, Figures 6(d)–6(f) correspond to the second propagation distance z_2 (=35.3 mm, 36 mm, and 37 mm, resp.) and Figures 6(g)–6(i) correspond to the propagation wavelength λ_p (=642.8 nm, 652.8 nm, and 672.8 nm, resp.). In each case, even the minor change in parameter value results in significant drop in the quality of the recovered images (Figures 6(a), 6(d), and 6(g)), whereas it is barely recognizable in the third column (Figures 6(c), 6(f), and 6(i)) which corresponds to slightly higher departure from the correct value. It clearly shows that the proposed scheme is highly sensitive to FrT parameters. Likewise, the sensitivity to DVTL parameters is observed in Figure 7 for wrong values of the focal length (Figures 7(a)–7(c)), the wavelength

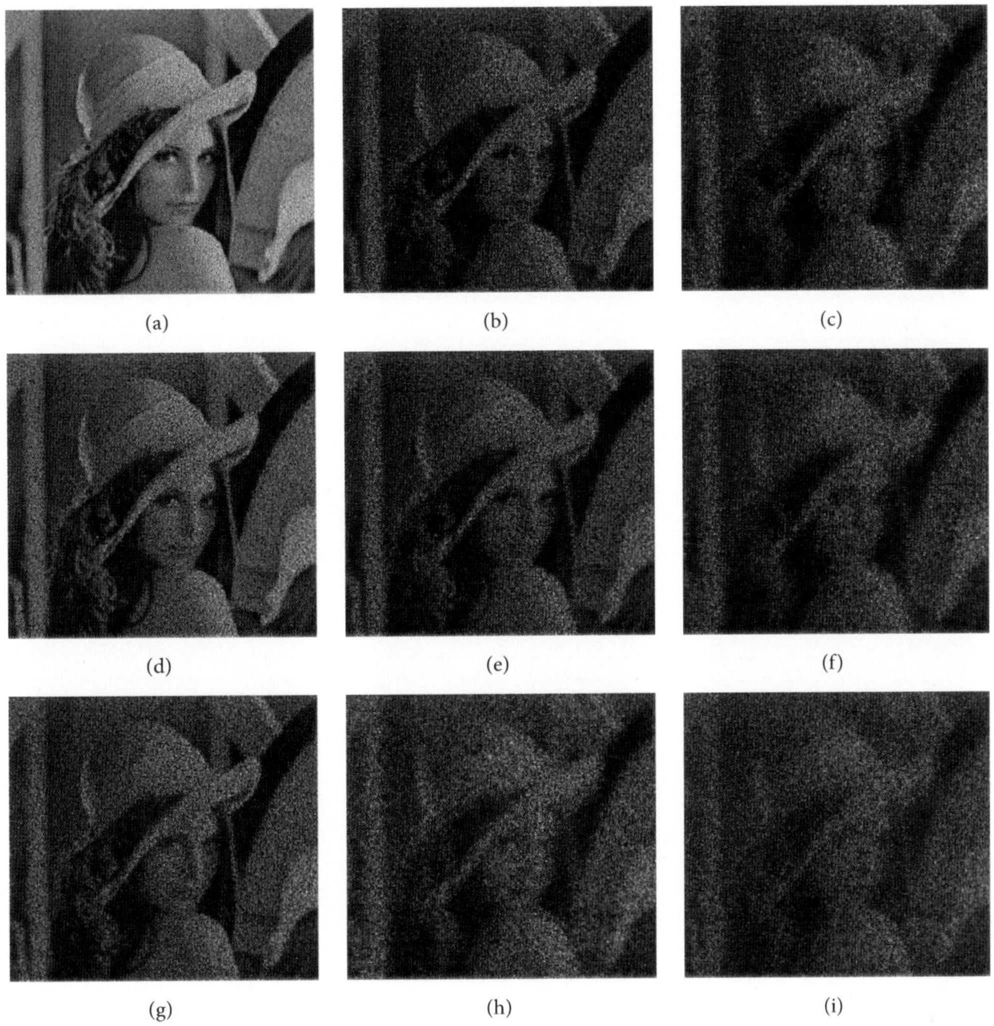

FIGURE 7: Results with incorrect parameters of DVTL: (a–c) decrypted images with incorrect focal lengths with deviation from correct value as 5, 10, and 50 mm, respectively, (d–f) decrypted images with incorrect λ with deviation from correct value as 10, 40, and 100 nm respectively, (g, h) decrypted images with incorrect topological charge with deviation from the correct value as 1 and 9, respectively, and (i) decrypted image with $S = 2$.

(Figures 7(d)–7(f)), the topological charge (Figures 7(g)-7(h)), and the devil's parameter S (Figure 7(i)).

In order to assess the efficacy of the proposed scheme, mean-squared-error (MSE) between the original input image and the decrypted image has been computed. If $I_0(x, y)$ and $I_d(x, y)$ denote, respectively, the pixel values of the original input image and the decrypted image, a mathematical expression for MSE can be written as

$$\text{MSE} = \sum_{x=0}^{255}\sum_{y=0}^{255}\frac{\left|I_0(x,y) - I_d(x,y)\right|^2}{256 \times 256}. \tag{13}$$

The computed values of MSE between the input and the recovered images for the grayscale and the binary images using the proposed scheme are 1.7087×10^{-20} and 2.3995×10^{-21}, respectively.

We have shown MSE plots for a wider range of values of FrT parameters in Figures 8(a)–8(c) and for DVTL parameters in Figures 8(d)–8(f). Each plot shows the MSE curves relative to the deviation from the correct parameter value for the grayscale and the binary inputs. It is clearly visible from the plots that the proposed algorithm is highly sensitive to the propagation parameters of FrT. Though the scheme is sensitive to DVTL parameters also, the variation in MSE is less steep. In all these subfigures, a comparison of the two curves indicates that the algorithm shows greater sensitivity for the grayscale as compared to the binary for each of the encryption parameters.

3.4. Occlusion Attack Analysis. We examine the robustness of the proposed algorithm against occlusion attacks on the encrypted image. The occluded images are shown in Figures 9(a)–9(d) for 10%, 25%, 50%, and 75% occlusion of the encrypted image of Lena. Figures 9(e)–9(h) show the corresponding decrypted images which are recovered fairly well even for occlusion up to 75%. Similar results are obtained when we tested the scheme's robustness for occlusion of the binary image (Figures 9(i)–9(l)). Additionally, we have

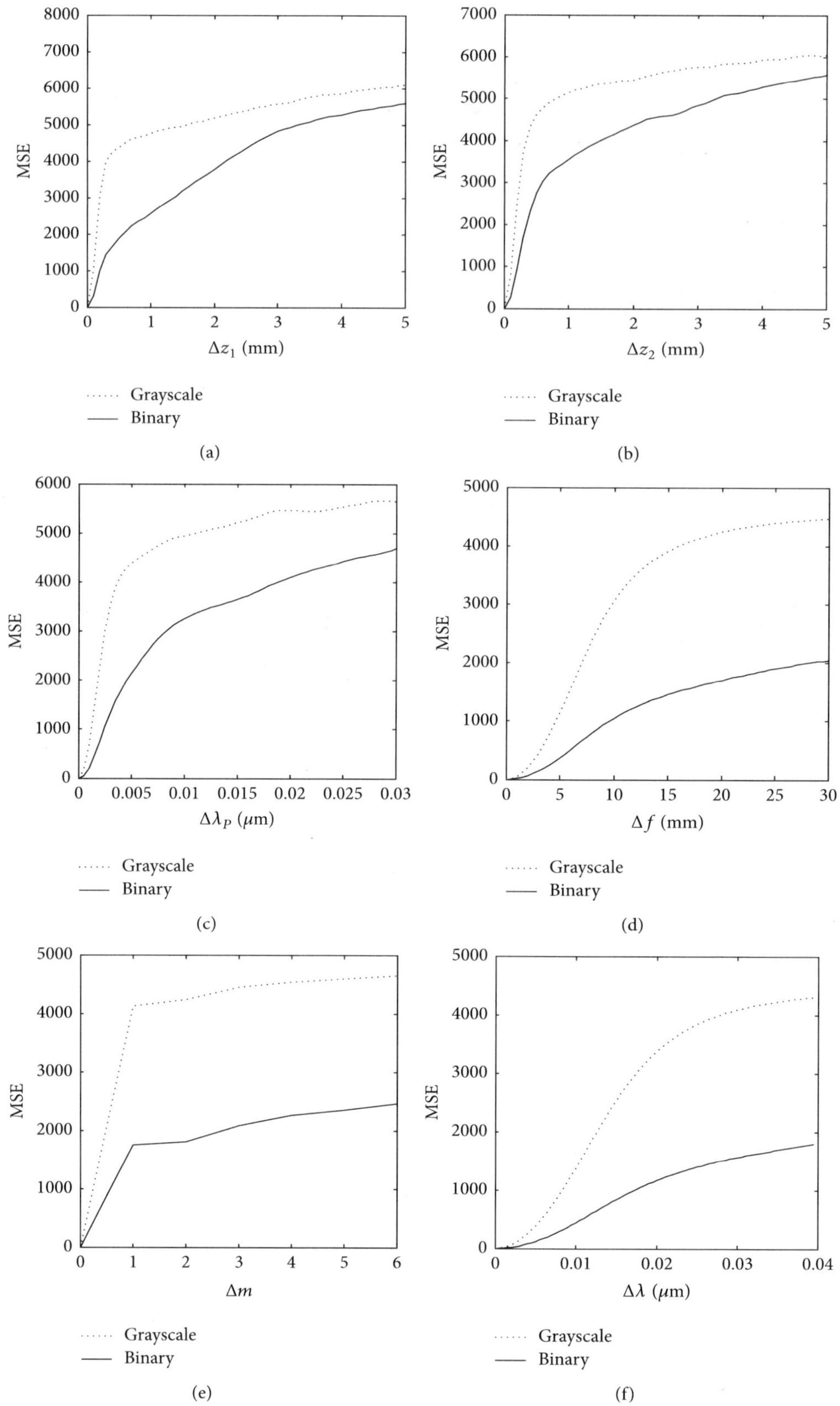

FIGURE 8: Sensitivity plots of MSE as a function of deviation from the correct values of various parameters of FrT and DVTL: (a) Δz_1 (mm), (b) Δz_2 (mm), (c) $\Delta \lambda_p$ (μm), (d) focal length f (mm), (e) topological charge m, and (f) wavelength λ (μm).

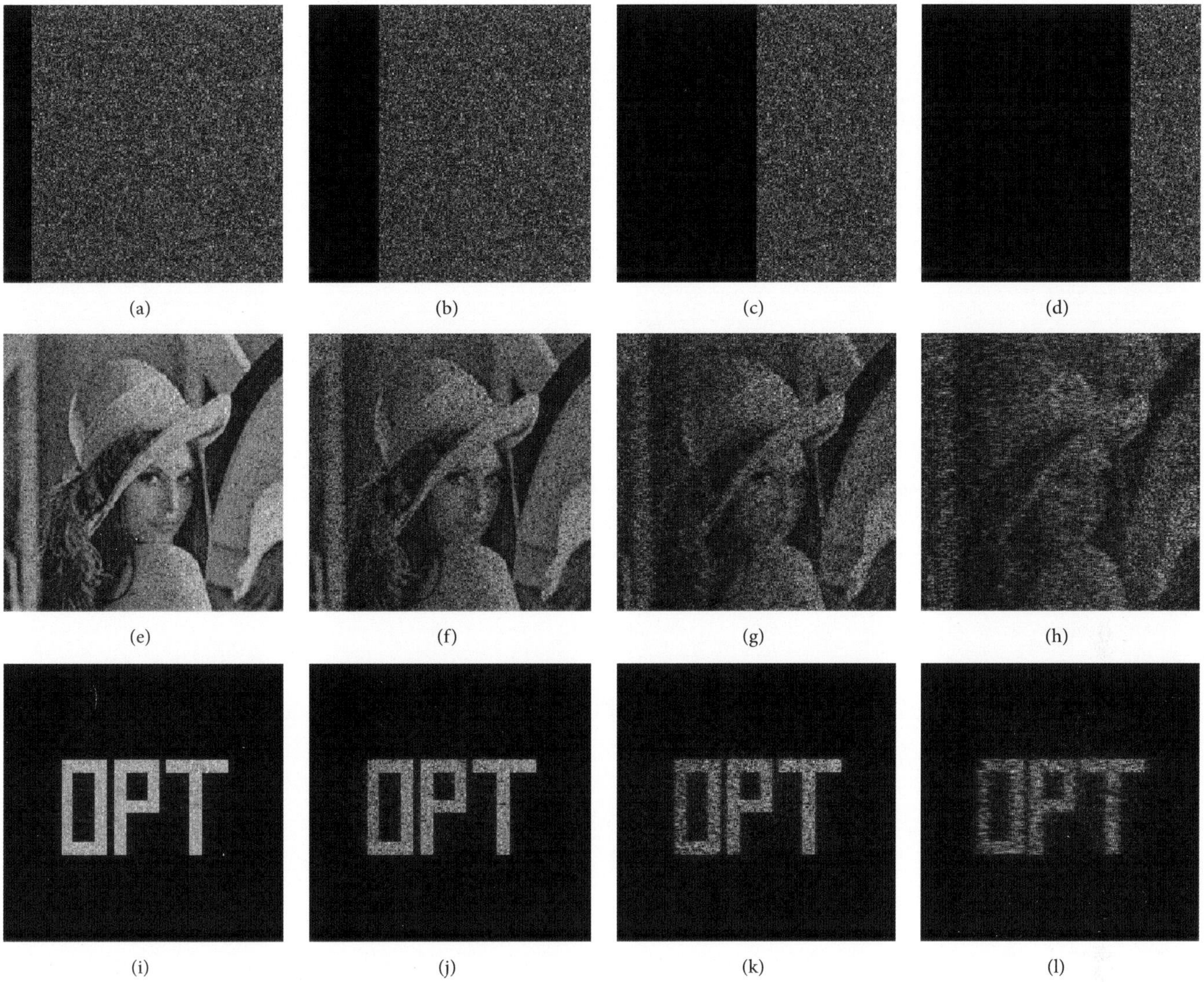

FIGURE 9: Occlusion results for the grayscale and the binary image for varying degrees of occlusion: encrypted image with (a) 10% occlusion, (b) 25% occlusion, (c) 50% occlusion, (d) 75% occlusion, (e–h) corresponding recovered grayscale images, and (i–l) corresponding recovered binary images.

plotted MSE and correlation coefficient (CC) against varying degrees of occlusion of the encrypted images (Figures 10(a)-10(b)). The variation of MSE and CC curves clearly indicates the scheme's robustness to occlusion attack. As expected, the binary image shows higher robustness as compared to the grayscale image.

3.5. Noise Attack Analysis. It is inevitable that the noise impacts directly the quality of the decrypted image. We have also tested the strength of the proposed scheme against noise attack [45–48] by considering additive noise (Gaussian, salt and pepper) and multiplicative noise (speckle) in the encrypted images. The multiplicative noise interferes with the encrypted images according to the following relation [48]:

$$e' = e(1 + kG), \tag{14}$$

where e and e' are, respectively, the encrypted and the noise-affected encrypted amplitude images, k is a coefficient which represents the noise strength, and G is a Gaussian random noise with zero-mean and unit standard deviation.

The input image of Lena (Figure 11(a)) is compared with the recovered images (Figures 11(b)–11(d)) when the encrypted image is affected by noise of the type salt and pepper (density = 0.05), additive Gaussian (variance = 0.05), and speckle (variance = 0.05), respectively. The corresponding images for the binary input are shown in Figures 11(e)–11(h). From the recovered images, we observe that the scheme is robust to noise attack, with maximum resistance to speckle noise. The drop in quality of the recovered images is comparable in the cases of additive noise. Figures 12(a)-12(b) show the plots of MSE curves against density/variance for grayscale and binary images, respectively. We see that there is a monotonic increase in MSE curves of both grayscale and

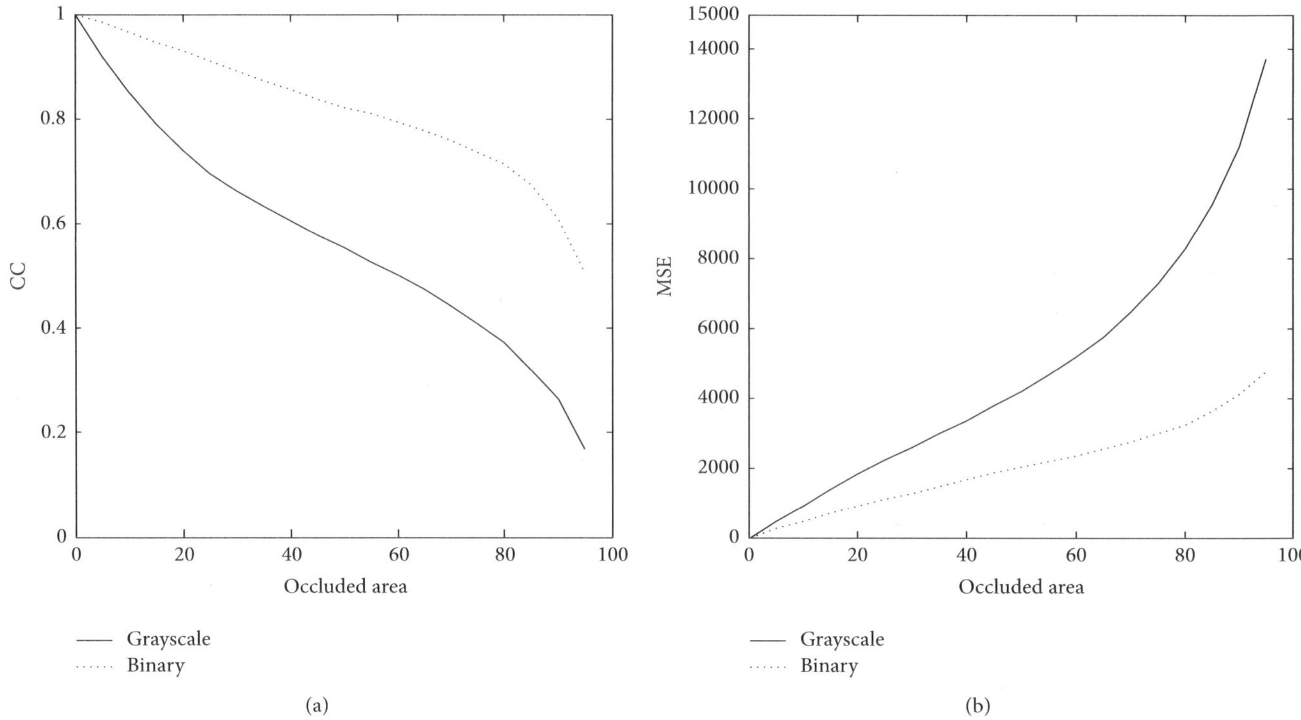

FIGURE 10: Plots of (a) correlation coefficient and (b) MSE for grayscale and binary images with varying occluded area.

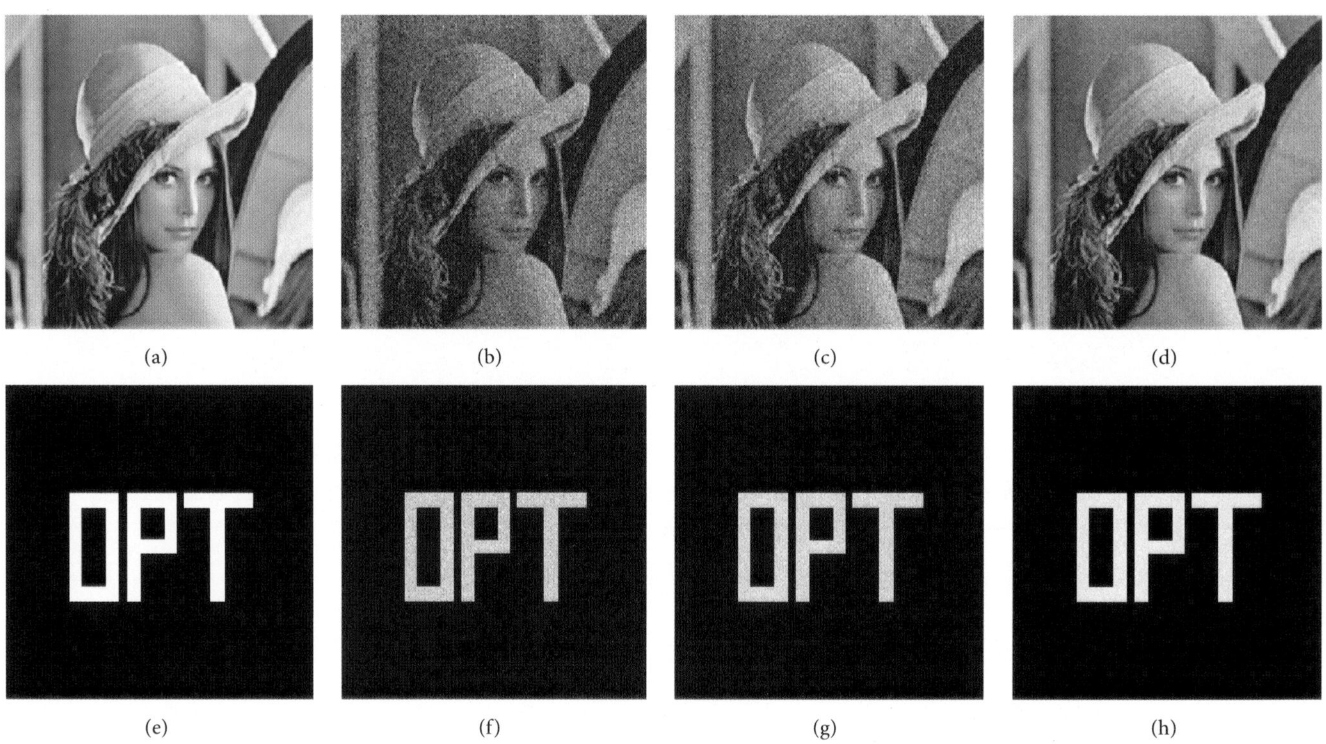

FIGURE 11: (a) Input image Lena; (b–d) recovered grayscale images corresponding to salt and pepper noise (density = 0.05), Gaussian noise (variance = 0.05), and speckle noise (variance = 0.05), respectively; (e) input binary image OPT; (f–h) recovered binary images corresponding to salt and pepper noise, Gaussian noise, and speckle noise, respectively.

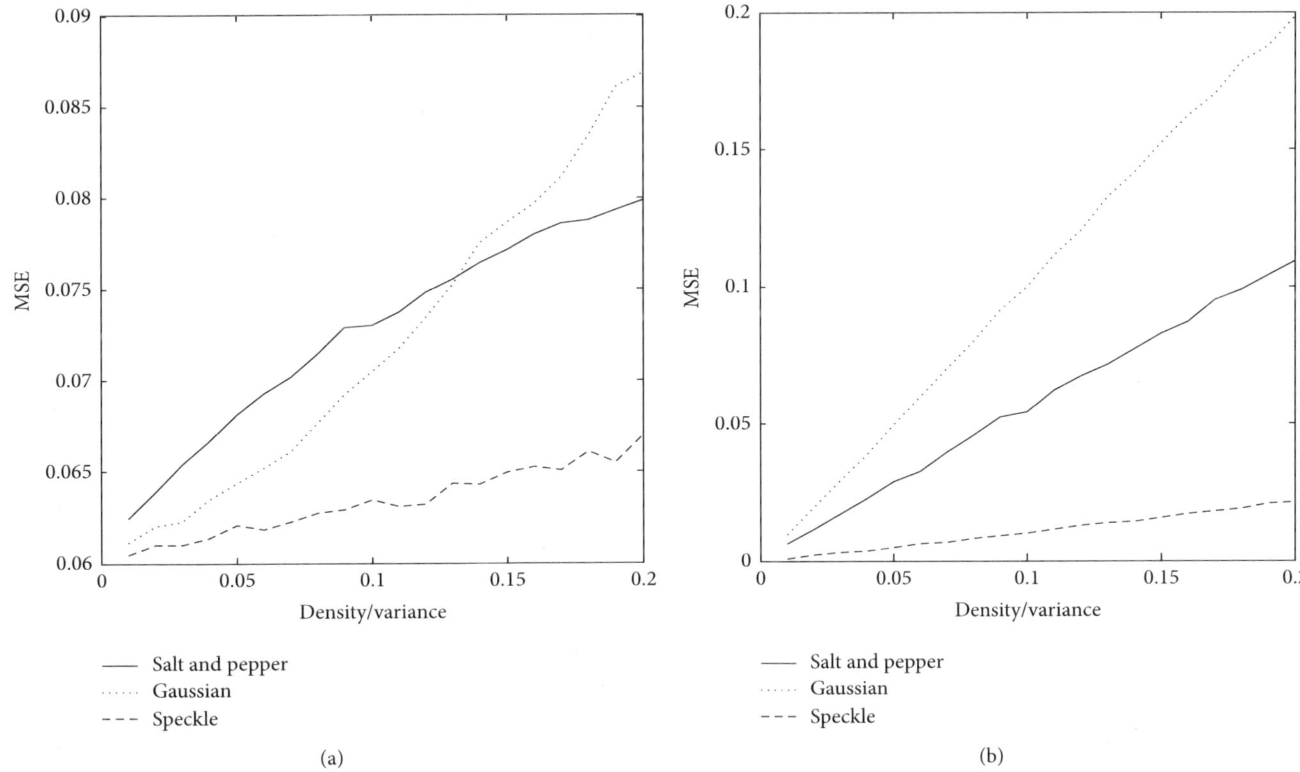

FIGURE 12: MSE plots for three types of noise against density/variance for (a) grayscale image Lena and (b) binary image OPT.

binary images, with the increase in density/variance. Unlike grayscale (Figure 12(a)), there is a distinct trend of MSE curves of binary image (Figure 12(b)) for the three types of noise. This establishes robustness of the proposed algorithm against commonly reported noise attacks.

4. Conclusions

A scheme for binary and grayscale images has been proposed, using RPM in the input plane and a DVTL phase mask in the frequency plane. The DVTL phase mask is preferred to introduce additional encryption parameters that enlarge the key space. This approach not only overcomes the problem of axis alignment of optical setup but also makes the proposed scheme more secure. The proposed scheme has been validated in the FrT domain. The entropy values and histograms show the validity of proposed scheme. Numerical results are presented to demonstrate the feasibility and security of the proposed system. The efficacy of the proposed scheme is seen from the computed values of MSE. The sensitivity of the scheme has also been studied for various parameters of FrT and DVTL. In addition, the results also demonstrate excellent robustness against noise and occlusion attacks.

Conflict of Interests

The authors declare that there is no conflict of interests regarding the publication of this paper.

Acknowledgment

The authors wish to thank the management of THE NORTHCAP UNIVERSITY for their encouragement.

References

[1] P. Refregier and B. Javidi, "Optical image encryption based on input plane and Fourier plane random encoding," *Optics Letters*, vol. 20, no. 7, pp. 767–769, 1995.

[2] B. Javidi, *Optical and Digital Techniques for Information Security*, Springer, New York, NY, USA, 2005.

[3] K. Singh, G. Unnikrishnan, and N. K. Nishchal, "Photorefractive optical processing for data security," in *Photorefractive Fiber and Crystal Devices: Materials, Optical Properties, and Applications VIII*, vol. 4803 of *Proceedings of SPIE*, pp. 205–219, Seattle, Wash, USA, July 2002.

[4] O. Matoba, T. Nomura, E. Pérez-Cabré, M. S. Millán, and B. Javidi, "Optical techniques for information security," *Proceedings of the IEEE*, vol. 97, no. 6, pp. 1128–1148, 2009.

[5] A. Alfalou and C. Brosseau, "Optical image compression and encryption methods," *Advances in Optics and Photonics*, vol. 1, no. 3, pp. 589–636, 2009.

[6] A. Kumar, M. Singh, and K. Singh, "Speckle coding for optical and digital data security applications," in *Advances in Speckle Metrology and Related Technique*, G. H. Kaufmann, Ed., chapter 6, pp. 239–299, Wiley-VCH, 2011.

[7] S. Liu, C. Guo, and J. T. Sheridan, "A review of optical image encryption techniques," *Optics & Laser Technology*, vol. 57, pp. 327–342, 2014.

[8] W. Chen, B. Javidi, and X.-D. Chen, "Advances in optical security systems," *Advances in Optics and Photonics*, vol. 6, no. 2, pp. 120–155, 2014.

[9] G. Unnikrishnan, J. Joseph, and K. Singh, "Optical encryption by double-random phase encoding in the fractional Fourier domain," *Optics Letters*, vol. 25, no. 12, pp. 887–889, 2000.

[10] B. M. Hennelly and J. T. Sheridan, "Image encryption and the fractional Fourier transform," *Optik*, vol. 114, no. 6, pp. 251–265, 2003.

[11] O. Matoba and B. Javidi, "Encrypted optical memory system using three-dimensional keys in the Fresnel domain," *Optics Letters*, vol. 24, no. 11, pp. 762–764, 1999.

[12] G. Situ and J. Zhang, "Double random-phase encoding in the Fresnel domain," *Optics Letters*, vol. 29, no. 14, pp. 1584–1586, 2004.

[13] B. M. Hennelly and J. T. Sheridan, "Random phase and jigsaw encryption in the Fresnel domain," *Optical Engineering*, vol. 43, no. 10, pp. 2239–2249, 2004.

[14] C.-H. Niu, Y. Zhang, and B.-Y. Gu, "Optical encryption and verification technique for information coding in multiple-wavelengths in Fresnel domain," *Optik*, vol. 117, no. 11, pp. 516–524, 2006.

[15] H. T. Chang, H.-E. Hwang, C.-L. Lee, and M.-T. Lee, "Wavelength multiplexing multiple-image encryption using cascaded phase-only masks in the Fresnel transform domain," *Applied Optics*, vol. 50, no. 5, pp. 710–716, 2011.

[16] J.-J. Huang, H.-E. Hwang, C.-Y. Chen, and C.-M. Chen, "Optical multiple-image encryption based on phase encoding algorithm in the Fresnel transform domain," *Optics & Laser Technology*, vol. 44, no. 7, pp. 2238–2244, 2012.

[17] S. Yuan, Y.-H. Xin, M.-T. Liu, S.-X. Yao, and X.-J. Sun, "An improved method to enhance the security of double random-phase encoding in the Fresnel domain," *Optics & Laser Technology*, vol. 44, no. 1, pp. 51–56, 2012.

[18] Q. Wang, Q. Guo, L. Lei, and J. Zhou, "Multiple-image encryption based on interference principle and phase-only mask multiplexing in Fresnel transform domain," *Applied Optics*, vol. 52, no. 28, pp. 6849–6857, 2013.

[19] S. K. Rajput and N. K. Nishchal, "Image encryption using polarized light encoding and amplitude and phase truncation in the Fresnel domain," *Applied Optics*, vol. 52, no. 18, pp. 4343–4352, 2013.

[20] S. K. Rajput and N. K. Nishchal, "Fresnel domain nonlinear optical image encryption scheme based on Gerchberg-Saxton phase-retrieval algorithm," *Applied Optics*, vol. 53, no. 3, pp. 418–425, 2014.

[21] J. M. Vilardy, M. S. Millán, and E. Pérez-Cabré, "Nonlinear optical security system based on a joint transform correlator in the Fresnel domain," *Applied Optics*, vol. 53, no. 8, pp. 1674–1682, 2014.

[22] Z. Liu, C. Guo, J. Tan et al., "Securing color image by using phase-only encoding in Fresnel domains," *Optics and Lasers in Engineering*, vol. 68, pp. 87–92, 2015.

[23] Y. Wang, C. Quan, and C. J. Tay, "Optical color image encryption without information disclosure using phase-truncated Fresnel transform and a random amplitude mask," *Optics Communications*, vol. 344, pp. 147–155, 2015.

[24] J. A. Rodrigo, T. Alieva, and M. L. Calvo, "Gyrator transform: properties and applications," *Optics Express*, vol. 15, no. 5, pp. 2190–2203, 2007.

[25] H. Singh, A. K. Yadav, S. Vashisth, and K. Singh, "Fully phase image encryption using double random-structured phase masks in gyrator domain," *Applied Optics*, vol. 53, no. 28, pp. 6472–6481, 2014.

[26] N.-R. Zhou, Y. Wang, and L. Gong, "Novel optical image encryption scheme based on fractional Mellin transform," *Optics Communications*, vol. 284, no. 13, pp. 3234–3242, 2011.

[27] S. Vashisth, H. Singh, A. K. Yadav, and K. Singh, "Devil's vortex phase structure as frequency plane mask for image encryption using the fractional Mellin transform," *International Journal of Optics*, vol. 2014, Article ID 728056, 9 pages, 2014.

[28] J. F. Barrera, R. Henao, and R. Torroba, "Optical encryption method using toroidal zone plates," *Optics Communications*, vol. 248, no. 1–3, pp. 35–40, 2005.

[29] J. F. Barrera, R. Henao, and R. Torroba, "Fault tolerances using toroidal zone plate encryption," *Optics Communications*, vol. 256, no. 4-6, pp. 489–494, 2005.

[30] S. K. Rajput and N. K. Nishchal, "Asymmetric color cryptosystem using polarization selective diffractive optical element and structured phase mask," *Applied Optics*, vol. 51, no. 22, pp. 5377–5386, 2012.

[31] M. R. Abuturab, "Color information security system using Arnold transform and double structured phase encoding in gyrator transform domain," *Optics & Laser Technology*, vol. 45, no. 1, pp. 525–532, 2013.

[32] J. W. Goodman, *Introduction to Fourier Optics*, McGraw-Hill, New York, NY, USA, 2nd edition, 1996.

[33] D. G. Voelz, *Computational Fourier Optics: A MATLAB Tutorial*, SPIE Press, Bellingham, Wash, USA, 2011.

[34] D. R. Chalice, "A characterization of the Cantor function," *The American Mathematical Monthly*, vol. 98, no. 3, pp. 255–258, 1991.

[35] W. D. Furlan, F. Giménez, A. Calatayud, and J. A. Monsoriu, "Devil's vortex-lenses," *Optics Express*, vol. 17, no. 24, pp. 21891–21896, 2009.

[36] A. Calatayud, J. A. Monsoriu, O. Mendoza-Yero, and W. D. Furlan, "Polyadic devil's lenses," *Journal of the Optical Society of America A*, vol. 26, no. 12, pp. 2532–2537, 2009.

[37] M. Mitry, D. C. Doughty, J. L. Chaloupka, and M. E. Anderson, "Experimental realization of the devil's vortex Fresnel lens with a programmable spatial light modulator," *Applied Optics*, vol. 51, no. 18, pp. 4103–4108, 2012.

[38] A. Calabuig, S. Sánchez-Ruiz, L. Martínez-León et al., "Generation of programmable 3D optical vortex structures through devil's vortex-lens arrays," *Applied Optics*, vol. 52, no. 23, pp. 5822–5829, 2013.

[39] H. Singh, A. K. Yadav, S. Vashisth, and K. Singh, "Double phase-image encryption using gyrator transforms, and structured phase mask in the frequency plane," *Optics and Lasers in Engineering*, vol. 67, pp. 145–156, 2015.

[40] A. K. Yadav, S. Vashisth, H. Singh, and K. Singh, "A phase-image watermarking scheme in gyrator domain using devil's vortex Fresnel lens as a phase mask," *Optics Communications*, vol. 344, pp. 172–180, 2015.

[41] J. A. Davis, D. E. McNamara, D. M. Cottrell, and J. Campos, "Image processing with the radial Hilbert transform: theory and experiments," *Optics Letters*, vol. 25, no. 2, pp. 99–101, 2000.

[42] M. Joshi, C. Shakher, and K. Singh, "Image encryption and decryption using fractional Fourier transform and radial Hilbert transform," *Optics and Lasers in Engineering*, vol. 46, no. 7, pp. 522–526, 2008.

[43] X. Tong, Y. Liu, M. Zhang, H. Xu, and Z. Wang, "An image encryption scheme based on hyperchaotic Rabinovich and exponential chaos maps," *Entropy*, vol. 17, no. 1, pp. 181–196, 2015.

[44] Y. Liang, G. Liu, N.-R. Zhou, and J. Wu, "Image encryption combining multiple generating sequences controlled fractional DCT with dependent scrambling and diffusion," *Journal of Modern Optics*, vol. 62, no. 4, pp. 251–264, 2014.

[45] X.-F. Meng, L.-Z. Cai, X.-L. Yang et al., "Digital color image watermarking based on phase-shifting interferometry and neighboring pixel value subtraction algorithm in the discrete-cosine-transform domain," *Applied Optics*, vol. 46, no. 21, pp. 4694–4701, 2007.

[46] M. Joshi, C. Shakher, and K. Singh, "Image encryption using radial Hilbert transform filter bank as an additional key in the modified double random fractional Fourier encoding architecture," *Optics and Lasers in Engineering*, vol. 48, no. 5, pp. 605–615, 2010.

[47] M. Joshi, C. Shakher, and K. Singh, "Fractional Fourier plane image encryption technique using radial Hilbert-, and Jigsaw transform," *Optics and Lasers in Engineering*, vol. 48, no. 7-8, pp. 754–759, 2010.

[48] L. Sui, M. Xin, A. Tian, and H. Jin, "Single-channel color image encryption using phase retrieve algorithm in fractional Fourier domain," *Optics and Lasers in Engineering*, vol. 51, no. 12, pp. 1297–1309, 2013.

2

Black Phosphorus Q-Switched Large-Mode-Area Tm-Doped Fiber Laser

Yan Ren,[1] Zhipeng Qin,[1] Guoqiang Xie,[1] Zhen Qiao,[1] Jingui Ma,[1] Peng Yuan,[1] Liejia Qian,[1] Shikai Wang,[2] Chunlei Yu,[2] and Lili Hu[2]

[1]*Key Laboratory for Laser Plasmas (Ministry of Education), Collaborative Innovation Center of IFSA (CICIFSA), School of Physics and Astronomy, Shanghai Jiao Tong University, Shanghai 200240, China*
[2]*Key Laboratory of High Power Laser Materials, Shanghai Institute of Optics and Fine Mechanics, Chinese Academy of Sciences, Shanghai 201800, China*

Correspondence should be addressed to Guoqiang Xie; xiegq@sjtu.edu.cn and Shikai Wang; woshiwsk@163.com

Academic Editor: Giulio Cerullo

We report on a passively Q-switched fiber laser with black phosphorus as saturable absorber. By employing the sol-gel fabricated large-mode-area Tm-doped fiber as gain medium, a high-energy Q-switched fiber laser has been demonstrated which delivers the maximum pulse energy of 11.72 μJ with the pulse width of 660 ns at the wavelength of 1954 nm. Our experimental results indicate that BP Q-switched large-mode-area Tm-doped fiber laser is an effective and reliable approach to generate high-energy pulses at 2 μm.

1. Introduction

High-power and high-energy pulsed lasers operating at 2 μm from the $^3F_4 \rightarrow {}^3H_6$ transition of Tm^{3+} ion have been extensively studied for their potential applications such as material processing, laser therapy, national defense security, free-space optical communication, and pumping for mid-infrared supercontinuum generation [1–5]. Passively Q-switched Tm-doped fiber lasers are favorable ways to generate high-energy pulses at 2 μm, where two-dimension (2D) materials such as graphene, topological insulators (TIs), and transition metal dichalcogenides (TMDCs) are usually adopted as saturable absorbers (SAs) instead of commercial semiconductor saturable absorber mirror (SESAM) due to their low cost and easy fabrication [6–13].

Black phosphorus (BP), as a newly emerged 2D material, has attracted wide attention recently for the common properties of 2D materials with wide bandwidth, ultrafast carrier dynamics, and planar characteristic [14]. More importantly, BP has a thickness-dependent direct energy bandgap from 0.3 eV (bulk) to 2.0 eV (single layer) [15], filling up the interval between the zero gap of graphene and large gap of TMDCs, which is of great importance to the optical applications. Up to now, the broadband BP saturable absorbers have been widely applied to pulsed lasers in a wide spectral range from visible to mid-infrared [16–25]. Specifically, in 2 μm Q-switched fiber lasers, Wang et al. reported a passively Q-switched Tm-doped fiber laser using a BP deposited microfiber, delivering the maximum pulse energy of 154 nJ [26]; Jiang et al. deposited the BP powders onto a side-polished fiber and achieved Q-switched operation with the maximum pulse energy of 276 nJ [27]. The output pulse energy was further improved up to 632 nJ with BP SA fabricated by optical deposition method [28]. Although BP has been confirmed to be a reliable and excellent SA for 2.0 μm Q-switched fiber lasers, the output pulse energy is limited below micro joule level due to the employment of single-mode fibers. Large-mode-area (LMA) fiber fabricated via sol-gel method combined with high temperature sintering will be a choice for generating high-energy laser pulses. The sol-gel method has a great merit of higher doping homogeneity for rare-earth ions, preventing the cluster of the rare-earth ions and the fluorescence quench effect. With the advance of LMA fiber fabrication, dual-cladding Tm-doped fiber with core diameter as large as 38 μm could be developed by sol-gel method combined with high temperature sintering [29],

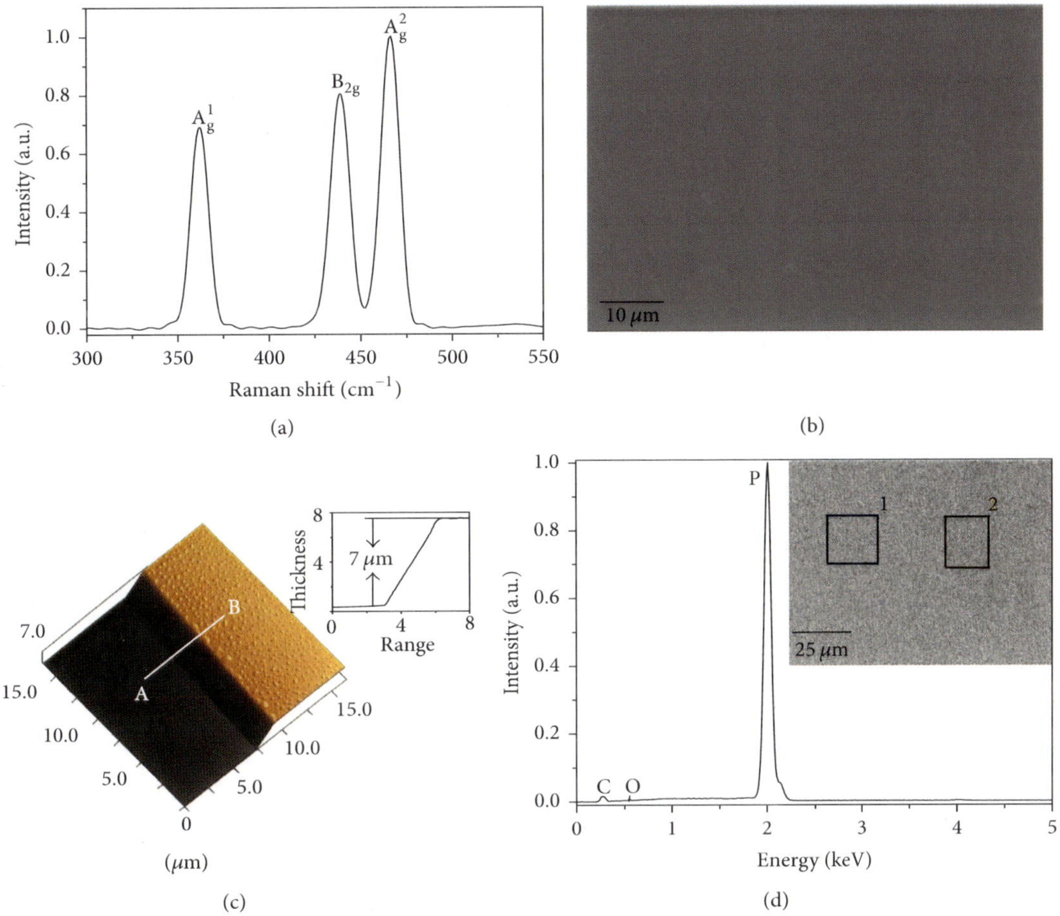

FIGURE 1: (a) Raman spectrum of the BP flake sample. (b) Morphology of BP flake measured by the scanning electron microscopy (SEM). (c) Three-dimension morphology of the BP flake scanned by the atomic force microscopy (AFM). Inset: height difference between mica sheet substrate (point A) and BP flake (point B). (d) Elemental components measured by the energy-dispersive X-ray spectroscopy (EDS). Inset: SEM image of BP flake.

offering a good alternative for higher pulse energy generation at 2 μm.

In this paper, we adopted the optimized LMA double-cladding Tm-doped fiber as gain medium with a core diameter as large as 30 μm. Based on a mechanically exfoliated BP SA, the high-energy Q-switching operation was demonstrated. The Q-switched fiber laser emitted pulses with the maximum average output power of 615 mW, the maximum pulse energy of 11.72 μJ, and the shortest pulse width of 660 ns at 1954 nm. Compared with the previously reported pulse energies in BP Q-switched Tm-doped fiber lasers, the pulse energy was improved by an order of magnitude, indicating that the sol-gel fabricated LMA Tm-doped fiber together with BP SA is an effective way to generate Q-switched fiber laser with high pulse energy at the wavelength of 2 μm.

2. Characteristics of BP Flake

Similar to the preparation approach of graphene, the layered BP flake could also be prepared via mechanical exfoliation method. By repeatedly exfoliating a bulk BP with scotch tape, the thin BP flake could be obtained. As displayed in Figure 1, the characteristics of as-prepared BP sample were analyzed. Excited by a 532 nm laser, Raman spectrum (Figure 1(a)) of the sample reveals three apparent peaks at the wavenumbers of 362 cm^{-1}, 439 cm^{-1}, and 466 cm^{-1}, corresponding to mode vibrations A_g^1, B_{2g}, and A_g^2 of phosphorus atoms in BP crystal lattice, respectively [30–32]. Using the scanning electron microscopy (SEM), the morphology of the as-prepared BP flake was obtained with an amplification rate of 1000. As Figure 1(b) shows, the smooth surface indicated that the mechanically exfoliated BP flake is uniform with a relatively high quality. To further confirm the thickness of the BP flake, the thickness measurement was performed with the atomic force microscopy (AFM). The three-dimensional (3D) AFM image (Figure 1(c)) shows the existence of some bubble-like bulges on the surface. The inset of Figure 1(c) shows that the height difference between the mica sheet substrate (point A) and the BP flake (point B) is 7.0 μm. Since the thickness of monolayer BP is 0.6 nm, the as-prepared BP flake could be regarded as bulk-like form with an energy gap of 0.3 eV, supporting the saturable absorption up to 4.1 μm wavelength. The elemental components of BP flake were confirmed by energy-dispersive X-ray spectroscopy (EDS). The analysis of the spectroscopic data in Figure 1(d) demonstrates the existence of carbon (C), oxygen (O), and phosphorus (P).

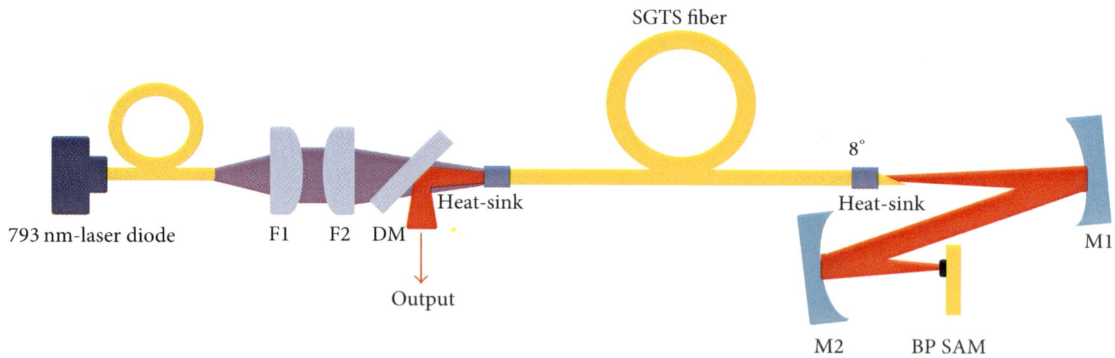

FIGURE 2: Experimental setup schematic of the BP Q-switched LMA Tm-doped fiber laser. DM: dichroic mirror; LMA TDF: large-mode-area Tm-doped fiber; BP SAM: black phosphorus saturable absorber mirror.

Two different rectangle areas with a size of about 400 μm^2 (inset of Figure 1(d)) were chosen for statistics. The average percentages of each component are 93.7%, 1.3%, and 5% in weight for P, O, and C, respectively. The existence of C and O might be attributed to defect in the fabrication process of bulk BP and oxidation in the atmosphere, respectively. The as-prepared BP flake was transferred onto a gold-coated mirror to fabricate a BP saturable absorber mirror (SAM). Compared with the BP dispersed in chemical solvent or polished into powder, the mechanically exfoliated BP flake is of high quality with flat surface, high purity, and large size which are very suitable to LMA Tm-doped fiber lasers for high-energy pulse generation.

3. Experimental Setup

The experimental setup schematic of BP passively Q-switched LMA Tm-doped fiber laser is depicted in Figure 2. The fiber-coupled 793 nm laser diode (LD) with a core diameter of 105 μm and a numerical aperture of 0.15 was adopted as the pump source. The pump laser from the LD was collimated and then focused onto the gain fiber by a pair of plano-convex lenses F1 and F2 ($f_1 = f_2 = 75$ mm). The 4 m length LMA double-cladding Tm-doped fiber was adopted as the gain fiber. The double-cladding Tm-doped fiber has a glass composition of $0.1Tm_2O_3$-$1.5Al_2O_3$-$98.4SiO_2$ in mol.% and a large core diameter of 30 μm with a NA of 0.102. The first cladding diameter and NA of the gain fiber were 250 μm and 0.366, respectively, guaranteeing the effective pump coupling. The normalized frequency of the fiber is about 4.9 for 1954 nm, indicating that the gain fiber supports multimode operation. In order to generate high-quality beam, the gain fiber was rolled up with a folding diameter of ~15 cm to suppress high-order transverse mode oscillation. The input fiber end was cleaved perpendicularly and acted as cavity feedback with Fresnel reflectivity of 4% and output coupler with a transmissivity of 96%. The other fiber end was cut with an angle of 8° to avoid parasitic oscillation. In order to remove the heat while pumping, both ends of fiber were mounted in an aluminum heat sink with a V-groove. The intracavity laser beam was collimated and focused onto BP SAM by a pair of concave mirrors M1 and M2 with a radius of curvature (ROC) of 150 mm and 100 mm, respectively. The residual pump light was filtered by the two concave mirrors (M1 and M2) with antireflection coating for pump wavelength, avoiding the adverse effect such as additional heat load on BP SAM. In order to separate the output laser beam from the pump beam, the 45°-placed dichroic mirror (DM) was coated with high transmissivity for pump wavelength ($T > 95\%$) and high reflectivity for laser wavelength ($R > 99\%$). For the convenience of measurement, the output laser beam was collimated by a plano-convex CaF_2 lens.

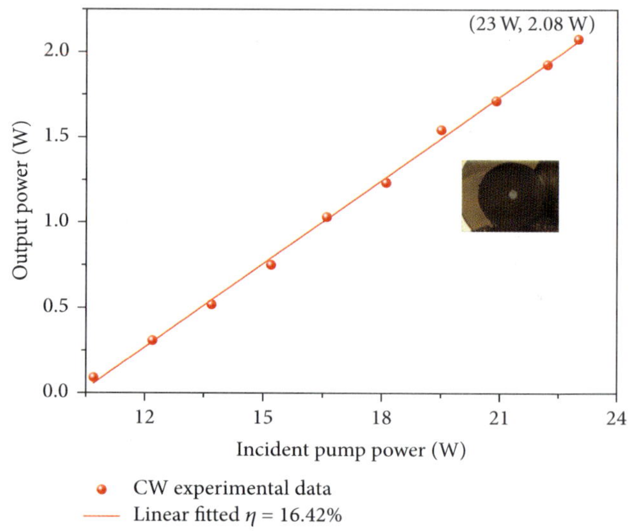

FIGURE 3: Continuous-wave output powers versus launched pump powers.

4. Results and Discussion

Using a highly reflective mirror instead of BP SAM, the continuous-wave (CW) laser performances with different-length Tm-doped fibers were tested and 4 m length LMA Tm-doped fiber demonstrated the best laser performance. CW operation with 4 m length gain fiber was achieved with a threshold of 10.7 W, as shown in Figure 3. With the increase of the launched pump power, the CW output power increased linearly with a slope efficiency of 16.42% and the maximum output power of 2.08 W was obtained at the launched pump power of 23 W. The output beam quality was measured by a commercial beam profiler (Thorlabs, M2MS, 400–2700 nm) with a M^2 factor of 1.3 [33]. Higher output power was only

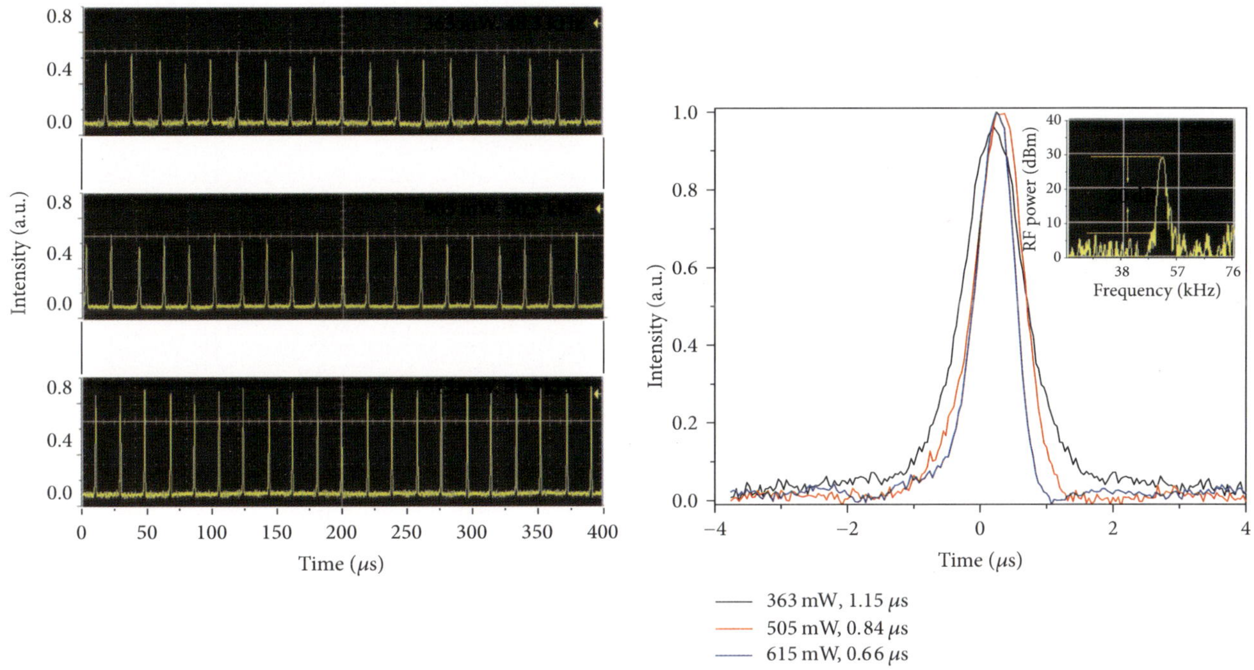

(a) (b)

Figure 4: (a) Typical pulse trains at the average output power of 363 mW, 505 mW, and 615 mW, respectively. (b) Corresponding pulse profiles. Inset: RF spectrum recorded at the average output power of 615 mW.

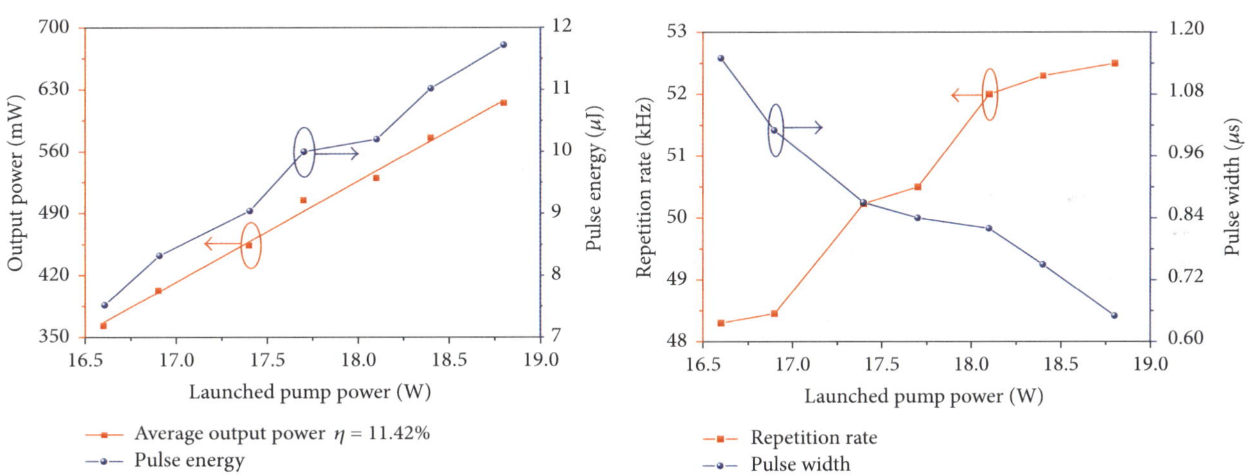

(a) (b)

Figure 5: (a) Average output power and pulse energy versus the launched pump power. (b) Repetition rate and pulse width versus the launched pump power.

limited by the available pump power in this experiment. In the whole pump power range from laser threshold to the maximum pump power, we did not observe the self-pulsed phenomenon. The inset of Figure 3 shows the cross section of the gain fiber. In order to achieve Q-switching operation, we replaced the highly reflective mirror with BP SAM.

By precisely adjusting the longitudinal and transverse position of BP SAM, self-started Q-switching operation was observed on a digital oscilloscope at the launched pump power of 16.6 W. Figure 4(a) shows three typical Q-switched pulse trains recorded at the output power of 363 mW, 505 mW, and 615 mW, respectively. Each output power corresponded to a repetition rate which increased with the launched pump power. The corresponding pulse profiles are shown in Figure 4(b), which were captured by a photoelectric detector (ALPHALAS, UPD-5N-IR2-P). The radio frequency (RF) spectrum (inset of Figure 4(b)), recorded at the average output power of 615 mW, shows a signal-to-noise ratio of 20 dB.

Figure 5(a) shows the average output power and pulse energy variation with the launched pump power in the passively Q-switched LMA Tm-doped fiber laser. Both average

Table 1: Passively Q-switched Tm-doped and Tm/Ho-co doped fiber lasers based on low dimensional SAs.

SA	Core diameter (μm)	Average power (mW)	Wavelength (nm)	Pulse width (μs)	Pulse energy (μJ)	Peak power (W)	Ref.
CNT	9	1.2	1890	7.2	8.6×10^{-2}	0.01	[11]
Graphene	6	4.5	2000	1.4	8.5×10^{-2}	0.06	[12]
Graphene	6	96	1957	0.76	0.38	0.5	[8]
Graphene	10	300	2032	3.8	6.7	1.8	[13]
Graphene	10	5200	2005	0.32	18	56	[34]
Bi_2Se_3	9	8.4	1980	4.2	0.31	0.07	[9]
Bi_2Te_3	9	0.68	1890	1.7	1.2×10^{-2}	6.7×10^{-3}	[10]
MoS_2	10	47	2032	1.8	1.0	0.56	[6]
$MoSe_2$	-	1.0	1924	5.5	4.2×10^{-2}	7.6×10^{-3}	[7]
BP	9	-	1890	2.5	0.28	0.1	[27]
BP	10	3.8	1948	5.6	0.15	0.27	[26]
BP	10	72	1912	0.73	0.63	0.86	[28]
BP	30	615	1954	0.66	11.72	18	This work

output power and pulse energy increased linearly with the launched pump power in the Q-switched regime. When the pump power increased to 18.8 W, the maximum pulse energy of 11.72 μJ was achieved. Compared with the previous reports of 2D SAs Q-switched fiber lasers at 2 μm (Table 1), there was a significant improvement of pulse energy in our experiment due to the employment of LMA fiber. With further increasing the pump power beyond 18.8 W, the pulse train of the passively Q-switched laser became unstable and then disappeared. This phenomenon might be attributed to performance degradation of BP flake due to excess heat. With the advance of fiber fabrication and optimization of BP SAM, we believe that higher pulse energy will be achieved.

Figure 5(b) shows the evolution of repetition rate and pulse width with the launched pump power. As expected, with the increase of launched pump power from 16.6 W to 18.8 W, the repetition rate increased from 48.3 kHz to 52.5 kHz and the pulse width decreased from 1.15 μs to 0.66 μs, respectively. During the experiment, no Q-switching mode-locked pulses were obtained.

At the maximum average output power of 615 mW, the spectrum of Q-switched fiber laser was measured by a mid-infrared spectrum analyzer with a resolution of 0.2 nm (Ocean Optics, SIR5000). As depicted in Figure 6, the Q-switched spectrum centered at the wavelength of 1954 nm with a full width of half maximum (FWHM) of 3.2 nm.

5. Conclusions

In conclusion, we have experimentally demonstrated a BP passively Q-switched fiber laser with LMA double-cladding Tm-doped fiber as gain medium for the first time. The Q-switched fiber laser emitted pulses with a maximum pulse energy of 11.72 μJ and the shortest pulse width of 0.66 μs at the wavelength of 1954 nm. The experimental results suggest that the sol-gel fabricated LMA dual-cladding Tm-doped fiber combined with BP SA is an effective way for generating high-energy laser pulses at the eye-safe wavelength of 2 μm.

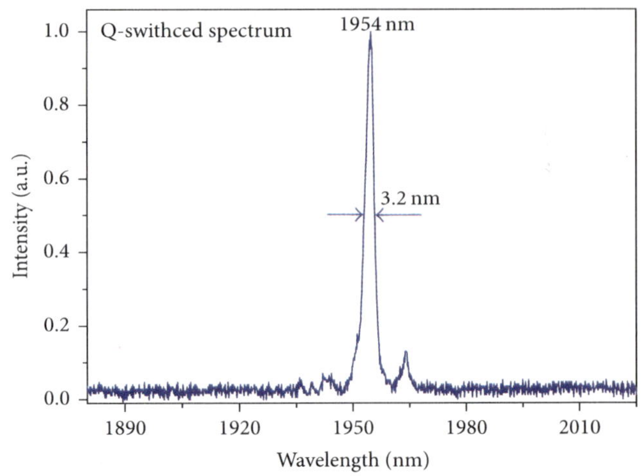

Figure 6: Q-switched pulse spectrum.

Conflicts of Interest

The authors declare that there are no conflicts of interest regarding the publication of this article.

Acknowledgments

The work is partially supported by the National Basic Research Program of China (Grant no. 2013CBA01505), the Shanghai Excellent Academic Leader Project (Grant no. 15XD1502100), and the National Natural Science Foundation of China (Grant nos. 61675130 and 11421064).

References

[1] T. Brabec and F. Krausz, "Intense few-cycle laser fields: frontiers of nonlinear optics," *Reviews of Modern Physics*, vol. 72, no. 2, pp. 545–591, 2000.

[2] W. Zeller, L. Naehle, P. Fuchs, F. Gerschuetz, L. Hildebrandt, and J. Koeth, "DFB lasers between 760 nm and 16 μm for sensing applications," *Sensors*, vol. 10, no. 4, pp. 2492–2510, 2010.

[3] C. R. Phillips, C. Langrock, J. S. Pelc et al., "Supercontinuum generation in quasi-phase-matched $LiNbO_3$ waveguide pumped by a Tm-doped fiber laser system," *Optics Letters*, vol. 36, no. 19, pp. 3912–3914, 2011.

[4] N. M. Fried and K. E. Murray, "High-power thulium fiber laser ablation of urinary tissues at 1.94 μm," *Journal of Endourology*, vol. 19, no. 1, pp. 25–31, 2005.

[5] H. Chen, S.-P. Chen, Z.-F. Jiang, and J. Hou, "Diversified pulse generation from frequency shifted feedback Tm-doped fibre lasers," *Scientific Reports*, vol. 6, Article ID 26431, 2016.

[6] Z. Luo, Y. Huang, M. Zhong et al., "1-, 1.5-, and 2-μm fiber lasers Q-switched by a broadband few-layer MoS_2 saturable absorber," *Journal of Lightwave Technology*, vol. 32, no. 24, Article ID 6920007, pp. 4077–4084, 2014.

[7] R. I. Woodward, R. C. T. Howe, T. H. Runcorn et al., "Wideband saturable absorption in few-layer molybdenum diselenide ($MoSe_2$) for Q-switching Yb-, Er- and Tm-doped fiber lasers," *Optics Express*, vol. 23, no. 15, pp. 20051–20061, 2015.

[8] M. Jiang, H. F. Ma, Z. Y. Ren et al., "A graphene Q-switched nanosecond Tm-doped fiber laser at 2 μm," *Laser Physics Letters*, vol. 10, no. 5, Article ID 055103, 2013.

[9] Z. Luo, C. Liu, Y. Huang et al., "Topological-insulator passively Q-switched double-clad fiber laser at 2 μm wavelength," *IEEE Journal of Selected Topics in Quantum Electronics*, vol. 20, no. 5, Article ID 0902708, 2014.

[10] J. Lee, M. Jung, J. Koo, C. Chi, and J. H. Lee, "Passively Q-switched 1.89-μm fiber laser using a bulk-structured Bi_2Te_3 topological insulator," *IEEE Journal of Selected Topics in Quantum Electronics*, vol. 21, no. 1, Article ID 0900206, 2015.

[11] M. Jung, J. Koo, Y. M. Chang, P. Debnath, Y.-W. Song, and J. H. Lee, "An all fiberized, 1.89-μm Q-switched laser employing carbon nanotube evanescent field interaction," *Laser Physics Letters*, vol. 9, no. 9, pp. 669–673, 2012.

[12] J. Liu, J. Xu, and P. Wang, "Graphene-based passively Q-switched 2 μm thulium-doped fiber laser," *Optics Communications*, vol. 285, no. 24, pp. 5319–5322, 2012.

[13] C. Liu, C. Ye, Z. Luo et al., "High-energy passively Q-switched 2 μm Tm^{3+}- doped double-clad fiber laser using grapheneoxide-deposited fiber taper," *Optics Express*, vol. 21, no. 1, pp. 204–209, 2013.

[14] J. Qiao, X. Kong, Z.-X. Hu, F. Yang, and W. Ji, "High-mobility transport anisotropy and linear dichroism in few-layer black phosphorus," *Nature Communications*, vol. 5, article no. 4475, 2014.

[15] V. Tran, R. Soklaski, Y. Liang, and L. Yang, "Layer-controlled band gap and anisotropic excitons in few-layer black phosphorus," *Physical Review B: Condensed Matter and Materials Physics*, vol. 89, no. 23, Article ID 235319, 2014.

[16] Y. Xie, L. Kong, Z. Qin, G. Xie, and J. Zhang, "Black phosphorus-based saturable absorber for Q-switched Tm: YAG ceramic laser," *Optical Engineering*, vol. 55, no. 8, Article ID 081307, 2016.

[17] J. Ma, S. Lu, Z. Guo et al., "Few-layer black phosphorus based saturable absorber mirror for pulsed solid-state lasers," *Optics Express*, vol. 23, no. 17, 2015.

[18] S. Zhang, X. Zhang, H. Wang et al., "Size-dependent saturable absorption and mode-locking of dispersed black phosphorus nanosheets," *Optical Materials Express*, vol. 6, no. 10, pp. 3159–3168, 2016.

[19] B. Zhang, F. Lou, R. Zhao et al., "Exfoliated layers of black phosphorus as saturable absorber for ultrafast solid-state laser," *Optics Expresss*, vol. 40, no. 16, pp. 3691–3694, 2015.

[20] Z. Qin, G. Xie, C. Zhao, S. Wen, P. Yuan, and L. Qian, "Mid-infrared mode-locked pulse generation with multilayer black phosphorus as saturable absorber," *Optics Letters*, vol. 41, no. 1, pp. 56–59, 2016.

[21] Z. Qin, G. Xie, H. Zhang et al., "Black phosphorus as saturable absorber for the Q-switched Er: ZBLAN fiber laser at 2.8 μm," *Optics Express*, vol. 23, no. 19, pp. 24713–24718, 2015.

[22] J. Li, H. Luo, B. Zhai et al., "Black phosphorus: a two-dimension saturable absorption material for mid-infrared Q-switched and mode-locked fiber lasers," *Scientific Reports*, vol. 6, Article ID 30361, 2016.

[23] L. Kong, Z. Qin, G. Xie et al., "Black phosphorus as broadband saturable absorber for pulsed lasers from 1 μm to 2.7 μm wavelength," *Laser Physics Letters*, vol. 13, no. 4, Article ID 045801, 2016.

[24] J. Sotor, G. Sobon, W. Macherzynski, P. Paletko, and K. M. Abramski, "Black phosphorus saturable absorber for ultrashort pulse generation," *Applied Physics Letters*, vol. 107, no. 5, Article ID 051108, 2015.

[25] J. Sotor, G. Sobon, M. Kowalczyk, W. Macherzynski, P. Paletko, and K. M. Abramski, "Ultrafast thulium-doped fiber laser mode locked with black phosphorus," *Optics Letters*, vol. 40, no. 16, pp. 3885–3888, 2015.

[26] Y. Wang, J. Li, L. Han et al., "Q-switched Tm^{3+}-doped fiber laser with a micro-fiber based black phosphorus saturable absorber," *Laser Physics*, vol. 26, no. 6, Article ID 065104, 2016.

[27] T. Jiang, K. Yin, X. Zheng, H. Yu, and X.-A. Cheng, *Black Phosphorus as A New Broadband Saturable Absorber for Infrared Passively Q-Switched Fiber Lasers*, 2015.

[28] H. Yu, X. Zheng, K. Yin, X. Cheng, and T. Jiang, "Nanosecond passively Q-switched thulium/holmium-doped fiber laser based on black phosphorus nanoplatelets," *Optical Materials Express*, vol. 6, no. 2, pp. 603–609, 2016.

[29] F. G. Lou, P. W. Kuan, L. Zhang et al., "2 μm laser properties of Tm^{3+}-doped large core sol-gel silica fiber," *Optical Materials Express*, vol. 4, no. 6, pp. 1267–1275, 2014.

[30] H. Liu, A. T. Neal, Z. Zhu et al., "Phosphorene: an unexplored 2D semiconductor with a high hole mobility," *ACS Nano*, vol. 8, no. 4, pp. 4033–4041, 2014.

[31] Y. Akahama, M. Kobayashi, and H. Kawamura, "Raman study of black phosphorus up to 13 GPa," *Solid State Communications*, vol. 104, no. 6, pp. 311–315, 1997.

[32] Z.-C. Luo, M. Liu, Z.-N. Guo et al., "Microfiber-based few-layer black phosphorus saturable absorber for ultra-fast fiber laser," *Optics Express*, vol. 23, no. 15, pp. 20030–20039, 2015.

[33] Y. Ren, Z. P. Qin, and G. Q. Xie, 2-μm mode-locking laser performances of sol-gel-fabricated large-core Tm-doped silica fiber, Chinese Optics Letter.

[34] Y. Tang, X. Yu, X. Li, Z. Yan, and Q. J. Wang, "High-power thulium fiber laser Q-switched with single-layer graphene," *Optics Letters*, vol. 39, no. 3, pp. 614–617, 2014.

3

Modified Three-Dimensional Multicarrier Optical Prime Codes

Rajesh Yadav and Gurjit Kaur

Department of ECE, School of ICT, Gautam Buddha University, Greater Noida, India

Correspondence should be addressed to Rajesh Yadav; raj_opyadav@yahoo.co.in

Academic Editor: Gang-Ding Peng

We propose a mathematical model for novel three-dimensional multicarrier optical codes in terms of wavelength/time/space based on the prime sequence algorithm. The proposed model has been extensively simulated on MATLAB for prime numbers (P) to analyze the performance of code in terms of autocorrelation and cross-correlation. The simulated outcome resembles the mathematical model and gives better results over other methods available in the literature as far as autocorrelation and cross-correlation are concerned. The proposed 3D optical codes are more efficient in terms of cardinality, improved security, and providing quality of services.

1. Introduction

Optical Code Division Multiple Access (OCDMA) network has a great potential to cater the needs of the fast access communication system. OCDMA is a suitable multiple access technique for local area network where traffic is bursty in nature [1]. Optical CDMA is an efficient multiplexing scheme, as it does not require synchronization between the transmitter and receiver and there is the efficient utilization of available bandwidth [2]. The performance of optical CDMA networks depends on the optimal selection of the optical code and coding configuration. It is desired to have code set that can correctly decode the desired user signal in the presence of other users [3]. The motivation for designing more dimensional code is to have increased code set and also have significant code weight to support quality of services for the given code length. The length of a code plays an important role in system performance and system complexity [4], although it is possible to improve the correlation property by using long code words. However, the code length cannot be increased much because of the power and processing time restrictions for encoder and decoder [5]. The weight of code represents the signal power. A code sequence with larger code weight is less sensitive to interference than that with a lower code weight. In this paper, novel modified three-dimensional (3D) multicarrier prime codes are proposed having sufficient code weight along with significant code length for supporting QoS in optical CDMA communication systems. The construction of proposed three-dimensional codes is conceptualized on the basis of prime sequence algorithm.

The rest of paper is organized as follows: Section 2 discusses the OCDMA coding theory that gives an insight of the optical codes for optical CDMA communication system. A concise introduction to optical prime codes for OCDMA systems is also discussed. Section 3 presents the mathematical modeling of proposed wavelength/time/space modified 3D multicarrier optical prime codes. Section 4 discusses the result of the proposed 3D optical prime codes and its performance estimation in terms of the autocorrelation and cross-correlation function. In Section 5 the current findings along with the future directions are concluded. The paper ends with the references studied and cited in the paper.

2. Optical Codes

A lot of researchers have studied optical codes like one-dimensional (1D), two-dimensional (2D), and now three-dimensional codes to provide efficient codes. Prucnal has proposed one-dimensional optical code based on the prime number P over a Galois field GF(P). The codes have an autocorrelation value of $P - 1$ and cross-correlation value of two [6]. The construction of prime codes is based on the congruence codes. Later on, one-dimensional prime code

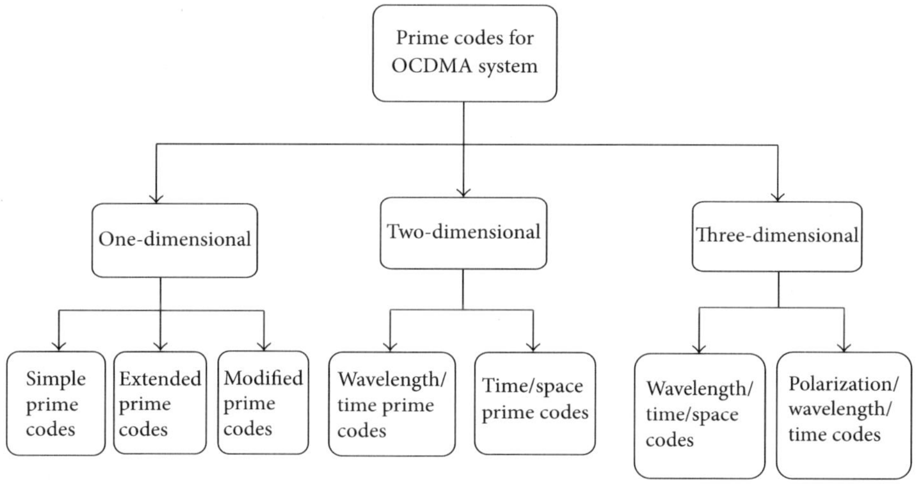

FIGURE 1: Prime code word family hierarchy [11].

with improved correlation property is proposed by many researchers [7–10]. In the one-dimensional code to increase the code set size, the code length is increased significantly to keep the correlation property satisfactory. To overcome this limitation two-dimensional codes are developed with better code set size and good correlation properties. Mendez et al. describe the 2D codes constructed from 1D code based on Golomb ruler to increase the number of code set sizes [7]. Yu and Park have proposed an algorithm to construct 2D code based on 1D prime sequence [8]. Yim et al. proposed a design for wavelength/time 2D code for O-CDMA and carried out the performance analysis of the 2D codes [9]. The performance analysis of the two-dimensional optical codes for bit error rate and a number of users supported carried out by Shivaleela and Kaur has revealed that there is a significant improvement in optical CDMA system using 2D codes over 1D code [10, 11].

The successful performance of two-dimensional optical codes over the one-dimensional codes has led the researcher to construct the three-dimensional codes with better code set size and improved system performance. McGeehan presented a time-wavelength-polarization 3D optical code for increasing the number of users in OCDMA LAN's application. This coding technique has limitation due to the polarization mode dispersion and complex polarization control at all the stages in the network [12, 13]. Kumar et al. present multiple pulse per plane codes using a row-wise orthogonal pairs (RWOP) algorithm for wavelength and spatial channel allocation [14]. The design offers the very low probability of error due to multiple access interference at lower cardinality. The RWOP algorithm based 3D code has a limitation imposed by multipath interference. The hierarchy of prime sequence codes is represented in Figure 1 [11]. Here we proposed wavelength/time/space spread 3D optical codes based on prime sequence algorithm derived from multicarrier prime codes for OCDMA [14, 15]. The comparative analysis of 1D, 2D, and 3D optical codes based on prime sequence is carried out by Yadav and Kaur [16]. The generation algorithm of 3D codes is discussed in detail and the orthogonality of the 3D optical codes is demonstrated mathematically and simulated on the MATLAB.

3. Generation of 3D Optical Code

In this research work, we have presented new modified three-dimensional multicarrier optical code for OCDMA network based on prime sequence algorithm. The 3D codes family can be represented as $(W * T * S, w, \lambda_a, \lambda_c)$ where W, T, and S denote the number of wavelengths, time, and spatial channels domain used, respectively. "w" signifies the weight of the code. λ_a and λ_c are the autocorrelation and cross-correlation of the optical code, respectively. Figure 2 illustrates the algorithm of modified three-dimensional multicarrier prime code constructions.

Three-dimensional optical code construction starts with the prime number P over the Galois field (GF). A prime sequence is constructed for a prime number P as [6, 17]

$$Z_r = \left(z_{r,0}, z_{r,1}, \ldots, z_{r,s}, \ldots, z_{r,(P-1)}\right), \quad r = 0, 1, \ldots, P-1, \tag{1}$$

where the element in this prime sequence is given by

$$Z_{r,s} = r \cdot s \pmod{P}, \tag{2}$$

where $r = (0, 1, \ldots, P-1)$ and $s = (0, 1, \ldots, P-1)$ are the elements over the Galois field GF(P). For prime number $P = 3$, the corresponding code set matrix $Z_{r,s}$ is shown by (3), where r and s are the elements over the Galois field GF(3). The following 3×3 matrix will be constructed:

$$Z_{r,s} = \begin{bmatrix} Z_{0,0} & Z_{0,1} & Z_{0,2} \\ Z_{1,0} & Z_{1,1} & Z_{1,2} \\ Z_{2,0} & Z_{2,1} & Z_{2,2} \end{bmatrix}. \tag{3}$$

This matrix represents a code set for prime number $P = 3$. In this 3D prime code set, every code word is represented by a matrix as given by

$$Z_{r,s} = \begin{bmatrix} C_{r,s,0,0} & C_{r,s,0,1} & \cdots & C_{r,s,0,P^2-1} \\ & \vdots & \ddots & \vdots \\ C_{r,s,P-1,0} & C_{r,s,P-1,1} & \cdots & C_{r,s,P-1,P^2-1} \end{bmatrix}. \quad (4)$$

This matrix represents a code word whose element can be derived as $C_{m,k} = C_{r,s,m,0} \cdots C_{r,s,m,k} \cdots C_{r,s,m,P^2-1}$, where $r = (0, 1, \ldots, P - 1)$, $s = (0, 1, \ldots, P - 1)$, $m = (0, 1, \ldots, P - 1)$, and $k = (0, 1, \ldots, P^2 - 1)$. With the prime sequence matrix $Z_{r,s}$ the value of $C_{m,k}$ will be mapped as

$$C_{m,k} = \begin{cases} W_{r\cdot n\oplus t\cdot s\cdot m, r\cdot m\oplus s\cdot n} & \text{if } k = ((r \cdot n \oplus t \cdot s \cdot m)\, P + n), \text{ for } m, n = \{0, 1, \ldots, P-1\} \\ 0 & \text{elsewere.} \end{cases} \quad (5)$$

This mapping will define the value of matrix elements or position of wavelength in the code word matrix of size 3×9 for $P = 3$. A matrix is constructed by setting each element according to the rule $C_{m,k} = W_{r\cdot n\oplus t\cdot s\cdot m, r\cdot m\oplus s\cdot n}$, where $n = (0, 1, \ldots, P - 1)$, $m = (0, 1, \ldots, P - 1)$, $t = \{0, 1, \ldots, P - 1\}/[B]$, and $\oplus$ denotes a modulo addition. $[B]$ is a set of numbers that contains every integer designed with $1/s^2 \pmod P$ for $s = \{1, 2, \ldots, P - 1\}/2$. The genuine value of t can be calculated using this formula and that can be used to obtain $Z_{r,s}$ for various prime numbers. Each element $W_{r\cdot n\oplus t\cdot s\cdot m, r\cdot m\oplus s\cdot n}$ represents a single transmitting wavelength, among P^2 different wavelengths in the mth space slot and nth time period position in $Z_{r,s}$. A binary 3D multicarrier prime code can be derived as having the exact position of wavelength in the space slot and time period.

In $Z_{r,s}$ code word matrix, each row will have one set of disjoint wavelength and in total, there are P distinct wavelengths per row. The space group is represented by $m = (0, 1, \ldots, P - 1)$. In this way, we have $P^2 - 1$ prime code words of length P^2 and weight P^2 and have P^2 different wavelengths where $m \leq P$. These P^2 distinct wavelengths are divided into "m" space groups with P different wavelengths per space/channel. Every time slot can have a minimum of one wavelength. The optical code words can then be transmitted over "m" different channels having P different wavelengths each distributed over P^2 time slots thus acting as a 3D code. In this way, the position of P different wavelengths can be calculated in every m spatial channel according to the formula given in (5). The cardinality of 3D optical prime code is given by $|C| = P^2 - 1$; therefore for $P = 3$ it will be 8 and $Z_{0,0}$ for $(a, b) = (0, 0)$, respectively, will not exist.

For prime number $P = 3$ the valid value of t we have is 2. In the codeword $Z_{0,1}$, $r = 0$ and $s = 1$ are in GF(3). The elements of the code word matrix are calculated according to the rule $C_{m,k} = W_{r\cdot n\oplus t\cdot s\cdot m, r\cdot m\oplus s\cdot n}$ as shown in

$$Z_{0,1} = \begin{bmatrix} W_{0,0}W_{0,1}W_{0,2} & 000 & 000 \\ 000 & 000 & W_{2,0}W_{2,1}W_{2,2} \\ 000 & W_{1,0}W_{1,1}W_{1,2} & 000 \end{bmatrix}. \quad (6)$$

The corresponding code matrix is of size 3×9 and weight 9. Each "W" in the code word matrix represents the position of "1" in the code word that will be transmitted using an optical pulse of certain wavelength over the space channel in the corresponding time slot. It can be seen that each code word matrix of a three-dimensional code is divided into P rows in the space domain having length P^2 for each row in the time domain; the P^2 column has been divided into P groups each having P subcolumns.

The code word has P^2 wavelengths distributed over the P^2 time slot with each column having only one wavelength and each space channel will have P wavelengths. The P^2 wavelengths are in the shape of 3 groups and 3 wavelengths per group. The length of the 3D optical code word is 9 for $P = 3$. Figure 3 shows the visualization of a wavelength/time/space spread 3D code word $Z_{0,1}$ for prime sequence $(P = 3)$ and $t = 2$. Similarly, for the code word $Z_{2,2}$ the elements are calculated according to the rule $C_{m,k} = W_{r\cdot m\oplus t\cdot s\cdot n, r\cdot n\oplus s\cdot m}$. The corresponding code matrices $Z_{r,s}$ for $r = 2$ and $s = 2$ are of size 3×9 and weight 9 as shown in Figure 4. Equation (7) shows the resulting code word matrix as follows:

$$Z_{2,2} = \begin{bmatrix} W_{0,0}00 & 00W_{1,1} & 0W_{2,2}0 \\ 0W_{0,1}0 & W_{1,2}00 & 00W_{2,0} \\ 00W_{0,2} & 0W_{1,0}0 & W_{2,1}00 \end{bmatrix}. \quad (7)$$

In OCDMA system using modified 3D multicarrier prime code, each subscriber is allocated a code matrix as its code word. To transmit information bit "1" subscriber sends out a sequence of the optical signal in the space, time, and wavelength domain as per the intended receiver code word matrix. At the receiving end, the optical signal in every spatial channel is segregated and routed to correlate the every wavelength separately in the time domain. After successfully correlating the received optical signals are combined together, thus resulting in a high peak optical signal for the proposed receiver. This is considered as autocorrelation function, whereas for all others, it gives a series of low peak optical signal. This is considered as cross-correlation function. The received peak optical pulse is further passed

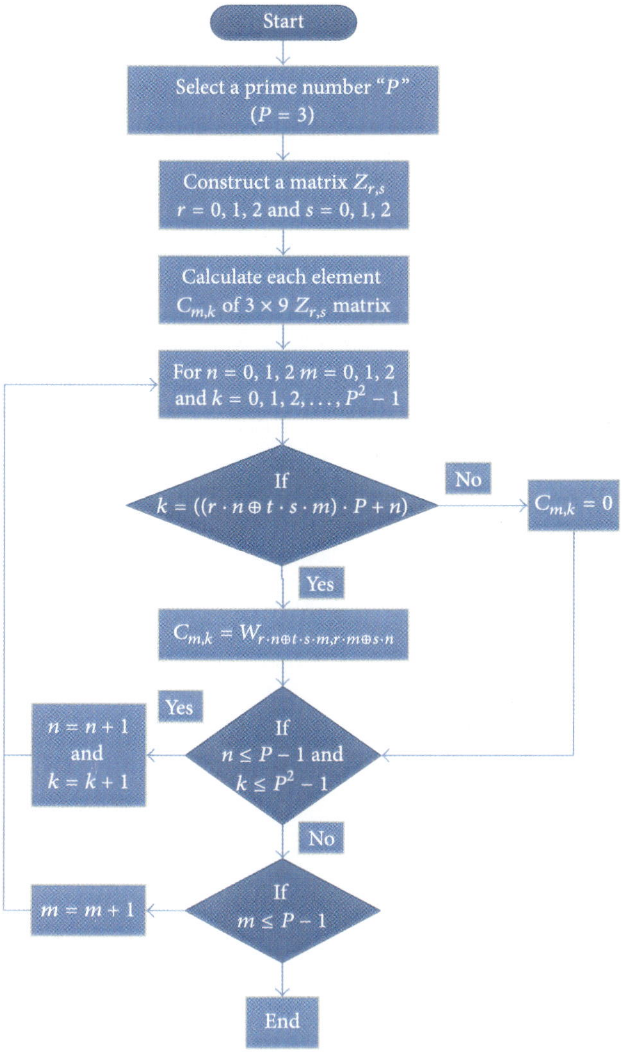

FIGURE 2: Flowchart of three-dimensional code construction algorithm.

Space	$m = 2$	W_{00}	W_{01}	W_{02}	0	0	0	0	0	0
Space	$m = 1$	0	0	0	0	0	0	W_{20}	W_{21}	W_{22}
Space	$m = 0$	0	0	0	W_{10}	W_{11}	W_{12}	0	0	0
	K	0	1	2	3	4	5	6	7	8
		$n = 0$	$n = 1$	$n = 2$	$n = 0$	$n = 1$	$n = 2$	$n = 0$	$n = 1$	$n = 2$
						Time ⟶				

FIGURE 3: Visualization of 3D optical prime code word $Z_{0,1}$ over GF($P = 3$) and $t = 2$.

to photodetector to detect the bit "1." For transmitting information bit "0" nothing is sent.

4. Result and Discussion

From optical CDMA network point of view, there is a need to design more secure codes with sufficient code weight, optimal code length, and higher code set size. The proposed code is compared with the 1D and 2D optical codes based on prime sequence algorithm for analysis as presented in Figure 5. It

Space	$m = 2$	W_{00}	0	0	0	0	W_{11}	0	W_{22}	0
Space	$m = 1$	0	W_{01}	0	W_{12}	0	0	0	0	W_{20}
Space	$m = 0$	0	0	W_{02}	0	W_{10}	0	W_{21}	0	0
	K	0	1	2	3	4	5	6	7	8
		$n = 0$	$n = 1$	$n = 2$	$n = 0$	$n = 1$	$n = 2$	$n = 0$	$n = 1$	$n = 2$
						Time ⟶				

FIGURE 4: Visualization of 3D optical prime code word $Z_{2,2}$ over GF($P = 3$) and $t = 2$.

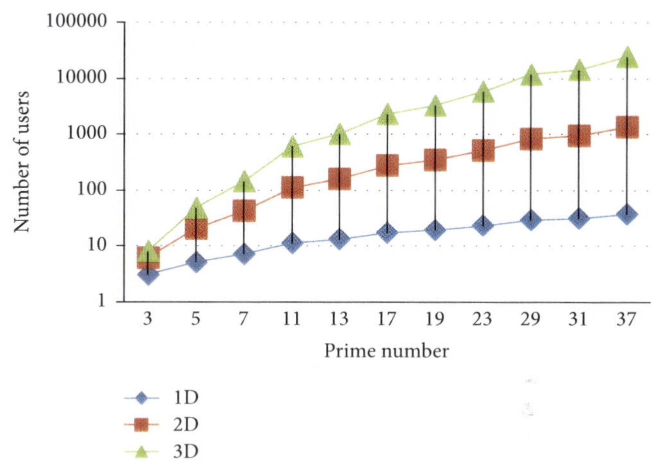

FIGURE 5: Number of users versus prime number for different optical prime codes.

can be observed that the proposed 3D codes are better in terms of a number of available codes (Cardinality) whereas they have comparable code length for a given prime number. However, the proposed 3D codes have higher code weight than 1D and 2D codes.

In optical CDMA, the interference among the users sharing the common fiber channel known as the multiple access interference is usually the dominant source of bit errors [6]. The performance of optical codes in OCDMA networks is mainly measured in terms of the autocorrelation and cross-correlation function values. The cross-correlation value depicts the interference between users. In contrast, the autocorrelation value depicts signal power and differentiates the users in the presence of other users. The intelligent design of the code word sequence is important to reduce the contribution of MAI to the total received signal. The address code words must satisfy two conditions; that is, all address code words should be distinguishable from shifted versions of it and all address code words should be distinguished from a possibly shifted version of every other code word in a code set [16].

The autocorrelation and cross-correlation value of the developed 3D optical code are evaluated on MATLAB tool. The wavelength/time/space spread 3D optical code is represented by ($W * T * S$, w, λ_a, λ_c). "W" signifies the wavelengths, "T" denotes the time slots, and "S" specifies the space channels. "w" signifies the weight of the code. λ_a and λ_c are the autocorrelation and cross-correlation of the code,

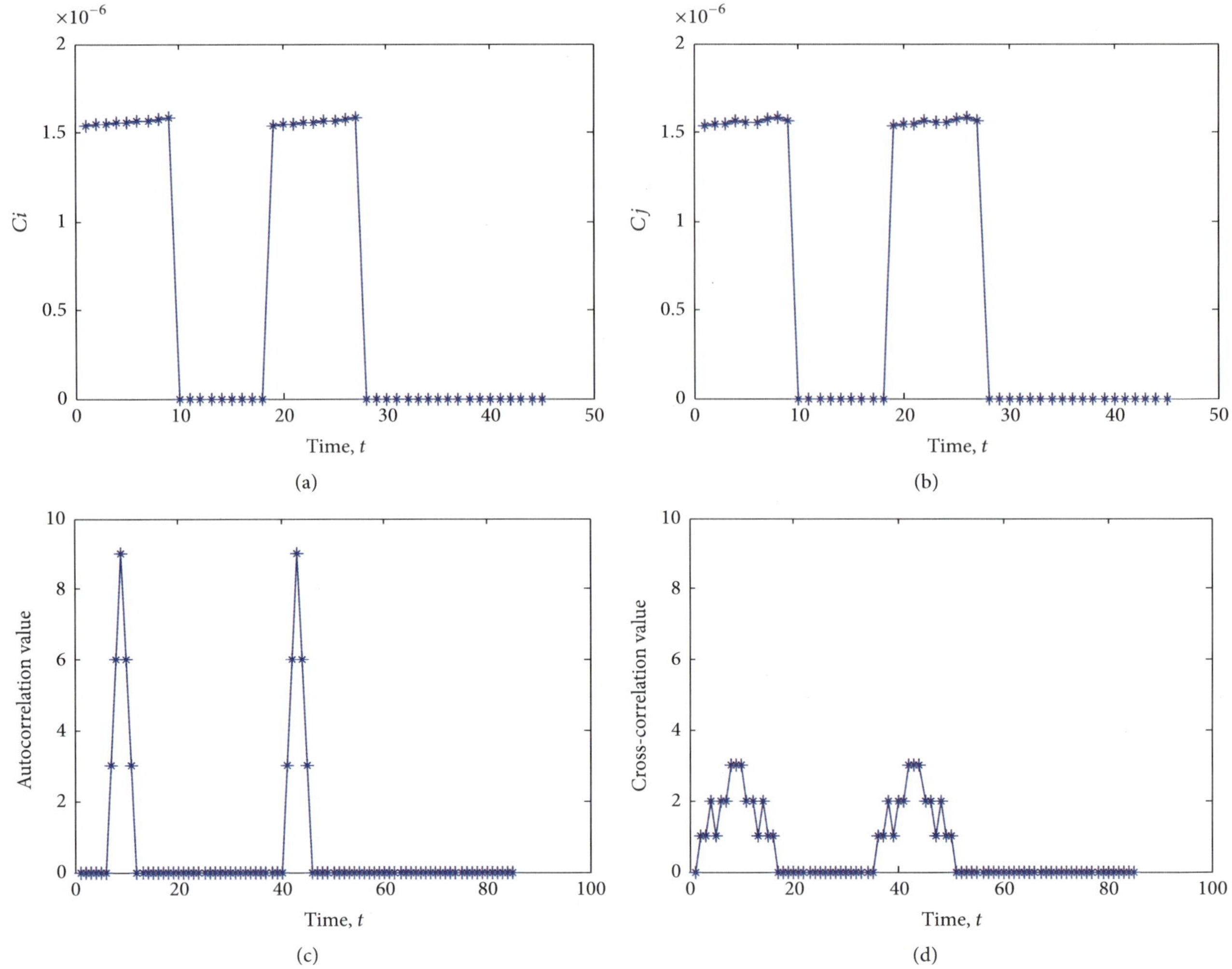

FIGURE 6: (a) Code word $Z_{0,1}$ of the code set of 3D optical code over GF(3). (b) Code word $Z_{2,2}$ of the code set of 3D optical code over GF(3). (c) Autocorrelation function of optical code $Z_{0,1}$ for the data bit stream 10100. (d) Cross-correlation function of optical codes $Z_{0,1}$ and $Z_{2,2}$, for the data bit stream 10100.

respectively. In (8a) and (8b) the correlation property for a pair of prime code words C_k and B_k with discrete data stream format are calculated.

$$\lambda_a = \sum_{k=0}^{P-1} \sum_{m=0}^{P-1} \sum_{n=0}^{P-1} C_{k,m,n} C_{k \cdot m \cdot n \oplus \tau} = P^2 \quad \text{for } 0 \le \tau \ge P^2, \quad (8a)$$

$$\lambda_c = \sum_{k=0}^{P-1} \sum_{m=0}^{P-1} \sum_{n=0}^{P-1} C_{k,m,n} B_{k \cdot m \cdot n \oplus \tau} \le 3 \quad \text{for } 0 \le \tau \ge P^2, \quad (8b)$$

where $C_{k,m,n}$ and $B_{k,m,n} \in (0,1)$ are an element of matrix Z, P is the prime number over Galois field, and $\oplus$ denotes the modulo addition.

It can be observed from the above correlation function that the peak autocorrelation is P^2 which occurs when $C_{k,m,n} = C_{k,m,n}$. Also, the cross-correlation value is "3" occurring at all synchronized time T when $C_{k,m,n} \neq B_{k,m,n}$. Figure 6 shows the two code words $Z_{0,1}$ and $Z_{2,2}$ of the code set of wavelength/time/space spread modified 3D multicarrier codes over GF(3) and of the length 9 and weight 9. It also illustrates that the autocorrelation property of 3D code sequence $Z_{0,1}$ having maximum peak value equals 9 at all synchronized times when the sequence follows the data stream 10100. Similarly, the cross-correlation property of 3D code sequence $Z_{0,1}$ with $Z_{2,2}$ is equal to 3 at all synchronized time T for the same data stream 10100. Similarly, Figure 7 represents the code words $Z_{2,0}$ and $Z_{1,1}$ and their autocorrelation and cross-correlation property, respectively, for the input data stream 10100.

Yen and Chen carried out the study of three-dimension Optical Code Division Multiple Access for optical fiber sensor networks. For efficient communication using OCDMA system, it is preferred to maximize the autocorrelation peak and minimize the cross-correlation function for sorting the correct signal and interference [18]. J. Singh and M. L. Singh present a new family of 3D wavelength/time/space codes named Golomb ruler-with-zero-insertions balanced codes for differential detection (GRZI-BCDD) using the unique interpulse distance property of Golomb rulers. The 3D code based on GRZI-BCDD requires fiber ribbons that

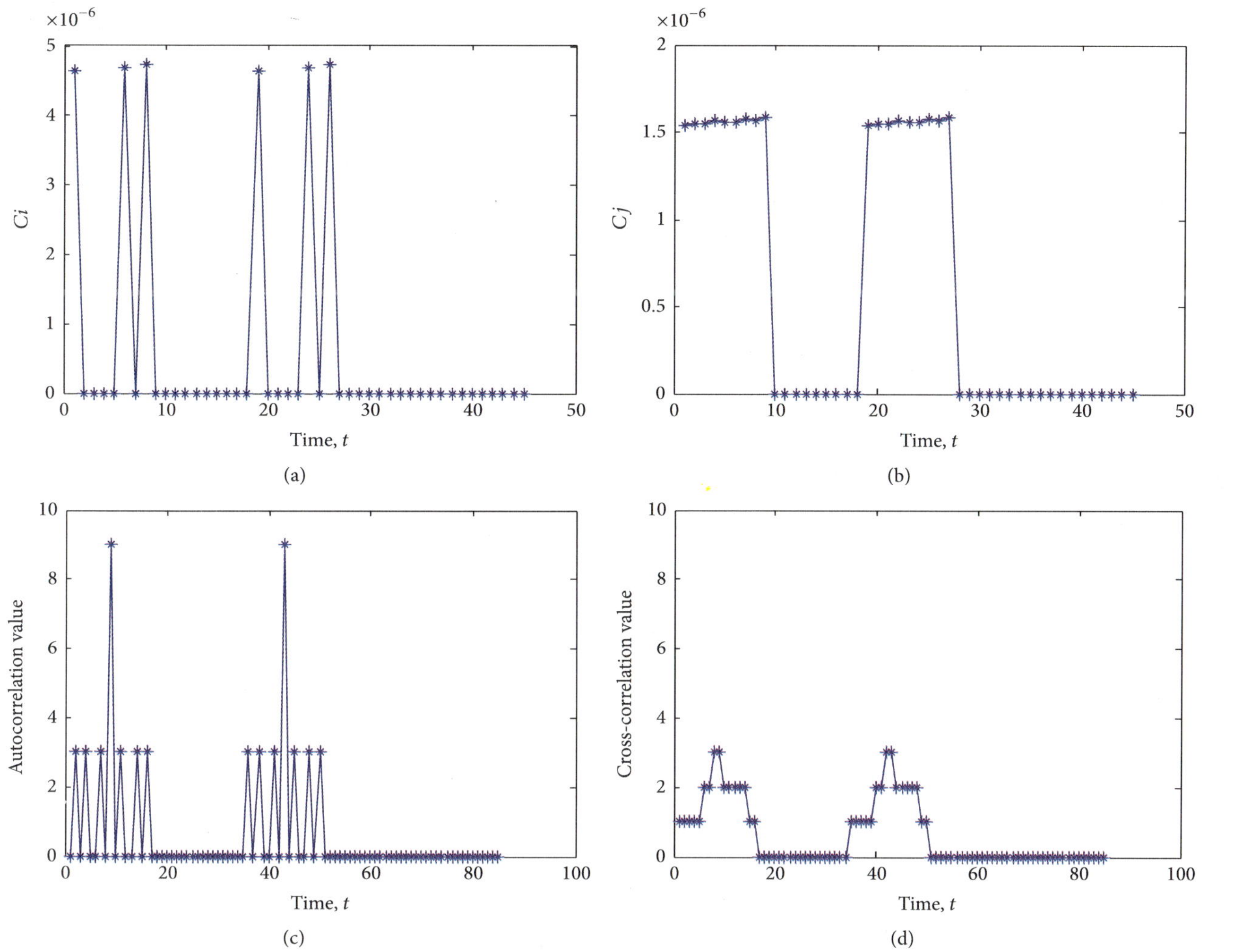

FIGURE 7: (a) Code word $Z_{2,0}$ of the code set of 3D optical code over GF(3). (b) Code word $Z_{1,1}$ of the code set of 3D optical code over GF(3). (c) Autocorrelation function of optical code $Z_{2,0}$ for the data bit stream 10100. (d) Cross-correlation function of optical codes $Z_{2,0}$ and $Z_{1,1}$, for the data bit stream 10100.

increase the system complexity [19]. To improve the ability of suppressing the multiple access interference, the 3D code set must have a sufficient code length of each code word. The proposed code incorporates the third dimension to improve the code cardinality, security, and coding flexibility that will help in providing multimedia services and quality of services. The limiting factor is its implementation because the complexity of system hardware is increased as the dimension increases. However do not require fiber ribbons for practical implementation.

5. Conclusion

This work presents the proposed wavelength/time/space spread modified 3D multicarrier codes using the prime sequence algorithm. The code discussed in the text has been designed for the prime number ($P = 3$) having code length 9; however, it can be increased for the assumed prime number. Thus, by introducing more dimensions, this coding approach not only overcomes the problem of a sufficient number of users for a given prime number but also makes the proposed code more secure and support quality of services. However, a marginal complexity is increased from an implementation point of view due to additional dimension as compared to 1D and 2D codes. In addition, the generation algorithm of code is simple and has optimum orthogonal properties. Therefore, it reduces the cost involved. The simulation results on MATLAB depict that these 3D optical codes have maximum autocorrelation peak ($\lambda_a = P^2$) with cross-correlation value ($\lambda_c = 3$).

Competing Interests

The authors declare that they have no competing interests.

References

[1] A. Stok and E. H. Sargent, "The role of optical CDMA in access networks," *IEEE Communications Magazine*, vol. 40, no. 9, pp. 83–87, 2002.

[2] J. A. Salehi, "Emerging optical CDMA techniques and applications," *International Journal of Optics and Photonics*, vol. 1, no. 1, pp. 15–32, 2007.

[3] G.-C. Yang and W. C. Kwong, "Performance analysis of optical CDMA with prime codes," *Electronics Letters*, vol. 31, no. 7, pp. 569–570, 1995.

[4] R. Yadav and G. Kaur, "Optical CDMA: technique, parameters and applications," in *Proceedings of the 4th International Conference on Emerging Trends in Engineering & Technology*, pp. 90–98, Kurukshetra, India, October 2013.

[5] S. Park, B. K. Kim, and B. W. Kim, "An OCDMA scheme to reduce multiple access interference and enhance performance for optical subscriber access networks," *ETRI Journal*, vol. 26, no. 1, pp. 13–20, 2004.

[6] P. R. Prucnal, M. A. Santoro, and T. R. Fan, "Spread spectrum fiber-optic local area network using optical processing," *Journal of Lightwave Technology*, vol. 4, no. 5, pp. 547–554, 1986.

[7] A. J. Mendez, R. M. Gagliardi, V. J. Hernandez, C. V. Bennett, and W. J. Lennon, "Design and performance analysis of wavelength/time (W/T) matrix codes for optical CDMA," *Journal of Lightwave Technology*, vol. 21, no. 11, pp. 2524–2533, 2003.

[8] K. Yu and N. Park, "Design of new family of two-dimensional wavelength-time spreading codes for optical code division multiple access networks," *Electronics Letters*, vol. 35, no. 10, pp. 830–831, 1999.

[9] R. M. H. Yim, L. R. Chen, and J. Bajcsy, "Design and performance of 2-D codes for wavelength-time optical CDMA," *IEEE Photonics Technology Letters*, vol. 14, no. 5, pp. 714–716, 2002.

[10] E. S. Shivaleela and T. Srinivas, "Construction of wavelength/time codes for fiber-optic CDMA networks," *IEEE Journal on Selected Topics in Quantum Electronics*, vol. 13, no. 5, pp. 1370–1377, 2007.

[11] R. Yadav and G. Kaur, "Optical CDMA codes: a review," in *Proceedings of the 7th International Conference Advanced Computing and Communication Technologies*, pp. 263–269, Panipat, India, 2013.

[12] J. E. McGeehan, S. M. R. Motaghian Nezam, P. Saghari et al., "3D time-wavelength-polarization OCDMA coding for increasing the number of users in OCDMA LANs," in *Proceedings of the Optical Fiber Communication Conference (OFC '04)*, pp. 444–446, February 2004.

[13] S. Amiralizadeh and K. Mehrany, "Analytical study of multiple access interference and beat noise in polarization-wavelength-time optical CDMA systems," in *Proceedings of the 18th International Conference on Telecommunications (ICT '11)*, pp. 206–210, IEEE, Ayia Napa, Cyprus, May 2011.

[14] M. Ravi Kumar, S. S. Pathak, and N. B. Chakrabarti, "Design and performance analysis of code families for multi-dimensional optical CDMA," *IET Communications*, vol. 3, no. 8, pp. 1311–1320, 2009.

[15] C.-F. Hong and G.-C. Yang, "Multicarrier FH codes for multicarrier FH-CDMA wireless systems," *IEEE Transactions on Communications*, vol. 48, no. 10, pp. 1626–1630, 2000.

[16] R. Yadav and G. Kaur, "Design and performance analysis of 1D, 2D and 3D prime sequence code family for optical CDMA network," *Journal of Optics*, 2016.

[17] G. C. Yang and W. C. Kwong, "Prime codes with applications to CDMA optical and wireless networks," in *Prime Codes with Applications to CDMA Optical and Wireless Networks*, pp. 43–108, Artech House, Norwood, Mass, USA, 2002.

[18] C.-T. Yen and C.-M. Chen, "A study of three-dimensional optical code-division multiple-access for optical fiber sensor networks," *Computers and Electrical Engineering*, vol. 49, pp. 136–145, 2016.

[19] J. Singh and M. L. Singh, "Design of 3-D wavelength/time/space codes for asynchronous fiber-optic CDMA systems," *IEEE Photonics Technology Letters*, vol. 22, no. 3, pp. 131–133, 2010.

4

Improving the Sensitivity of Humidity Sensor based on Mach-Zehnder Interferometer Coated with a Methylcellulose

Jun Huang,[1] Bowen Wang,[2] and Kai Ni[2]

[1]*College of Modern Science and Technology, China Jiliang University, Hangzhou, Zhejiang 310018, China*
[2]*Institute of Optoelectronic Technology, China Jiliang University, Hangzhou, Zhejiang 310018, China*

Correspondence should be addressed to Jun Huang; hjun@cjlu.edu.cn

Academic Editor: Sulaiman W. Harun

A novel humidity sensor based on Mach-Zehnder interferometer (MZI) with the single-mode fiber (SMF) coated with methylcellulose (MC) is proposed and experimentally demonstrated. The MZI consists of two waist enlarged structures. Such an all-fiber MZI incorporates an intermodal interference between the core mode and cladding modes. The MC is coated on the surface of the SMF. External humidity changes the refractive index of MC, causing the intensity changes of the interference pattern. The proposed sensor is linearly responsive to refractive humidity (RH) within the range from 45% to 85% RH, with sensitivity of 0.094 dB/%RH. Moreover the insensitivity of the fiber to the temperature makes this structure more suitable for practical measurement.

1. Introduction

The optical fiber humidity sensor has the advantages of high measuring accuracy, high relative speed, simple fabrication, and compact size. Different types of humidity sensors based on fiber optical sensing have been proposed, such as polymer-coated Bragg grating fiber [1], long period fiber gratings (LPFG) [2], agarose-infiltrated photonic crystal fiber (PCF) interferometers [3], nonadiabatic tapered fiber coated with PDDA/Poly-R-478 film [4], U-shaped fiber coated by phenol red-doped PMMA [5], and interference structures [6]. Above all, the interference structures, particularly MZI, are more practical because of their compact structure, low cost, and easy fabrication and are widely used for temperature, strain, and refractive index measurement [7]. The humidity sensor based on MZI with waist enlarged structure is constructed by two waist enlarged fusion bitapers sandwiched between the SMFs. The core mode and the cladding mode have been coupled and recombined, the latter sensitive to outside environment refractive index (RI) with some humidity sensitive material coated on the SMF-MZI [8].

In the research of optical fiber humidity sensor, although optical fiber humidity sensor type and structure design are very important, humidity sensitive material also plays a key role in improving measurement performance of optical fiber humidity sensor [9]. In order to improve humidity sensor sensitivity, researchers tried to adopt various materials as humidity sensitive material, such as polyvinyl alcohol [10], metal oxide films [11], and chitosan [8]. However, all of these sensors have their own disadvantages. MC, as a special kind of high polymer chemical products, has good dispersion, thickening, film forming, and so on and is becoming a feasible material for biological sensing studies [12]. Recently, the layer-by-layer self-assembly technique has attracted significant attention because of its precise control of thickness at the nanometer level and construction on nonflat surfaces [8, 13]. Owing to its unique properties, it has been widely used for the construction of ultrathin multilayer films.

In this article, a methylcellulose coated humidity sensor based on MZI with waist enlarged structure is demonstrated. It can be monitored by intensity, which makes this sensor production of low cost and easy to make.

2. Design and Principle

A schematic diagram of a proposed humidity sensor structure is shown in Figure 1. It consists of two waist enlarged structures. The light which comes from the optical power source

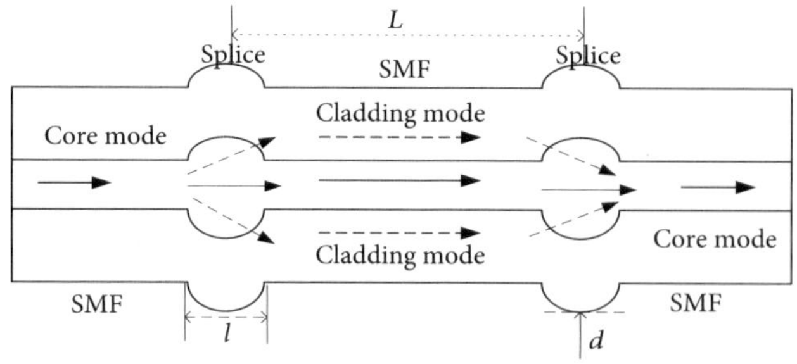

FIGURE 1: The schematic of a proposed humidity sensor structure; the inset shows the image of the MC material.

passes through the SMF to the first waist enlarged region; the cladding mode is excited and propagates inside the SMF cladding region [8]. When the two lights reach the second waist enlarged structure region, since two lights of optical path are different, interference will occur in the second waist enlarged structure region. The first waist enlarged structure is equivalent to the beam splitter, and the second waist enlarged structure plays the role of coupler.

The transmission of the interferometer can be expressed in the equation stated below [14]:

$$I = I_{\text{core}} + I_{\text{cladding}} + 2\sqrt{I_{\text{core}} I_{\text{cladding}}} \cos\left(\Phi + \Phi_0\right), \quad (1)$$

where I_{core} and I_{cladding} are the intensities of the core and cladding modes, Φ_0 is the initial phase, and Φ is the phase difference between the core mode and the cladding mode, defined as

$$\Phi = \frac{2\pi \Delta n_{\text{eff}} L}{\lambda}, \quad (2)$$

where λ is the wavelength and Δn_{eff} is the effective refractive index difference between the core and the cladding and L is the length of interference.

The change in the RI of the external environment affects the effective RI of the cladding mode and results in a variety in Δn_{eff}. Δn_{eff} leads to changes in Φ, which eventually results in variety in I.

The MC is coated on the surface of single-mode fiber, and it acts as the external environment converter of RI. The external environment will change the humidity sensitive film RI; humidity sensitive film material will affect the effective index of the cladding mode and make the different refractive index between the core and the cladding.

3. Experimental Results and Discussions

Fiber optic sensor based on waist enlarged structure is an ordinary single-mode fiber (SMF G.652). The production process was as follows: using stripping pliers to take off two single-mode fiber coating layers; cutting the end together with a cutter (EFC-6S); placing two segments of optical fiber on the FSM-60; changing the procedure of the welding machine; and setting overlap at 120 μm. The program was used to fuse two single-mode optical fibers. In the stripped-off processing before the process of splicing, the fiber ends of the SMF must be cut by a high-precision cleaver and spliced to the SMF, by a conventional fiber splicer, in a manual operation program (Fujikura FSM-40S). With precise manual adjustments, two prepared SMF stubs were butt-coupled and aligned along axis. To form a waist enlarged fiber bitaper, the splicing parameters in the manual mode must be default, but a large overlapping distance in which the two fibers are pushed further together was 80 μm for fusion splicing correspondingly [14].

By the above method, a waist enlarged structure can be obtained. Figure 2 shows microscope photograph of the sensor structure with 8x magnification. The waist enlarged structure diameter is 12.1 μm and length is 139.8 μm. The distance L between the two waists enlarged structures is 252.5 μm, which is the length of the interference.

MC has certain moisture absorption and film forming ability and can be quickly dispersed in hot water and expanded to form a definite or cloudy colloidal solution. The real image of MC is shown in Figure 3. In the experiment, 2 g MC material was dissolved in 100 ml deionized water and dissolved in a magnetic mixer for 1 hour at room temperature. After the preparation of MC solution, the deposition was achieved by dipping the sensor into the MC solution for 15 min. To create a reproducible and reliable approach for the coating process, repeating trials had to be conducted to achieve smooth and uniform coating on the SMF. Then the SMF-MZI was drawn from the solution at the rate of 30 mm/min by a stepper motor. This drawing process should be repeated for at least three times, upon the completion of which the coated SMF-MZI is placed in for 1 hour to get dried.

The experimental system of this proposed humidity sensor is shown in Figure 4. The humidity chamber is fixed, in which the MC coated SMF-MZI fiber sensing probe has been held by two clamps and referential humidity meter (TRACEABLE VWR L207080). The fixed chamber could avoid sensor movement and fiber refixing in measurements, which is very important in decreasing the influence of bending and twisting. The corresponding transmission spectra of the sensor at different humidity levels are measured by a combination of a broadband light source (BBS, Asef15296410) and an optical spectrum analyzer (OSA, Yokogawa AQ6370) with a resolution of 0.02 nm. In the humidity measurement experiment, the sensing head is put into a closed humidity control box; the temperature of the humidity control box is kept at a room temperature of about 25°C. In order to measure

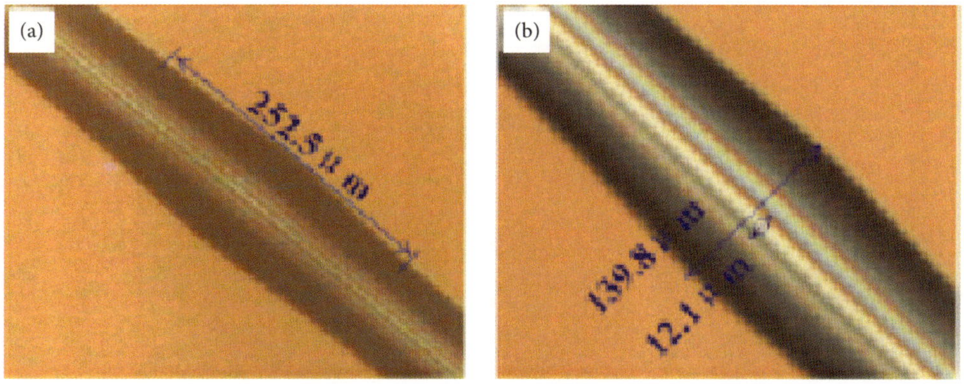

(a) (b)

Figure 2: Microscope photograph of the sensor structure with 8x magnification.

Figure 3: Real image of MC.

the response of the sensor head to humidity, the humidity in the humidity control box is controlled by pumping dry or moist gas into it, and the humidity gauge in the humidity control box is used to monitor the change of humidity in the humidity control box. The change of interference spectrum with the change of humidity in the humidity control box can be observed by the OSA.

In the humidity test, both ends of the sensor are fixed. The wavelength shift of interference spectrum is monitored between the range from 35% to 85% RH and the measured data is recorded when the humidity is kept stable in the humidity control box. Figure 5 shows that the interference intensity increases with the increasing of the humidity. The result shows the intensity dip shift from −11.36 to −7.32 dB with a total shift of 4.04 dB. As showed in Figure 6, the wavelength shift of the dip has a linear relationship with the refractive index changing, for which the sensitivity of the sensor is 0.094 dB/%RH.

Because of the change in the refractive index of MC coating, the change rate of the wavelength is higher than that of the intensity, which is completely consistent with Lorenz theoretical analysis [15]. The MC refractive index will rise with the increase of humidity at the temperature below 50°C. When MC is expanded in the water, its refractive index will reach the cladding one, which leads to the fact that most higher-order modes in the cladding will not be guided by the reducing intensity of transmitted light. The shift of the spectra is caused by the change of MC in water swelling, which will affect the effective refractive indices of the fast and slow axis of the SMF-MZI; interference is modulated by the RH and brought in the shifting of the spectra.

The experimental result showed that when the ambient humidity is increased, MC will absorb water and the RI will decrease. When the moisture sensitive film absorbs more water molecules, the RI of the humidity sensitive film itself will be reduced and the intensity of the interference spectrum will be increased, which is completely consistent with [8] analysis. According to [8], the change of RH will alter the effective RI of both bare cladding and MC coating. With the increase of RH, more water molecules will diffuse into MC coating, which leads to the MC swelling and increase in the thickness of coating. Same as other swelling polymers, the increase of polymer in water will decrease the RI of MC coating. Meanwhile, the thickness of the coating also will improve the effective RI of the cladding modes. When RH is above 70% RH, the changes of the MC thickness will increase effective RI and result in the increase in extinction ratio of the spectrum. When the MC coating RI is equal to the RI of the fiber cladding, the sensitivity increases greatly [16].

As shown in Figure 7, whether the humidity increases or decreases, the relative difference between the intensities is less than 0.2 dB, which proved that the proposed sensor has a good stability. In order to further verify the stability of the proposed humidity sensor, the experiment coated with MC film formed by 10 mg/ml MC solution was done under 55% and 75% fixed RH level, respectively. The experiment results in Figure 8 showed that only small fluctuations (maximum 0.3% at 55% RH and 0.7% at 75% RH) were recorded in the two cases over a period of 60 min. As we know, MC is one kind of the high polymer and the space structure is long chain, which has curve revolving ability and many active groups in MC [17]. By adsorbing colloidal particles with active groups, the polymer has different configurations, thus forming a network structure. Just this main unique property in MC makes MC coated humidity sensor have better stability and offer perfect film forming capability.

The temperature is a vital parameter in the measurement of fiber humidity. Figure 9 shows that when the temperature was increased from 20°C to 90°C, the intensity only had

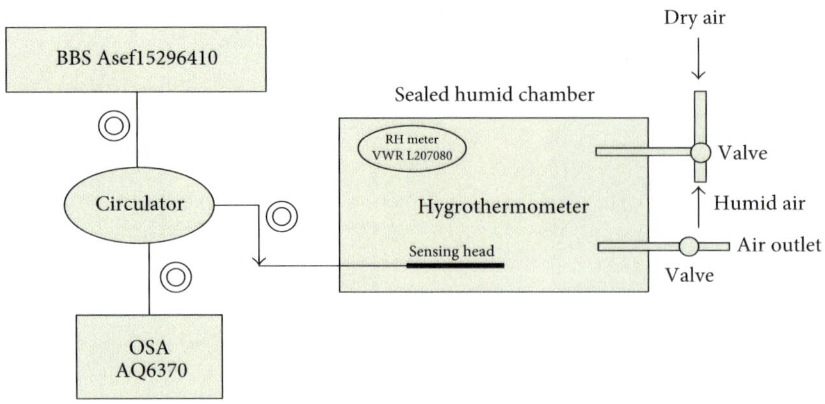

FIGURE 4: Experimental setup for the proposed humidity sensor.

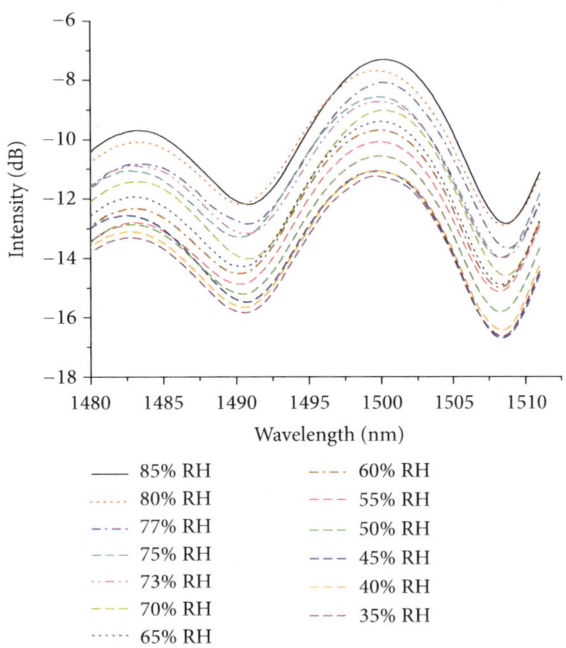

FIGURE 5: The interference spectrum with the change of humidity in the external environment sensor.

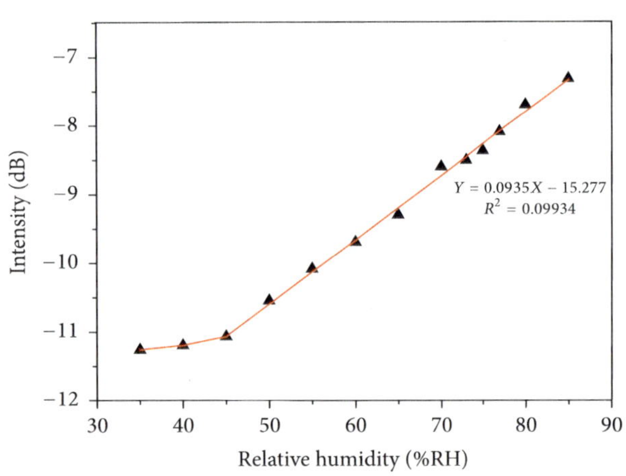

FIGURE 6: The relationship between the intensity change and humidity.

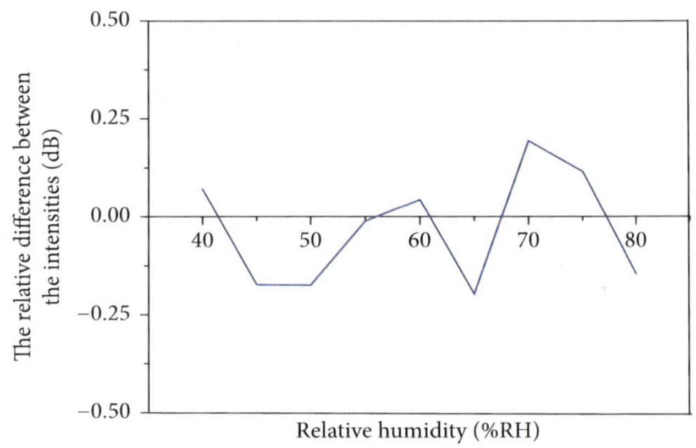

FIGURE 7: Relation of average relative difference of intensity and humidity.

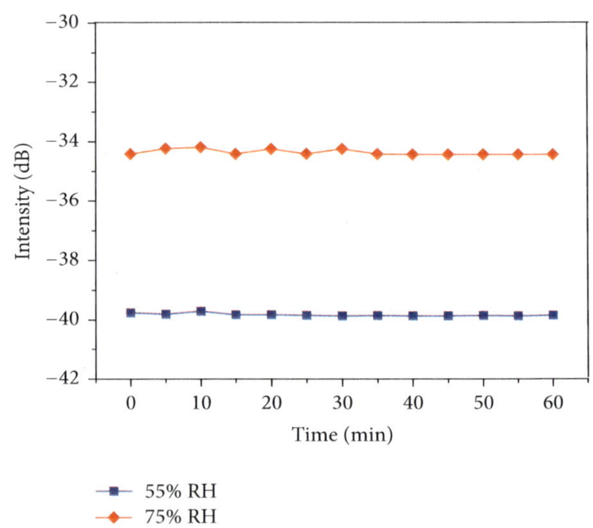

FIGURE 8: Stability testing results.

a 0.03 dB change; the maximum intensity change was only 0.16 dB, which testified that the sensor is insensitive to the temperature.

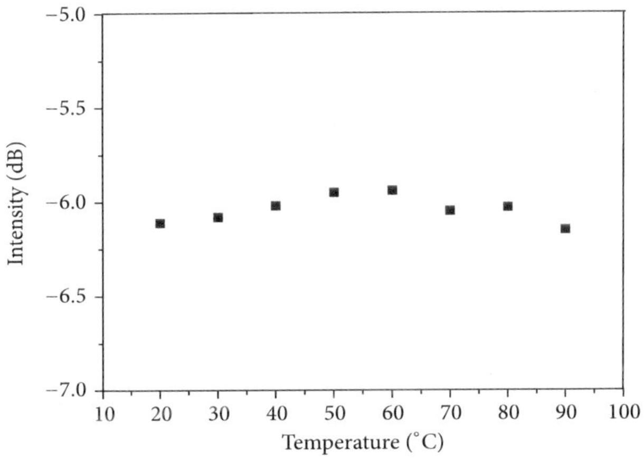

FIGURE 9: The variation of temperature and wavelength in the external environment.

4. Conclusion

A MC coated humidity sensor based on Mach-Zehnder interferometer with waist enlarged structure has been proposed and experimentally demonstrated. The waist enlarged structure could effectively couple guided light from the fiber core mode to the high-order cladding modes. When humidity is increased or decreased, the average relative difference between the intensities is 0.03 dB. There is a good linear relationship between humidity and intensity; the linear fit is as high as 0.9934 and has a good stability. The maximum sensitivity is up to 0.094 dB/%RH, over a range of 45% to 85% RH. In the meantime, low cost, compact size, easy fabrication, and the insensitivity to temperature make it suitable for the practical applications.

Conflicts of Interest

The authors declare that they have no conflicts of interest.

Acknowledgments

This work has been supported by the National Natural Science Foundation of China (no. 11404310).

References

[1] T. L. Yeo, T. Sun, K. T. V. Grattan, D. Parry, R. Lade, and B. D. Powell, "Characterisation of a polymer-coated fibre Bragg grating sensor for relative humidity sensing," *Sensors and Actuators B: Chemical*, vol. 110, no. 1, pp. 148–156, 2005.

[2] T. Venugopalan, T. L. Yeo, T. Sun, and K. T. V. Grattan, "LPG-based PVA coated sensor for relative humidity measurement," *IEEE Sensors Journal*, vol. 8, no. 7, pp. 1093–1098, 2008.

[3] J. Mathew, Y. Semenova, and G. Farrell, "Relative humidity sensor based on an agarose-infiltrated photonic crystal fiber interferometer," *IEEE Journal of Selected Topics in Quantum Electronics*, vol. 18, no. 5, pp. 1553–1559, 2012.

[4] T. Venugopalan, T. Sun, and K. T. V. Grattan, "Long-period grating-based humidity sensor for potential structural health monitoring," *Sensors and Actuators A: Physical*, vol. 148, no. 1, pp. 57–62, 2008.

[5] B. D. Gupta and Ratnanjali, "A novel probe for a fiber optic humidity sensor," *Sensors and Actuators B: Chemical*, vol. 80, no. 2, pp. 132–135, 2001.

[6] M. Shao, X. Qiao, H. Fu, N. Zhao, Q. Liu, and H. Gao, "An In-Fiber Mach–Zehnder Interferometer Based on Arc-Induced Tapers for High Sensitivity Humidity Sensing," *IEEE Sensors Journal*, vol. 13, no. 5, pp. 2026–2031, 2013.

[7] K. Ni, C. C. Chan, L. Chen, X. Dong, R. Huang, and Q. Ma, "A chitosan-coated humidity sensor based on Mach-Zehnder interferometer with waist-enlarged fusion bitapers," *Optical Fiber Technology*, vol. 33, pp. 56–59, 2017.

[8] Q. Ma, K. Ni, and R. Huang, "A carboxy-methyl cellulose coated humidity sensor based on Mach-Zehnder interferometer with waist-enlarged bi-tapers," *Optical Fiber Technology*, vol. 33, pp. 60–63, 2017.

[9] W. C. Wong, C. C. Chan, L. H. Chen, T. Li, K. X. Lee, and K. C. Leong, "Polyvinyl alcohol coated photonic crystal optical fiber sensor for humidity measurement," *Sensors and Actuators B: Chemical*, vol. 174, pp. 563–569, 2012.

[10] R. K. Verma and B. D. Gupta, "Surface plasmon resonance based fiber optic sensor for the IR region using a conducting metal oxide film," *Journal of the Optical Society of America A: Optics and Image Science, and Vision*, vol. 27, no. 4, pp. 846–851, 2010.

[11] L. I. Ya-Juan, G. Y. Dang, and J. D. Fei, "Review of light intensity modulated humidity sensor," *Transducer & Microsystem Technologies*, vol. 28, pp. 5–12, 2009.

[12] F. J. Arregui, Y. Liu, I. R. Matias, and R. O. Claus, "Optical fiber humidity sensor using a nano Fabry-Perot cavity formed by the ionic self-assembly method," *Sensors and Actuators B: Chemical*, vol. 59, no. 1, pp. 54–59, 1999.

[13] Y. Wang, S. Wang, L. Jiang et al., "Temperature-insensitive refractive index sensor based on Mach-Zehnder interferometer with two microcavities," *Chinese Optics Letters*, vol. 15, no. 2, Article ID 020603, 2017.

[14] Y. Geng, X. Li, X. Tan, Y. Deng, and Y. Yu, "High-sensitivity Mach-Zehnder interferometric temperature fiber sensor based on a Waist-Enlarged fusion bitaper," *IEEE Sensors Journal*, vol. 11, no. 11, pp. 2891–2894, 2011.

[15] L. H. Chen, T. Li, C. C. Chan et al., "Chitosan based fiber-optic Fabry-Perot humidity sensor," *Sensors and Actuators B: Chemical*, vol. 169, pp. 167–172, 2012.

[16] P. Hu, X. Dong, K. Ni, L. H. Chen, W. C. Wong, and C. C. Chan, "Sensitivity-enhanced Michelson interferometric humidity sensor with waist-enlarged fiber bitaper," *Sensors and Actuators B: Chemical*, vol. 194, pp. 180–184, 2014.

[17] Z. Tong and H. Dandan, "Effect of methylcellulose on the properties of silica membrane," *Journal of Membrane Science & Technology*, vol. 33, no. 1, pp. 39–43, 2013.

5

Optical Bistability in a Controllable Giant Self-Kerr Nonlinear Gaseous Medium under Electromagnetically Induced Transparency and Doppler Broadening

Phuong Le Thi Minh,[1,2] Doai Le Van,[1] Khoa Dinh Xuan,[1] and Bang Nguyen Huy[1]

[1] *Vinh University, Le Duan 182, Vinh City, Vietnam*
[2] *Saigon University, 273 An Duong Vuong, Ho Chi Minh City, Vietnam*

Correspondence should be addressed to Bang Nguyen Huy; bangnh@vinhuni.edu.vn

Academic Editor: Jiayang Wu

We study optical bistability (OB) in a controllable giant self-Kerr nonlinear atomic gaseous medium placed in a unidirectional ring cavity. The medium is coherently excited by strong controlling field and a weak probe laser field under electromagnetically induced transparency (EIT) and Doppler broadening. In a weak field limit of the probe light, an analytic OB equation for the probe light field is derived as an analytic function of parameters of the controlling field and temperature of the medium. It is shown that OB characters can be manipulated with the parameters due to the controllable properties of the self-Kerr nonlinearity. Furthermore, enhancement of the Kerr nonlinearity reduces the switching intensity threshold and width of the OB.

1. Introduction

The OB is one of the most interesting fields of research in nonlinear optics because it has a large number of potential applications in both optical sciences and photonic technology, for example, all-optical switches, all-optical memories, optical transistors, and all-optical logic gates [1–3]. For the OB systems using Kerr nonlinear materials, a large nonlinearity is a fundamental need to gain sensitivity. In the early years of the OB research for atomic media, a great interest was focused on the two-level atomic system due to resonantly enhanced nonlinearity [1, 2]. However, applications of the two-level atomic system are limited due to strong resonant absorption and only one optical field being employed for both applying and switching, thus lacking control for switching intensity thresholds.

The advent of EIT [4] offers coherent media of vanished resonant absorption [5–8] and controllable giant Kerr nonlinearity [9–11]. Wang et al. performed a direct measurement of self-Kerr nonlinear coefficient in a three-level lambda EIT medium in the presence of Doppler broadening by using a cavity scanning technique. The measurement shows a great enhancement of the Kerr nonlinear index of refraction in several orders of magnitude around atomic resonance [12]. Recently, Doai et al. [13] have developed an analytical model of EIT enhanced self-Kerr nonlinearity with controllable parameters of a light field and temperature of atomic sample which is in good agreement with the experimental observations [12]. The studies showed that the Kerr nonlinearity is greatly enhanced and depends on temperature of the medium. Furthermore, the magnitude, slop, and sign of the Kerr nonlinear coefficient can be controlled by tuning frequency and/or intensity of the coupling light or by changing temperature of the medium. Such controllable giant Kerr nonlinear media are used for controllable optical bistability [14], generating four-wave mixing beams [15] which exploit new ways in designing devices for optical switching in optical communication and all-optical signal processing.

Therefore, owing to the controllable giant-nonlinearity of the EIT medium, switching intensity thresholds and width of the OB can be manipulated. For the first time, Joshi et al. [14] demonstrated experimentally the controllable atomic optical bistability in the three-level EIT system inside an optical ring cavity in the presence of Doppler broadening. Since then, aspects that influence OB hysteresis loop have been extensively investigated [16–26]. Similar optical bistability and vacuum Rabi splitting of ring cavity in EIT medium are well known theoretically and experimentally [27–29]. Several reviews [1, 2] and monographs [3] on this field of the OB are available, which give a deeper insight into the topic and provide lists of original references.

In the OB studies related to the EIT atomic gaseous medium, it can be categorized into two types of OB which concern the dispersive and absorptive properties of the medium, respectively [3]. The dispersive OB type arises from the linear and nonlinear contributions, among which the giant Kerr nonlinearity often dominates because it is enhanced by atomic coherence [12, 13]. However, study of OB concerning Kerr nonlinearity and temperature of such medium is still quite modest. Furthermore, in several cases, a precise knowledge of the relationship of the input and output intensities as a function of the controlling light and temperature of the medium is crucially needed to optimize the operating processes and to choose experimental parameters. With the growth of this interest, in this work, we studied controlling OB of the three-level EIT atomic gaseous medium under presence of giant self-Kerr nonlinearity and Doppler broadening by using an analytic method. Using the weak field approximation of the probe light, a relationship of output and input intensities of OB is derived in the presence of Doppler broadening. The influences of Doppler broadening and the controlling light on switching intensity thresholds and width of the OB are investigated.

2. The Mater Equation of OB

We consider an atomic medium (with length L) composed of three-level Λ-type atoms with density N placed in unidirectional ring cavity, as shown in Figure 1(a). A weak probe laser field with frequency ω_p and electric field amplitude E_p drives the transition $|1\rangle \leftrightarrow |2\rangle$, whereas an intense coupling laser field with frequency ω_c and electric field amplitude E_c couples a transition between the states $|2\rangle$ and $|3\rangle$.

For simplicity, we assumed both mirrors 1 and 2 are perfect reflectors; each has a reflectivity R and transmittivity T, with $R + T = 1$. In the unidirectional ring cavity, a part of the probe field E_p is circulated in the cavity but not the coupling field E_c. The coherent field E_p^I enters through mirror M_1, interacts with the atomic medium of the length L, circulates in the cavity, and partially comes out of the mirror M_2 as E_p^T. Part of the output intensity reflected back into the medium provides the feedback which is essential for generating OB.

The relationship of the output and input intensities of the unidirectional ring cavity or master equation of OB due to the self-Kerr nonlinearity is given by [30]:

$$I_{\text{out}} = \left[\frac{1}{2} + \frac{1}{2}\cos\left(\frac{2\pi L}{\lambda}n_2 I_{\text{out}} + \varphi\right)\right] I_{\text{in}}, \tag{1}$$

where n_2 is the Kerr coefficient, λ is wavelength of the probe light, $I_{\text{in}} \propto (E_p^I)^2$ and $I_{\text{out}} \propto (E_p^T)^2$ are incident and transmitted intensities of the probe light, and

$$\varphi = \frac{2\pi L}{\lambda}n_0 + \varphi_0 \tag{2}$$

is the round-trip phase of the probe light, φ_0 is the initial phase, and n_0 is the linear index of refraction.

The self-Kerr coefficient n_2 can be determined by using the density matrix formalism. Under the framework of the semiclassical theory and dipole and rotating wave approximations, the evolution of the three-level system can be represented by [13]

$$\begin{aligned}
\dot{\rho}_{11} &= \gamma_{31}\left(\rho_{33} - \rho_{11}\right) + \gamma_{21}\rho_{22} - \frac{i}{2}\Omega_p\rho_{21} + \frac{i}{2}\Omega_p\rho_{12}, \\
\dot{\rho}_{22} &= -\left(\gamma_{23} + \gamma_{21}\right)\rho_{22} - \frac{i}{2}\Omega_p\rho_{12} + \frac{i}{2}\Omega_p\rho_{21} - \frac{i}{2}\Omega_c\rho_{32} \\
&\quad + \frac{i}{2}\Omega_c\rho_{23}, \\
\dot{\rho}_{33} &= \gamma_{31}\left(\rho_{11} - \rho_{33}\right) + \gamma_{23}\rho_{22} + \frac{i}{2}\Omega_c\rho_{32} - \frac{i}{2}\Omega_c\rho_{23}, \\
\dot{\rho}_{21} &= -\left(\gamma - i\Delta_p\right)\rho_{21} + \frac{i}{2}\Omega_p\left(\rho_{22} - \rho_{11}\right) - \frac{i}{2}\Omega_c\rho_{31}, \\
\dot{\rho}_{23} &= -\left(\gamma - i\Delta_c\right)\rho_{23} + \frac{i}{2}\Omega_c\left(\rho_{22} - \rho_{33}\right) - \frac{i}{2}\Omega_p\rho_{13}, \\
\dot{\rho}_{31} &= -\left[\gamma_{31} - i\left(\Delta_p - \Delta_c\right)\right]\rho_{31} + \frac{i}{2}\Omega_p\rho_{32} - \frac{i}{2}\Omega_c\rho_{21},
\end{aligned} \tag{3}$$

where

$$\begin{aligned}
\Delta_p &= \omega_p - \omega_{21}, \\
\Delta_c &= \omega_c - \omega_{23},
\end{aligned} \tag{4}$$

is frequency detunings of the probe coupling lasers, respectively; $\Omega_p = d_{21}E_p/\hbar$ and $\Omega_c = d_{23}E_c/\hbar$ are the Rabi frequencies associated with the probe and coupling beams, respectively; d_{lm} represents the dipole moment of the $|l\rangle - |m\rangle$ transition; γ_{lm} is represented with the decay rates Γ_{lm} from $|l\rangle$ to $|m\rangle$ as follows [13]:

$$\begin{aligned}
\gamma_{lm} &= \frac{1}{2}\left(\sum_{k<l}\Gamma_{lk} + \sum_{j<m}\Gamma_{mj}\right), \\
\gamma &= \frac{\gamma_{21} + \gamma_{31} + \gamma_{23}}{2}.
\end{aligned} \tag{5}$$

In (3), the diagonal matrix elements ρ_{mm} ($m = 1 \div 3$) represent population in the state $|m\rangle$, whereas the off-diagonal matrix elements ρ_{lm} ($l \neq m$) represent the coherence between the states $|l\rangle$ and $|m\rangle$. Using perturbation theory, the matrix element ρ_{21} is calculated up to third order as [13]

$$\begin{aligned}
\rho_{21} &= \frac{-i\Omega_p}{2F}\left(1 - \frac{2\gamma_{31}}{2\gamma + \gamma_{21}}\right) \\
&\quad + \frac{i\Omega_p}{2F}\frac{\Omega_p^2}{2\gamma + \gamma_{21}}\left(\frac{1}{F} + \frac{1}{F^*}\right),
\end{aligned} \tag{6}$$

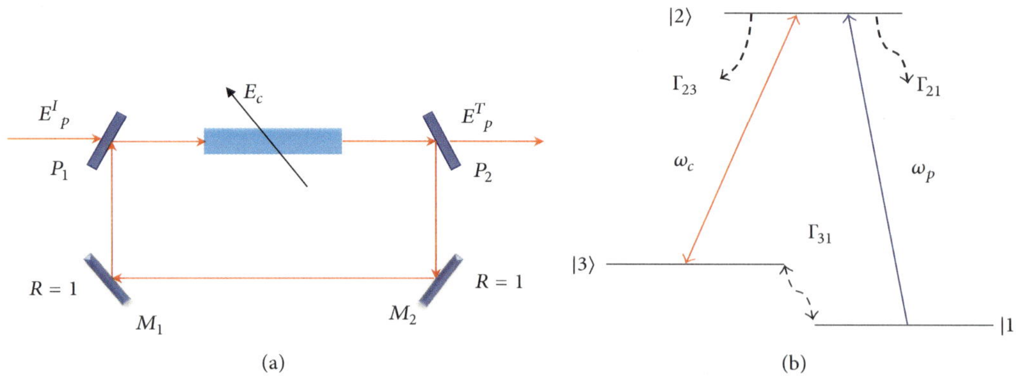

FIGURE 1: Schematic setup of a unidirectional ring cavity (a) containing a three-level atomic sample (b), where E_p^I and E_p^T denote the incident and transmitted probe field, respectively, and E_c represents the coupling field that is not circulated inside the cavity.

where

$$F = \gamma - i\Delta_p + \frac{\Omega_c^2/4}{\gamma_{31} - i\left(\Delta_p - \Delta_c\right)}, \quad (7)$$

and F^* is the complex conjugation of F.

For matrix element ρ_{21}, the total susceptibility χ for the probe light is determined by

$$\chi = -2\frac{Nd_{21}}{\varepsilon_0 E_p}\rho_{21}. \quad (8)$$

In alternative form, the total susceptibility χ can be represented with the first-order and the third-order susceptibilities ($\chi^{(1)}$ and $\chi^{(3)}$) by the following way:

$$\chi = \chi^{(1)} + 3E_p^2\chi^{(3)}. \quad (9)$$

For a gaseous phase, it is necessary to take into account Doppler broadening. In such case, the linear and third-order nonlinear susceptibilities can be derived as [13]

$$\chi^{(1)} = \frac{iN_0 d_{21}^2\sqrt{\pi}}{\varepsilon_0\hbar\left(\omega_p u/c\right)}\left(1 - \frac{2\gamma_{31}}{2\gamma + \gamma_{21}}\right)e^{z^2}\left[1 - \text{erf}\left(z\right)\right],$$

$$\chi^{(3)} = -\frac{iN_0 d_{21}^4}{3\sqrt{\pi}\varepsilon_0\hbar^3\left(\omega_p u/c\right)^2}\left(\frac{1}{2\gamma + \gamma_{21}}\right) \times \left\{2\sqrt{\pi}\left(-1 + \sqrt{\pi}ze^{z^2}\left[1 - \text{erf}\left(z\right)\right]\right) + \frac{\pi\left(e^{z^2}\left[1 - \text{erf}\left(z\right)\right] + e^{z^{*2}}\left[1 - \text{erf}\left(z^*\right)\right]\right)}{z + z^*}\right\}, \quad (10)$$

where $u = \sqrt{2k_B T/m}$ is the root mean square velocity, N_0 is the total atomic density of the vapor,

$$z = \frac{c}{\omega_p u}\left(\gamma - i\Delta_p + \frac{\Omega_c^2/4}{\gamma_{31} - i\left(\Delta_p - \Delta_c\right)}\right) = \frac{c}{\omega_p u}F, \quad (11)$$

z^* is the complex conjugation of z, and erf is the error function.

Having the linear and third-order nonlinear susceptibilities, a nonlinear Kerr index of refraction n_2 is determined by the following relations:

$$n_2 = \frac{3\text{Re}\left(\chi^{(3)}\right)}{4\varepsilon_0 n_0^2 c}, \quad (12)$$

where $n_0 = \sqrt{1 + \text{Re}(\chi^{(1)})}$ is the linear index of refraction.

The self-Kerr nonlinear coefficient given be (12) represents a function of the intensity and frequency detuning of the coupling light and temperature of the medium. By substituting expression (12) into master equation (1) the OB behaviors can be investigated with the parameters.

3. Analysis of OB

In order to analyze the analytic result of OB, we consider the ^{87}Rb atomic gaseous medium. In this case, the states $|1\rangle$, $|2\rangle$, and $|3\rangle$ are chosen as $5S_{1/2}$ ($F = 1$), $5P_{1/2}$ ($F = 2$), and $5S_{1/2}$ ($F = 2$), respectively. We used values for the parameters as follows (in which the frequencies and decay rates are given in the unit of 2π) [12, 31]: $\lambda = 780 \times 10^{-9}$ m, $\omega_p = 377 \times 10^{12}$ Hz, $\gamma_{21} = 3$ MHz, $\gamma_{31} = 1.1$ MHz, $\gamma = 3.5$ MHz, and $N = 4.5 \times 10^{17}$ atoms/m^3, $d_{21} = 1.6 \times 10^{-29}$ C·m.

In the first step, we plotted a surface of OB at $\Delta_c = 0$, $\Omega_c = 70$ MHz and $T = 300$ K and plotted OB curves at particular values of the probe frequency, as shown in Figures 2(a) and 2(b). It is shown that the threshold intensity and width of OB decrease as the probe frequency moves to the blue side. To explain this case, we plotted the Kerr nonlinear coefficient n_2 with the respect to the probe frequency detuning Δ_p (Figure 2(c)).

Indeed, the value of n_2 is given at $\Delta_p = -10$ MHz, -15 MHz, and -20 MHz is 7.2×10^{-6} cm^2/W, 5.1×10^{-6} cm^2/W, and 3.2×10^{-6} cm^2/W, respectively. The results show that, in the region of Δ_p from -20 MHz to 10 MHz, amplitude of the Kerr nonlinearity increases as frequency of the probe field moves to the blue and thus lowers the OB threshold

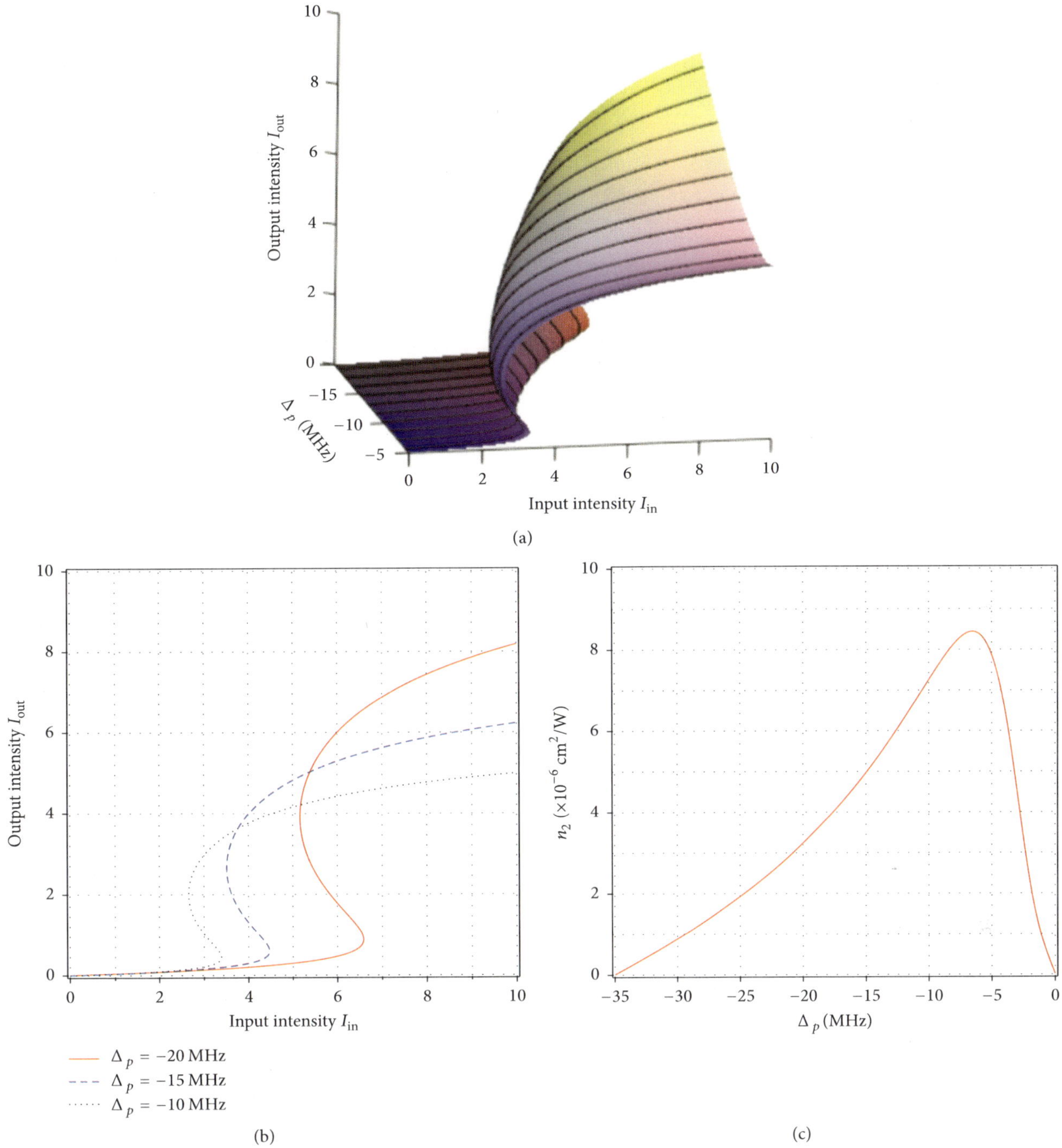

FIGURE 2: (a) Surface plot of the OB; (b) OB curves at particular values of the probe frequency detuning; (c) variation of the Kerr nonlinear coefficient n_2 versus the probe detuning. All the plots given at the fixed values $\Delta_c = 0$, $\Omega_c = 70$ MHz, and $T = 300$ K.

and OB width. Furthermore, for the given parameters of the coupling light and temperature of the medium, one can find a frequency region from $\Delta_p = -35$ MHz to 0 MHz with positive value of the self-Kerr nonlinearity which can be used for OB (see Figure 2(c)). It is worth emphasizing that the magnitude of self-Kerr coefficient here reaches order 10^{-5} cm^2/W that is much larger than that of the conventional Kerr materials [26].

Next, we investigated influences of the frequency and intensity of the coupling field by making plots of OB curves at particular values of these parameters, as shown in Figures 3 and 4, respectively. In the first case, the threshold and width of the OB change with variation of frequency of the coupling field. This can be attributed (see Figure 3(b)) to changing the Kerr nonlinearity. Indeed, the values of n_2 at $\Delta_c = 3$ MHz, 7 MHz, and 15 MHz are determined as 3.7×10^{-6} cm^2/W,

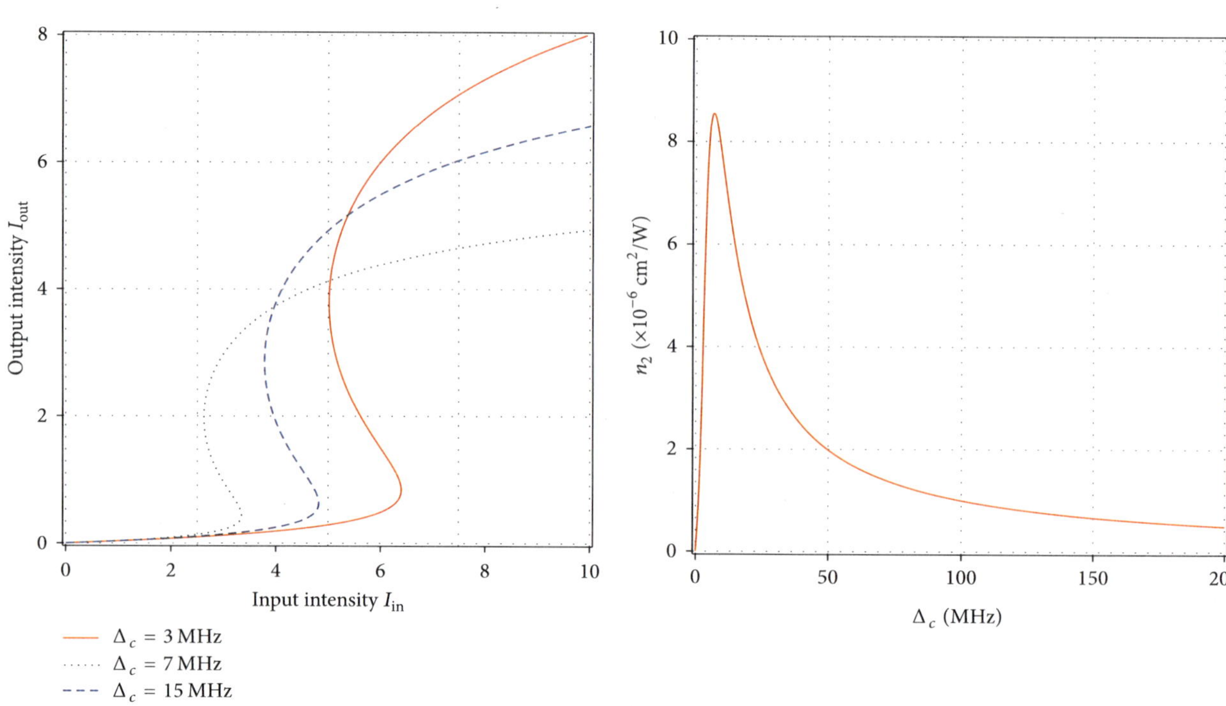

(a) (b)

FIGURE 3: OB curves at particular values of frequency detuning of the coupling field (a) and variation of Kerr coefficient n_2 versus the coupling detuning (b). The parameters are $\Omega_c = 70$ MHz, $\Delta_p = 0$, and $T = 300$ K.

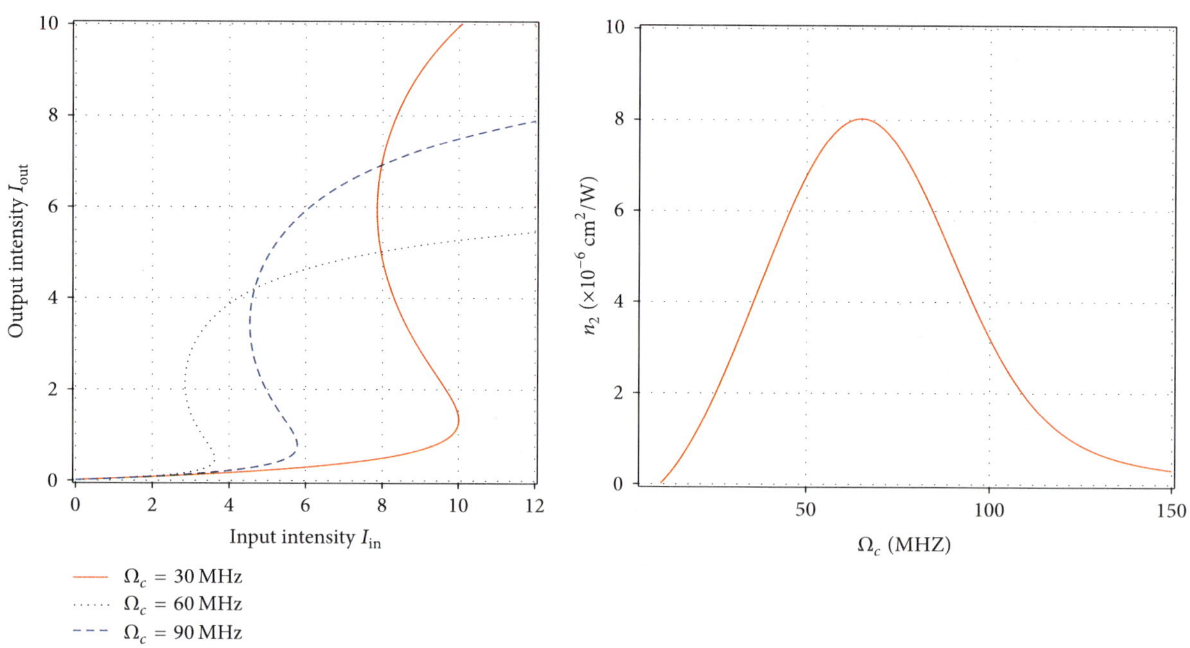

(a) (b)

FIGURE 4: OB curves at particular values of intensity of the coupling field (a) and variation of Kerr coefficient n_2 versus intensity of the coupling field (b). The parameters are $\Delta_p = -5$ MHz, $\Delta_c = 0$, and $T = 300$ K.

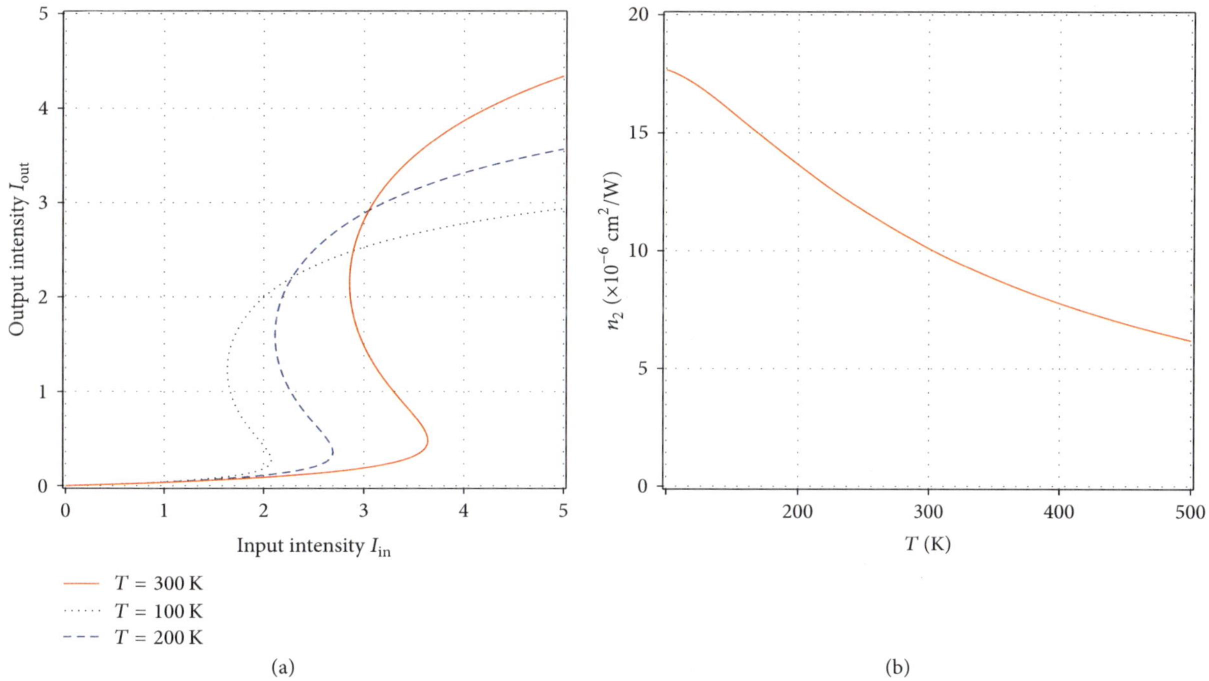

FIGURE 5: The OB curves at particular values of the temperature (a) and variation of Kerr nonlinear coefficient n_2 versus the temperature (b). The parameters are $\Delta_p = -5$ MHz, $\Delta_c = 0$, and $\Omega_c = 70$ MHz.

8.5×10^{-6} cm^2/W, and 5.9×10^{-6} cm^2/W, respectively. The case of maximum value of the Kerr nonlinearity corresponds to the lowest threshold and width of the OB. Consequently, the threshold and width of the OB decrease as growing magnitude of Kerr nonlinearity. This behavior is as same as the second case which is indicated in Figure 4. Here, the lowest threshold and width of the OB are obtained at $\Omega_c = 60$ MHz corresponding to the largest value of the Kerr nonlinearity.

Finally, we studied influence of Doppler broadening by plotting OB curves at particular values of temperature and plotted variation of the Kerr nonlinearity, as shown in Figure 5. Similar to the cases in Figures 2–4, the results here show that growing temperature leads to reduction of Kerr nonlinearity and thus increases the threshold and width of the OB.

4. Conclusion

We have studied the OB behaviors of a unidirectional ring cavity containing an EIT enhanced self-Kerr nonlinear medium under Doppler broadening. In a weak field limit of the probe light, an analytic OB equation was derived as a function of controllable parameters. It is shown that, owing to the controllable Kerr nonlinearity, the characters of the OB can be manipulated with intensity and frequency of the coupling field and temperature of the medium. Furthermore, enhancement of the Kerr nonlinearity leads to reduction of the switching intensity thresholds and width of the OB. The analytic model is helpful to understand OB phenomena using Kerr nonlinear gaseous material, and it is suitable for finding the related OB applications at low light intensity.

Conflicts of Interest

The authors declare that they have no conflicts of interest.

Acknowledgments

The financial support from the Vietnam Ministry of Science and Technology through the Grant no. ĐTĐLCN.17/17 is acknowledged.

References

[1] L. A. Lugiato, "Theory of optical bistability," in *Progress in Optics*, E. Wolf, Ed., vol. 21, pp. 71–216, 1984.

[2] H. M. Gibbs, *Optical Bistability: Controlling Light with Light*, Academic Press, New York, NY, USA, 1985.

[3] A. Joshi and M. Xiao, *Controlling Steady-State and Dynamical Properties of Atomic Optical Bistability*, World Scientific Publishing, 2012.

[4] K.-J. Boller, A. Imamolu, and S. E. Harris, "Observation of electromagnetically induced transparency," *Physical Review Letters*, vol. 66, no. 20, pp. 2593–2596, 1991.

[5] M. Fleischhauer, A. Imamoglu, and P. J. Marangos, "Electromagnetically induced transparency," *Reviews of Modern Physics*, vol. 77, no. 2, pp. 633–673, 2005.

[6] L. V. Doai, P. V. Trong, D. X. Khoa, and N. H. Bang, "Electromagnetically induced transparency in five-level cascade scheme of 85Rb atoms: an analytical approach," *International Journal for Light and Electron Optics*, vol. 125, no. 14, pp. 3666–3669, 2014.

[7] D. X. Khoa, P. V. Trong, L. V. Doai, and N. H. Bang, "Electromagnetically induced transparency in a five-level cascade system under Doppler broadening: an analytical approach," *Physica Scripta*, vol. 91, no. 3, Article ID 035401, 2016.

[8] D. X. Khoa, L. C. Trung, P. Van Thuan, L. Van Doai, and A. N. H. Bang, "Measurement of dispersive profile of a multiwindow electromagnetically induced transparency spectrum in a Doppler-broadened atomic medium," *Journal of the Optical Society of America B: Optical Physics*, vol. 34, no. 6, pp. 1255–1263, 2017.

[9] H. Schmidt and A. Imamoğlu, "Giant Kerr nonlinearities obtained by electromagnetically induced transparency," *Optics Expresss*, vol. 21, no. 23, pp. 1936–1938, 1996.

[10] S. E. Harris, "Nonlinear optics at low light levels," *Physical Review Letters*, vol. 82, no. 23, pp. 4611–4614, 1999.

[11] D. X. Khoa, L. Van Doai, D. H. Son, and N. H. Bang, "Enhancement of self-Kerr nonlinearity via electromagnetically induced transparency in a five-level cascade system: an analytical approach," *Journal of the Optical Society of America B: Optical Physics*, vol. 31, no. 6, pp. 1330–1334, 2014.

[12] H. Wang, D. Goorskey, and M. Xiao, "Enhanced Kerr nonlinearity via atomic coherence in a three-level atomic system," *Physical Review Letters*, vol. 87, no. 7, Article ID 073601, p. 073601/4, 2001.

[13] L. V. Doai, D. X. Khoa, and N. H. Bang, "EIT enhanced self-Kerr nonlinearity in the three-level lambda system under Doppler broadening," *Physica Scripta*, vol. 90, no. 4, Article ID 045502, 2015.

[14] A. Joshi, A. Brown, H. Wang, and M. Xiao, "Controlling optical bistability in a three-level atomic system," *Physical Review A: Atomic, Molecular and Optical Physics*, vol. 67, no. 4, p. 4, 2003.

[15] Y. Zhang, Z. Wang, Z. Nie et al., "Four-wave mixing dipole soliton in laser-induced atomic gratings," *Physical Review Letters*, vol. 106, no. 9, Article ID 093904, 2011.

[16] A. Joshi, W. Yang, and M. Xiao, "Effect of quantum interference on optical bistability in the three-level V-type atomic system," *Physical Review A: Atomic, Molecular and Optical Physics*, vol. 68, no. 1, p. 015806, 2003.

[17] J. Li, "Coherent control of optical bistability in a microwave-driven V-type atomic system," *Physica D: Nonlinear Phenomena*, vol. 228, no. 2, pp. 148–152, 2007.

[18] Z. Wang, A.-X. Chen, Y. Bai, W.-X. Yang, and R.-K. Lee, "Coherent control of optical bistability in an open Λ-type three-level atomic system," *Journal of the Optical Society of America B: Optical Physics*, vol. 29, no. 10, pp. 2891–2896, 2012.

[19] J.-H. Li, X.-Y. Lü, J.-M. Luo, and Q.-J. Huang, "Optical bistability and multistability via atomic coherence in an N -type atomic medium," *Physical Review A: Atomic, Molecular and Optical Physics*, vol. 74, no. 3, Article ID 035801, 2006.

[20] X.-Y. Lü, J.-H. Li, J.-B. Liu, and J.-M. Luo, "Optical bistability via quantum interference in a four-level atomic medium," *Journal of Physics B: Atomic, Molecular and Optical Physics*, vol. 39, no. 24, article 013, pp. 5161–5171, 2006.

[21] M. Sahrai, H. R. Hamedi, and M. Memarzadeh, "Kerr nonlinearity and optical multi-stability in a four-level Y-type atomic system," *Journal of Modern Optics*, vol. 59, no. 11, pp. 980–987, 2012.

[22] H. R. Hamedi, S. H. Asadpour, M. Sahrai, B. Arzhang, and D. Taherkhani, "Optical bistability and multi-stability in a four-level atomic scheme," *Optical and Quantum Electronics*, vol. 45, no. 3, pp. 295–306, 2013.

[23] J.-H. Li, "Controllable optical bistability in a four-subband semiconductor quantum well system," *Physical Review B: Condensed Matter and Materials Physics*, vol. 75, no. 15, Article ID 155329, 2007.

[24] L. Ebrahimi Zohravi, R. Doostkam, S. M. Mousavi, and M. Mahmoudi, "Controlling the optical bistability in a kobrak-rice 5-level quantum system," *Progress in Electromagnetics Research M*, vol. 25, pp. 1–11, 2012.

[25] D. X. Khoa, L. V. Doai, L. N. M. Anh et al., "Optical bistability in a five-level cascade EIT medium: an analytical approach," *Journal of the Optical Society of America B: Optical Physics*, vol. 33, no. 4, pp. 735–740, 2016.

[26] Z. Zhu, A.-X. Chen, Y. Bai, W.-X. Yang, and R.-K. Lee, "Controllable optical steady behavior from nonradiative coherence in GaAs quantum well driven by a single elliptically polarized field," *Modern Physics Letters B*, vol. 28, no. 15, Article ID 1450117, 2014.

[27] J. Yuan, W. Feng, P. Li et al., "Controllable vacuum Rabi splitting and optical bistability of multi-wave-mixing signal inside a ring cavity," *Physical Review A: Atomic, Molecular and Optical Physics*, vol. 86, no. 6, Article ID 063820, 2012.

[28] Z. Zhang, D. Ma, J. Liu et al., "Comparison between optical bistabilities versus power and frequency in a composite cavity-atom system," *Optics Express*, vol. 25, no. 8, pp. 8916–8925, 2017.

[29] Z. Zhang, H. Chen, L. Zhang et al., "Unveiling the relationship between optical bistability and vacuum Rabi splitting," *EPL (Europhysics Letters)*, vol. 117, no. 5, Article ID 53001, 2017.

[30] B. E. A. Saleh and M. C. Teich, *Fundamentals of photonics*, John Wiley & Sons, 2nd edition, 2007.

[31] Daniel Adam Steck, " ^{87}Rb D Line Data," http://steck.us/alkalidata.

6

Impact of Temperature and Relative Humidity on PMD in Directly Buried Optical Fibre Cables in Semi-Arid and Tropical Highlands in Kenya

Moses Ndunda,[1] **Alix Dehayem-Kamadjeu**,[2] **and David Waswa**[3]

[1] *University of Nairobi Kenya, Chiromo Campus, OSA Student Member and Optics Transmission Engineer in Liquid Telecom Kenya, Nairobi, Kenya*
[2] *Department of Physics and Researcher in Applied Optics Research Group in University of Nairobi, Chiromo Campus, Nairobi, Kenya*
[3] *Fibre Optics and Laser Research Group in Department of Physics, University of Eldoret, Eldoret, Kenya*

Correspondence should be addressed to Moses Ndunda; ndundamm@students.uonbi.ac.ke

Academic Editor: Sulaiman W. Harun

The telecommunication industry has implemented fibre deployment guidelines that reliably safeguard cable health during installation in the field. While installed fibre cables remain buried in the field, temperature and moisture in the locality subject them to mechanical expansions and corrosion. Directly buried fibre cables experience accelerated degradation that results from exposure to harsh environments. This increases pulse spreading and overlaps, with a mean time duration, known as Differential Group Delay (DGD), on the signal, as it propagates along the cable. DGD is stochastic; thus, Mean DGD is determined and presented as Polarization Mode Dispersion (PMD). This work undertook a real life assessment of how fluctuations in temperature and relative humidity influence PMD in directly buried fibre optical links, in a case study that focused on the fibre cable network owned by Liquid Telecom Kenya. The network spans across two key climatic ecosystems, namely, rift valley highlands and northern lowlands. The analysis revealed that fibre cables experience higher PMD in semi-arid areas by a factor of 2.6, compared to highland areas.

1. Introduction

Fibre cables that are directly buried in the ground often encounter operating conditions that impact negatively their performance, particularly with regard to high capacity data transmission. Temperature and humidity variations in the directly buried cables have been cited as significant influences on Polarization Mode Dispersion (PMD) [1, 2]. This can be very challenging to economies, including Kenya, which operate liberal policies on fibre cable infrastructure. Under such environments, individual fibre cable providers independently deploy and maintain their own infrastructure. Given that fresh excavation of areas where fibre cables are laid is an expensive exercise, it is not economically viable to upgrade the already deployed cables to duct standards. Existing fibre plants therefore require managing until they attain their end-life. This implies that telecommunication providers must keep tracking the PMD fluctuations associated with the cables in the field so as to determine the appropriate period to replace them. This has to be undertaken in a timely manner so as to minimize the traffic disruptions to customers, underpinned by the emerging regard for consumer rights.

The variants can be monitored for their impact on data transmission in the cables. This could be undertaken in different localities where the cables pass, so as to inform associated network designs and engineering. Such monitoring is valuable because the behavior of fibre cables is best analyzed using data that is specific to cables buried in the field at given localities [3]. PMD is nonlinear. It is therefore deterministic or stochastic due to changes in weather, mechanical stress, and/or strain while the cables remain installed in the field [4]. All data transmission designs therefore ought to factor PMD impacts in any fibre cable links. Dead cables point to their inability to transmit data through them without loss of traffic, which calls for their replacement.

2. Literature Review

PMD analyses in different climatic regions have exhibited varied Maxwellian Distribution Functions (PDF). This means that PMD is influenced by the environment and varied for different climatic zones [3]. Environmental changes like temperature and moisture lead to mechanical stain and stress, that in turn influence PMD fluctuations in fibre cables [5]. Since the cables are exposed to strain and stress, they develop cross-sectional deformations in their core cross sections. The cores tend to lose the circular cross section to elliptical or worse that leads to high birefringence and signal mode coupling [6]. Besides the mechanical stress and strain on the cables, temperature and humidity also affect fibre cables.

Distributed Temperature Sensing (DTS) technique for testing the inside of fibre cable cores and use of weather stations to test the outside temperatures in aerial cables have both revealed that either approach is reliable in analyzing the impact of temperature on PMD in fibre cables [1, 2, 7].

For buried cables, algorithms have been applied on metrological weather data to determine temperature at the depth where the cable is buried [7]. This method has been used to analyze frequency-shifts and chromatic dispersion. Factors like depths where the fibre cable is buried, specific heat capacity, thermal conductivity, and density of the soil have been taken into account by use of an algorithm [7, 8].

Analysis of temperature fluctuations inside the cable using DTS and outside the cable using metrological data has been applied alongside various PMD tests techniques. An example is Scrambled State-of-polarization Analysis (SSA) and Generalized InterferomeTrY (GINTY) that have been compared and, with consistency, demonstrated that temperature fluctuations have an impact on PMD [9]. Algorithms [7] have been applied to metrological weather data for buried cables and so far been only applied to chromatic dispersion (CD) analyses, but not on PMD.

PMD analyses based on laboratory simulation where the cable was placed in temperature controlled chambers have also been undertaken [2]. For this, a temperature range was set to an upper and a lower limit; then the variable (temperature) is adjusted in steps, so as to simulate an adiabatic environment. The same setup is repeated, with temperature being randomly varied in order to simulate fluctuations in an isothermal environment [2, 10]. A simulation in both adiabatic and isothermal ensures that all possible temperature fluctuation patterns experienced in the field are exposed to the fibre cable in the laboratory. Such simulations have also been used alongside PMD tests using Jones Matrix Eigen Analysis (JME) [2, 10] and have concurred with the field-based methods that temperature has an impact on PMD fluctuations [3].

All tests on temperature vs PMD are normally conducted at uniformly spaced time intervals (typically between 1 hour and 2 hours) to allow distinct transitions in both variables. Various PMD analysis techniques have been conducted using the said intervals and have demonstrated consistent trend showing rise and drop for both PMD and temperature [2, 7–9]. The trend has been very closely related, which confirms the strong dependence of PMD on temperature.

Studies on the response of PMD to fluctuations caused by relative humidity in telecommunication fibre optics have not attracted much attention. This could partly be attributed to the observation that fluctuations in relative humidity have impact on temperature [11]. Another assumption is that the impact of relative humidity is effectively managed in telecommunication fibre cables by filling the gaps between fibre cores and protective jacket with water-blocking agents [12]. That does not however mean that fibre optical cables are not affected by fluctuations in relative humidity.

Studies that have used the zero-stress aging technique to simulate relative humidity in laboratories have concentrated on the impact of cables' mechanical strength and connections with the rate of aging of the cables [2, 13, 14]. Other studies have focused on using the FOCs high response to relative humidity to produce humidity sensors [15].

Since relative humidity and temperature are inseparable in the field, this study therefore acquired relative humidity data alongside temperature fluctuations in order to examine the nature of their combined contributions to PMD in the field.

3. Experimental Setup

The locations for the Fibre under Test (FUT) were identified and selected distinctively to represent lowland and highland climates in Kenya [8] as presented in Figure 1. Analyzed sections for highland climate are labelled in blue while sections for lowlands are labelled in red. The FUT was part of jacketed direct buried cable that was not deployed in ducts and was serving a live network since 2007.

Two locations selected for tests were as follows: a section of about 60 km in highland area (Mau-Summit) located at coordinates -0.149087N, 35.687885E at the highest altitude peak (2850 m) within cable route and a similar length located in a low altitude area (Sultan-Hamud (1500 m)) located at coordinates -1.906248N, 37.279186E. Selection of the study sites excluded the Kenyan coast because of the region's proximity to the Indian Ocean, which makes it have higher alkalinity that results in higher cable corrosion compared to other regions in Kenya. Even so, the Kenya coastal region is only a small percentage of about 4% of the whole fibre cable that was to be analyzed in this study. PMD tests were randomly undertaken in 2 other locations so as to verify consistency of the tests.

A capacitive soil sensor ML3 [16, 17] was buried at a depth where the fibre cable under test was located. This was typically between 1 m and 1.2 m below earth's surface. Humidity tests sought to establish the effect of a mixture of soil particles, air, and water on the performance of the fibre optic cable. Damp soil is essentially a mixture of soil particles, air, and water. These components together determine the dielectric properties of a given soil, including the refractive index ($\sqrt{\varepsilon}$). The refractive index is evaluated from [17].

$$\sqrt{\varepsilon} = \alpha_0 + \alpha_1 \cdot \theta \tag{1}$$

where $\sqrt{\varepsilon} = 1.0 + 6.175V + 6.303V^2 - 73.578V^3 + 183.44V^4 - 184.78V^5 + 68.017V^6$ and V represents the sensor output in

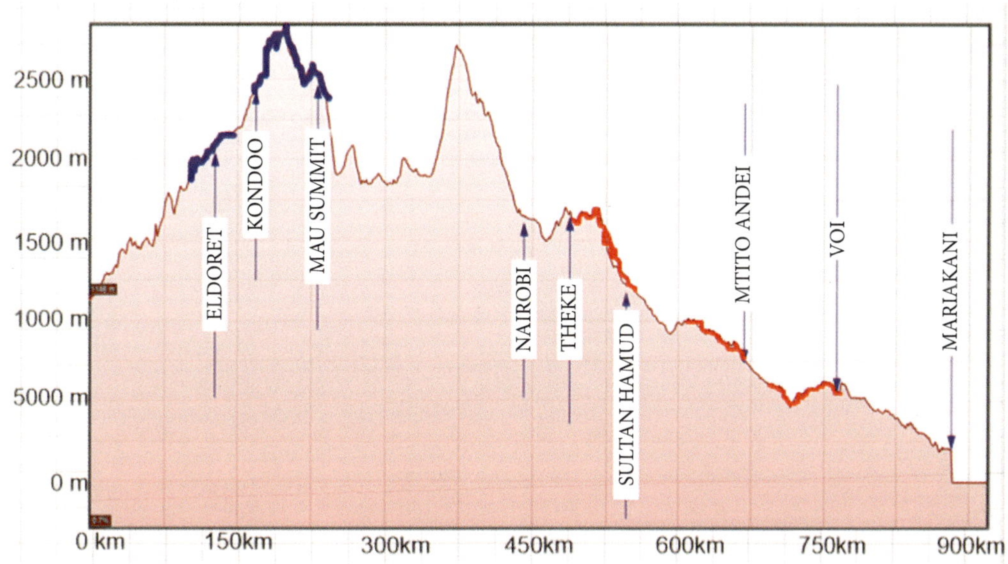

FIGURE 1: 900 km cross section from Kenya's sea level (coast) to rift valley highlands through northern lowlands. Vertical plane represents fibre route altitude while horizontal plane represents fibre length.

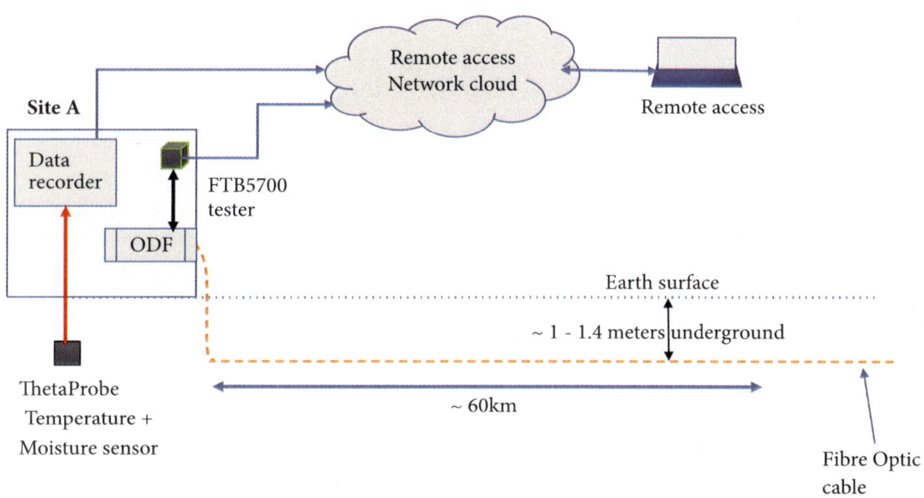

FIGURE 2: Setup of the field tests that were conducted using a single ended tester, moisture, and temperature sensor.

volts. α_0 and α_1 are coefficients that conveniently characterize the dielectric properties of a dry and wet soil, respectively. θ represents moisture.

If $\sqrt{\varepsilon}$ is known, (2) can be rearranged to assist in evaluating the moisture in the soil, as indicated in the equation below [17].

$$\theta = \frac{\sqrt{\varepsilon} - \alpha_0}{\alpha_1} \tag{2}$$

A data logger was configured to query and save sensor readings on an hourly basis, in a bid to register moisture content fluctuations in the entire test period. Deployment of the equipment was undertaken in a manner that ensured that the moisture data relating to different seasons was captured. A ML3 Theta-probe was used to acquire temperature-related data. The probe had inbuilt circuitry for measuring temperature alongside soil moisture. The soil temperature was measured using a 10 KΩ thermistor, which was mounted on the ML3 and connected to a data logger. The data logger had a look-up chart that contained readings of resistances that corresponded to the temperature of the locality where the thermistor was buried.

Field PMD tests were conducted on the buried fibre cable using the Scrambled State-of-polarization Analysis (SSA) (EXFO FTB-5700) [18, 19]. Figure 2 is the schematic diagram of the experiment setup in the field. Strands in 60 km fibre spans were identified in both semi-arid and highland climatic regions. The analyzer is a single ended device so it was placed at one end of the fibre and the other end was spliced to a connector to provide a proper reflection at the end of the 60 kms in a manhole. A mouse bot was then installed in the analyzer to run PMD data acquisition every end of hour.

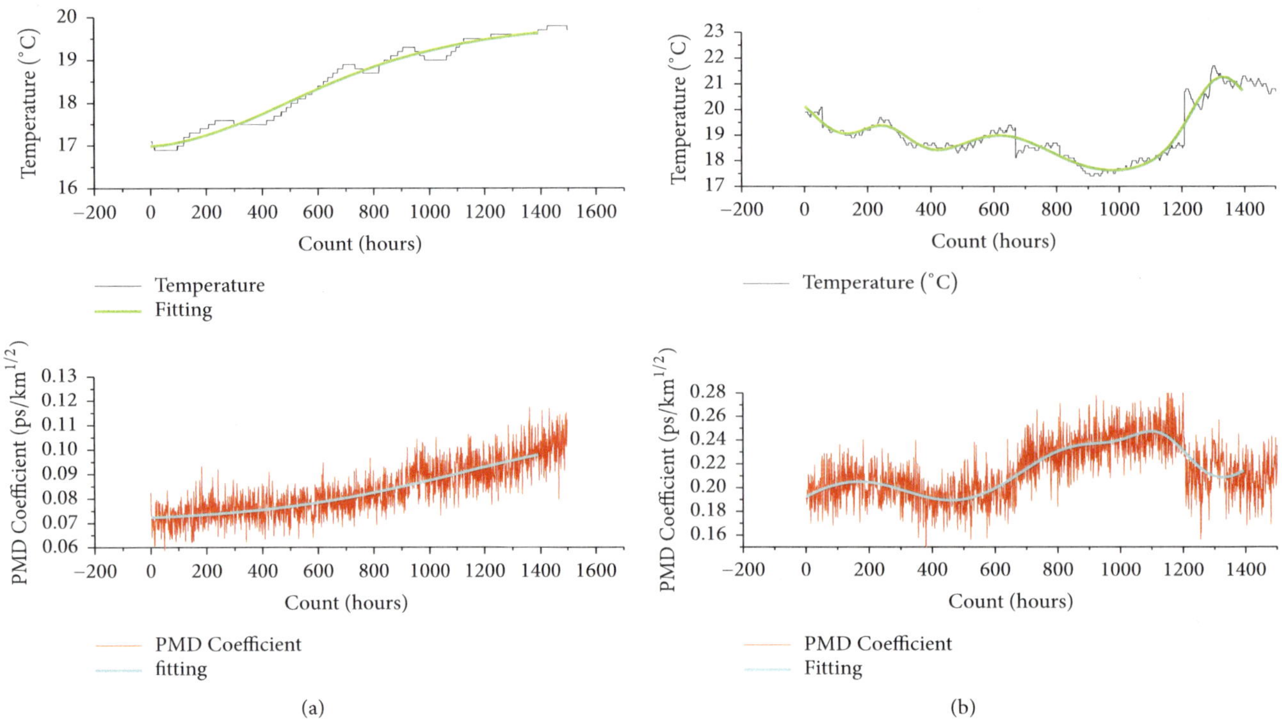

FIGURE 3: (a) Temperature (black) with fitting (green) and PMD (red) with fitting (cyan) for highland climate in Mau-Summit area. (b) Temperature (black) with fitting (green) and PMD trend-line (red) with fitting (cyan) for semi-arid climate in Sultan Hamud area.

Each PMD test-run was configured to coincide with the humidity and temperature data acquisition and ran for the same duration for purposes of correlation. PMD testing in each season also provided the study with the annual fluctuation ranges. PMD values are better expressed in terms of PMD coefficient that is calculated as PMD per square root km of the fibre length under measurement.

4. Results and Discussion

In highland area, highest recorded PMD coefficient was 0.1175 $ps\sqrt{km}$, while the lowest was 0.0588 $ps\sqrt{km}$. A PMD coefficient mean of 0.0836 $ps\sqrt{km}$, a standard deviation of 0.05113 $ps\sqrt{km}$, and a variance of 0.13 were observed. On the other hand, the highest recorded PMD coefficient in semi-arid area was 0.2956 $ps\sqrt{km}$, and the lowest was 01559 $ps\sqrt{km}$, while the mean and standard deviation of the PMD were 0.21524 $ps\sqrt{km}$ and 0.02177 $ps\sqrt{km}$, respectively. This effectively led to a variance of 0.1 in semi-arid areas.

The PMD coefficient in semi-arid compared to highland area was high by a factor of 2.6 leading to indications of higher rate of cable PMD aging in semi-arid areas. The various correlations for PMD coefficient alongside temperature and relative humidity in the two climatic zones follow as shown below.

4.1. Temperature Correlation with PMD. In highland area, both temperature and PMD coefficient had an upward trend as seen in Figure 3(a) until test hour 1,500 when both tests ended. Since the cable was buried, the trend in daily temperature fluctuations was stable. This was because the cable did not experience day-night transitions. Clear PMD trends that were in response to distinct temperature transitions can be attributed to general seasonal fluctuations. Change in temperature fluctuations corresponded to changes in PMD values. The correlation demonstrated that changes in temperature had influence on PMD fluctuations, as has also been noted in other studies [2, 7, 8, 10]. The graph in Figure 3(a) presents compiled PMD and temperature correlation for highlands (Mau-Summit area)

And the graphs in Figure 3(b) represent temperature and PMD trend-line correlations for semi-arid climate that is located in Sultan Hamud area.

In the semi-arid region, namely, Sultan Hamud, lowest PMD concentrations were between test hours 0 and 500, though the lowest temperature was recorded between test hours 800 and 1000. Highest PMD fluctuations were recorded around test hour 1100 although the period between test hours 1,300 and 1,500 was the hottest, as presented in Figure 3(b).

Temperature fluctuations in the semi-arid area were high compared to those in the highland area. Higher PMD was observed in the semi-arid test area compared to the highland area. It was noted in the period between test counts 450 and 1500 that temperature downward trend corresponded to upward trend in PMD coefficient. The consistency of PMD vs temperature trend-line had an inverse relationship and was a contrast observation compared to temperature vs PMD coefficient relationship in highlands. Given the observations in this analysis and the observations that were made by Haro

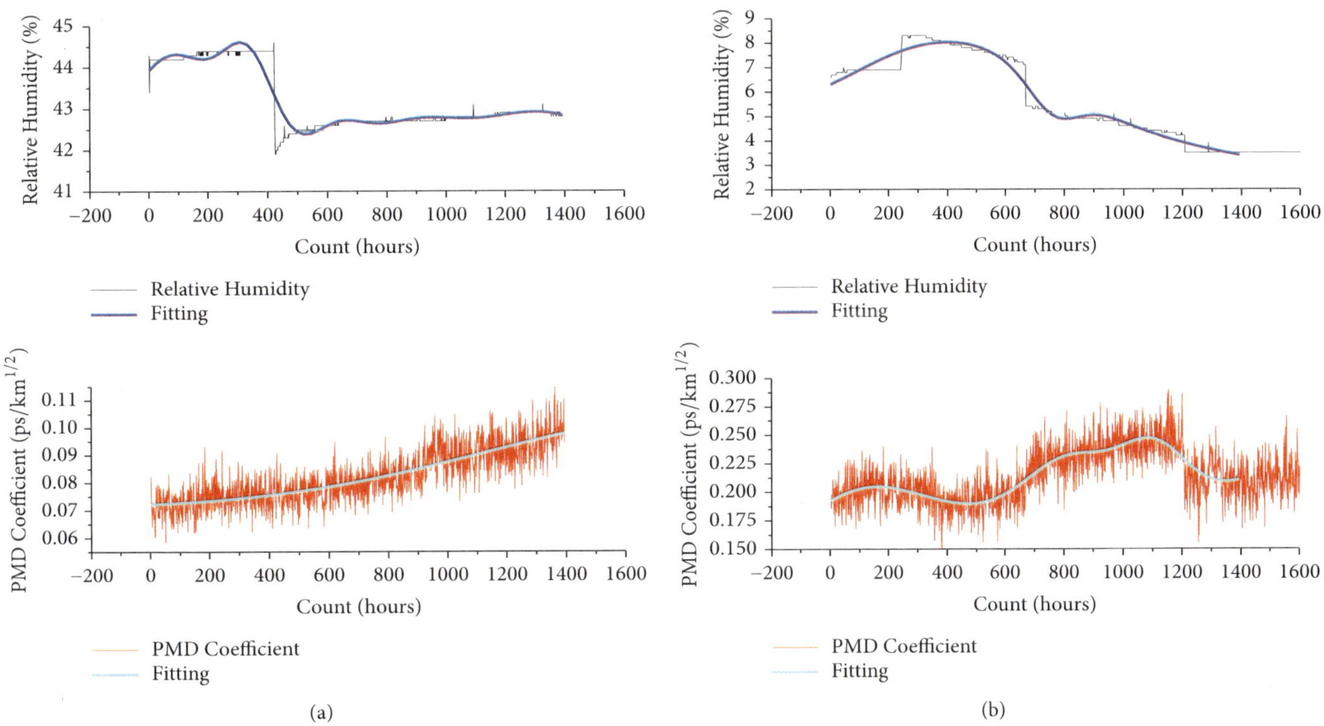

FIGURE 4: (a) Relative humidity (black) with fitting (blue) and PMD coefficient (red) with fitting (cyan) for highland climate in Mau-Summit area. (b) Relative humidity (black) with fitting (blue) and PMD coefficient (red) with fitting (cyan) for semi-arid climate in Sultan Hamud area.

et al., there is an indication that PMD in fibre cables exhibits a clear inverse relationship in environments where temperature exceeds 20 degrees Celsius [20].

4.2. Relative Humidity Correlation with PMD. Relative humidity (RH) in highland area, namely, Mau-Summit, had an average of 42.8% and a fluctuation range of 2.7%. Relative humidity vs PMD results are presented in Figure 4(a). Major RH fluctuation incidences were compared with corresponding PMD fluctuations. A key incident was the 2.5% drop within a duration of 4 hours, which was recorded between test hours 400 and 500. This drop in RH did not trigger a PMD drop but, instead, PMD fluctuations remained uninterrupted. Although RH in highland areas had very minimal fluctuations as seen in Figure 4(a), it was observed that the only notable incidence in the entire 1,600 hours did not have any impact on PMD. This implies that RH has negligible impact, if any, on PMD in a highland area.

RH in the semi-arid area where the study was undertaken (Sultan Hamud) was very low, compared to that of the highland area (Mau-Summit), but with a higher fluctuation range of 5%. The fluctuations in semi-arid area were analyzed based on two major incidences: one being a RH rise incidence and the downward trend, as presented in Figure 4(b).

The first incident was a 1.4% sharp rise in RH, recorded at test hour 200 and which took 5 hours at the peak, followed by a downward trend until test hour 1,250. During this test period, a direct relationship between PMD and RH was observed. This was followed by a stable nonvarying RH period (second incident) that lasted until test hour 1600. During this time, relative humidity remained the same for a period of 400 hours, while the PMD fluctuated with a downward trend and then floored about 100hrs after RH floored. A fairly direct relationship between PMD and RH was observed as seen in Figure 4(b) traces.

RH did not have impact on PMD in highlands but it was observed that it had a notable influence in semi-arid areas. The reduced influence of PMD in highlands compared to semi-arid environments could be attributed to the small fluctuation in RH range, of 2.3% that was recorded, compared to 5% that was recorded in semi-arid areas. Although the temperature and relative humidity variations do not have wide range of fluctuations, it should be noted that the variations represent real data that was obtained by use of sensors that were buried 1.2 meters underground. The depth is the same that the cable was buried.

5. Conclusion

This is the first documented real life study of PMD of fibres deployed in a field, where correlations with T and RH% have been explored. It is noteworthy to stress that the impacts of RH on PMD cannot be ignored in semi-arid regions, where they correspond to direct relationship for the better part of test durations. Findings of this study further apprise that RH alone is not sufficient towards investigating the impact caused by fluctuations of temperature and humidity on PMD. This is because, as the results of this study demonstrate, PMD fluctuates in both semi-arid areas and highlands and at times when RH was stable and nonvarying. In the highlands, however, RH experienced minimal variations that were less likely to lead to meaningful impact on PMD fluctuations.

Findings of this work in the highlands concur with those of other investigations that have been conducted elsewhere on the connection between temperature and PMD fluctuations [2, 3, 10]. The findings of the results in semi-arid area present a unique observation and a contrast for results obtained in other investigations. Variances of 0.1 and 0.13 were observed in respect of the semi-arid and highland areas, respectively. This indicates reduced PMD fluctuation-spread from the mean attributable to semi-arid areas, compared to that for highland areas. It demonstrates reduced PMD fluctuations in buried cables, compared to the 0.42 obtained in laboratory simulations [2] and 0.39 in field tests for those that were conducted in aerial cables in different climatic zones in South Africa [3].

The nature of PMD fluctuations vs temperature and relative humidity in semi-arid area was an interesting finding. This presents a need to further analyze PMD in many other regions that have the same climatic conditions or higher temperature and lower relative humidity to establish the repeatability of outcome of the unique climatic relationship with PMD.

Conflicts of Interest

The authors declare that they have no conflicts of interest.

Acknowledgments

We would like to acknowledge our supervisors, Dr. Alix D. Massop and Dr. David W. Waswa, for their invaluable inputs throughout the period of this research. Special thanks are due to Hans Haerdtle, the Chief Technical Officer (C.T.O) of Liquid Telecom East Africa, for giving us an opportunity to analyze Liquid Telecom infrastructure, as well as the facilitation in acquiring test and analysis tools. The entire Liquid Telecom fraternity is also gratefully acknowledged for the support that they individually and collaterally provided us during the study.

References

[1] J. Bohata, S. Zvanovec, and M. Pisarik, "Outdoor atmospheric influence on polarization mode dispersion in optical cables," pp. 1–4, General Assembly and Scientific Symposium, Beijing, China, 2014.

[2] S. Karlik and G. Yilmaz, *Effects Of Temperature On Polarization Mode Dispersion Of G.652 Optical Fibers*, Uludag University, Faculty of Engineering and Architecture, Department of Electronics Engineering, Bursa, Turkey, 2006.

[3] T. Gibbon, W. Lorinda, and D. Waswa, "Polarization mode dispersion compensation for the South African optical-fibre telecommunication network," *South Afr. J. Sci*, vol. 104, pp. 3-4, 2008.

[4] W. T. Ireeta, V. Musara, and A. Leitch, "Monitoring FO-PMD and SO-PMD over time with respect to Environmental conditions," *Can. Cent. Sci. Educ*, vol. 6, no. 4, pp. 10–15, 2014.

[5] K. R. Lefebvre, "Environmental Effects on Chromatic and Polarization Mode Dispersion," *NetTest*, 2001.

[6] E. A. Lindholm, H. G. Limberger, M. J. Matthewson et al., "Aging behavior of optical fibers in aqueous environments," in *Reliability of Optical Fiber Components, Devices, Systems, and Networks II*, 2004.

[7] T. J. Pinkert, O. Böll, L. Willmann et al., "Effect of soil temperature on optical frequency transfer through unidirectional dense-wavelength-division-multiplexing fiber-optic links," *Applied Optics*, vol. 54, no. 4, pp. 728–738, 2015.

[8] H. C. Ji, J. H. Lee, and Y. C. Chung, "System Outage Probability Due to Dispersion Variation Caused by Seasonal and Regional Temperature Variations," *J. Fiber Opt. Commun*, vol. 60, no. 2330, pp. 1–3, 2005.

[9] J. Bohata, J. Jaros, S. Pisarik, S. Zvanovec, and M. Komanec, "Long-Term Polarization Mode Dispersion Evolution and Accelerated Aging in Old Optical Cables," *IEEE Photonics Technology Letters*, vol. 29, no. 6, pp. 519–522, 2017.

[10] P. Kondamuri and C. Allen, *PMD Characterization of Installed Fiber*, The University of Kansas Center for Research, Lawrence, Kan, USA, 2004.

[11] A. K. Betts, R. Desjardins, D. Worth, and B. Beckage, "Climate coupling between temperature, humidity, precipitation, and cloud cover over the Canadian Prairies," *Journal of Geophysical Research: Atmospheres*, vol. 119, no. 23, pp. 13305–13326, 2014.

[12] J. Peters, *A Comparison of Dry Versus Gel Filled Optical Cables*, Sterlight Technologies, 2012.

[13] B. Risch, S. Fox, and R. Delden, *Lifetime Prediction of Fiber Optic Cable Materials for Nuclear Power Applications: Evaluation of Failure Mechanism, End of Life Criteria, and Test Methodology*, International wire and cable symposium, Rode Island, Pa, USA, 2009.

[14] A. Dwivedi and S. Glaesemann, "Optical Fiber Strength, Fatigue and Handleability After Aging in a Cable," *Corning Incorporated*, 2011.

[15] M. Batumalay, S. W. Harun, N. Irawati, H. Ahmad, and H. Arof, "A study of relative humidity fiber-optic sensors," *IEEE Sensors Journal*, vol. 15, no. 3, pp. 1945–1950, 2015.

[16] Delta T devices, *ML3 ThetaProbe moisture sensor*, Cambridge, 2014.

[17] Delta-T Devices Ltd, *User manual for the ML3 ThetaProbe*, Delta-T Devices Ltd, 2013.

[18] EXFO Inc, *FTB-5700 Single-Ended Dispersion Analyzer*, EXFO Inc., 2014.

[19] G. André, C. Hongxin, A. Francis et al., *A new test method for the quantitative determination of bad pmd sections along an installed fiber link*, EXFO Inc., 2009.

[20] J. Haro and P. R. Horche, "Evolution of PMD with the temperature on installed fiber," *Optical Fiber Technology*, vol. 14, no. 3, pp. 203–213, 2008.

7

Alternative Approach of Developing Optical Binary Adder using Reversible Peres Gates

Dhoumendra Mandal,[1] Sumana Mandal,[2] Mrinal Kanti Mandal,[2] and Sisir Kumar Garai[3]

[1]*Department of Physics, Saldiha College, Saldiha, Bankura, West Bengal 722173, India*
[2]*Department of Physics, NIT Durgapur, West Bengal 713209, India*
[3]*Department of Physics, MUC Women's College, Burdwan, West Bengal 713104, India*

Correspondence should be addressed to Sisir Kumar Garai; skgarai@gmail.com

Academic Editor: Giulio Cerullo

All-optical devices will play a very significant and crucial role in the modern all-optical network by eliminating the bottleneck of opto-electro-opto- (O-E-O-) conversion. Unfortunately, the conventional logic gates lose information at the output, and the states of the outputs cannot give any credible impressions of the states of the inputs. In this article, at first, the authors have proposed a method of designing an optical three-input-three-output reversible Peres gate. Authors have deployed polarization switching characteristic of Semiconductor Optical Amplifier (SOA) for designing this circuit. The authors have also proposed a method of designing an optical reversible full adder, using two such Peres gates and subsequently a data recovery circuit which can recover the input data of the adder. The authors have chosen frequency encoded data for processing the operation. The proposed scheme has been verified by simulation results.

1. Introduction

In the modern communication network, where the system deals with a large number of data, the reversible logic circuit is preferable to conventional logic circuit due to low power consumption. The conventional logic gates like AND, OR, EX-OR, and so on commonly generate less output information than the inputs. Thus, information loss occurs at the output. The information loss leads to energy loss. Landauer's principle exposes that the system generates $kT * \ln 2$ joules of heat energy for every bit loss [1]. Though, at room temperature, the heat energy dissipated due to single bit loss is very small, when a system deals with a large amount of data and there exist larger numbers of logic processors, then the loss of bits is larger. Then, the heat generated due to bit loss is larger, which in turn increases the temperature of the system. Consequently, this cumulative process dissipates more energy. Bennett has illustrated that $kT * \ln 2$ joules of energy dissipation in an irreversible circuit can be avoided if the computation is done in a reversible manner [2]. In case of the reversible logic gate, there exists a direct mapping between inputs and outputs, and therefore ideally there is no chance of bit loss. This is not the be-all and end-all here. The output information of the conventional logic gates does not give any clear idea about the input information. In case of reversible logic gate, the inputs are individually retrievable from the outputs. Hence, the reversible logic gate may be the bliss to overcome such shortcomings. In the electronic communication network, various types of arithmetic and logical units have been designed using reversible logic gates [3–5]. Van Rentergem and De Vos designed reversible full adder, implemented into electronic circuitry based on CMOS technology and pass-transistor design [3]. Aradhya et al. designed control unit for low power AU using reversible logic for the implementation in electronics [4]. But, in the 21st century, the scientists are dealing with the optical signals in the communication network. To develop an all-optical network, optical devices are the basic requirement. In case

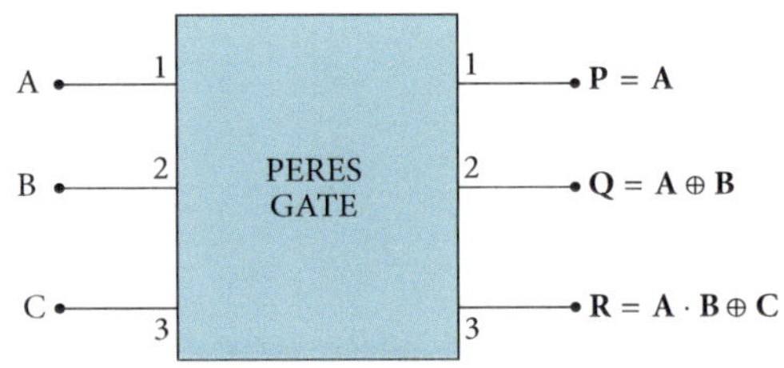

FIGURE 1: Block diagram of Peres gate.

of optical devices, the signals remain in the photonic state, thereby removing the "electronic bottleneck." Consequently, the speed of the system can reach up to THz.

Peres gate is a three-input three-output, reversible logic gate with quantum cost "4." The block diagram of Peres gate is shown in Figure 1. Three input signals "A," "B," and "C" are fed into the Peres gate through three input terminals "1," "2," and "3," respectively, and three output signals "P," "Q," and "R" are received from three output terminals "1," "2," and "3," respectively. Logic expressions of the outputs in terms of inputs are as follows: $\mathbf{P} = \mathbf{A}$, $\mathbf{Q} = \mathbf{A} \oplus \mathbf{B}$, $\mathbf{R} = \mathbf{A} \cdot \mathbf{B} \oplus \mathbf{C}$; that is, output "P" always gives the result, same as "A," and "Q" gives "A" EX-OR "B." If both the inputs "A" and "B" are in binary "1" state, then "R" will be "$\mathbf{1} \oplus \mathbf{C}$," that is, nothing but $\overline{\mathbf{C}}$. When at least one of the inputs "A" and "B" is in binary "0" state, then "R" will be "$\mathbf{0} \oplus \mathbf{C}$"; that is, "R = C". Here, it is easy to determine the input "A" from "P," input "B" from "P" and "Q," input "C" from "P," and "Q" and "R." In this article, authors at first have designed an optical Peres gate using polarization switch (PSW), and then they have also designed optical full adder circuit using two such Peres gates and subsequently a data recovery circuit which can recover the input data of the adder. The principle of nonlinear polarization rotation of the probe beam in presence of pump beam in a Semiconductor Optical Amplifier (SOA) is exploited here to explain the switching characteristic of PSW. Simulated results of the switching action of the PSW are shown in the figures in the respective sections. The authors deal with frequency encoded data as the frequency is the fundamental character of the signal and it does not change due to reflection, refraction, or attenuation. To execute the operation, binary state "1" is represented by the optical beam of frequency ν_2 and binary state "0" is represented by the optical beam of frequency ν_1. Simulated power spectrums of input-output of Peres gate are shown in the figures in respective sections which confirm the feasibility of the proposed schemes. Some optical Multiplexers (MUXs) and Demultiplexers (DMUXs) have been used to construct the optical circuits. MUX and DMUX work on frequency encoded data also. DMUX routes the optical signal of frequency ν_1 to channel 1 of it. If the frequency of the optical signal is ν_2, then it is transmitted via channel 2. MUXs are used to connect the optical channels.

So many researchers have designed Peres gate using various techniques. Kotiyal et al. developed Peres gate using four Mach-Zehnder interferometers (MZI) [6]. Maity et al. also designed the Peres gate using Mach-Zehnder interferometers (MZI) [7]. Kotiyal et al. designed binary adder using MZI [8]. Mandal et al. have also designed all-optical full adder circuit using five reversible Toffoli gates [9]. But in this article the circuit complexity of the proposed full adder is reduced as here the authors have designed a full adder using two Peres gates only. Quantum cost of Toffoli gate is greater than that of Peres gate. Full adder circuit using reversible logic gate is developed based on CMOS technology and pass-transistor design [3, 10] in the electronic domain, but in this article the authors have used SOA as the key element and the whole circuit is optical and frequency encoded. SOA is a worthy choice for designing various types of optical devices due to the fact of compactness, integration potential, high gain, and large operating bandwidth. From the outputs of the proposed full adder circuit, one can easily retrieve the inputs and this makes the scheme novel one. The switching speed of the PSW is very high and it is near 0.01 ns. Thus, the proposed full adder will be very helpful for designing various types of optical processors with less energy dissipation. The simulated power spectrums of the input-outputs of the Peres gate and full adder circuit make the scheme more reliable and these are shown in respective section of the article. The article is arranged as follows: Section 1 is the introductory part of the article. The working principle of the proposed scheme is explained in Section 2. The scheme of optical Peres gate is discussed in Section 3. Proposed optical full adder using Peres gates is designed and explained in Section 4. Sections 5 and 6 are the discussion part and the conclusion part of the article, respectively.

2. Working Principle

In this article, the proposed work focuses on the polarization switches (PSW). PSW works on the principle of nonlinear polarization rotation of the Semiconductor Optical Amplifier (SOA). So many researchers have already explained the nonlinear polarization rotation characteristic of SOA in different ways [11–13], but in this article the basic theory of nonlinear polarization rotation is explained with the help of simple rate-equation model as stated by Dorren et al. [14].

To explain the basic theory, two different optical beams are considered. One is less intense probe beam, and another is a highly intense pump beam, and these are fed into the SOA from the two opposite directions of the SOA as shown in Figure 2. The probe beam is injected into SOA via a polarization controller (PC) which controls the polarization direction of the incident probe beam at 45° to the orientation of the SOA layer. An attenuator is placed in the path of the probe beam and this attenuator attenuates the power of the probe beam to the desired value so that the SOA works in a linear way in the absence of the pump beam. In the presence of the highly intense pump beam, SOA produces nonlinearity and this leads to polarization dependent gain at the output. The output probe beam of SOA is fed into a polarization beam splitter (PBS). PBS transmits the output beam to port 1 or port 2 according to the polarization direction of the output beam. PBS and PC are set in such a way that in the absence of

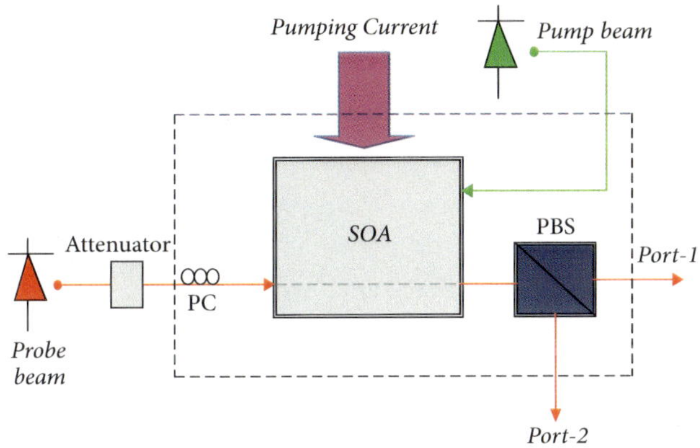

FIGURE 2: SOA acting as a PSW.

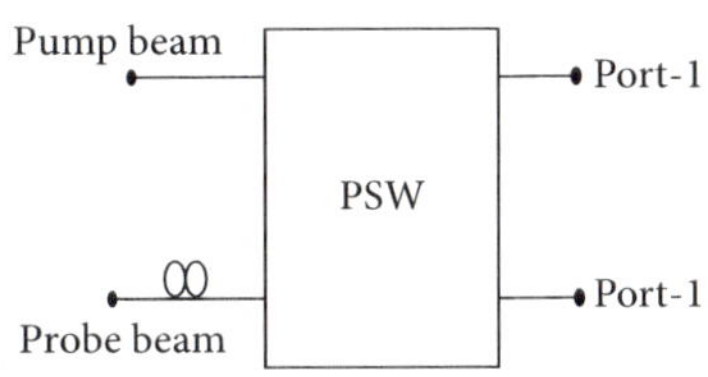

FIGURE 3: Block diagram of PSW.

the pump beam total power is developed at port 2, but in the presence of the pump beam, the output beam reaches port 1. Block diagram of PSW is shown in Figure 3.

Detailed theory of nonlinear polarization rotation in SOA is demonstrated using the simple rate-equation model [14–17]. If the optical probe beam in SOA is decomposed into two mutually perpendicular components, called TE mode and TM mode, then these two components experience different optical properties in the presence of the pump beam. According to Dorren's model, the gain of the SOA for TE mode is [14]

$$g^{\mathrm{TE}}(z,t) = \xi^{\mathrm{TE}}\left[n_c(z,t) + n_x(z,t) - N_0\right]. \quad (1)$$

Similarly for TM mode gain will be

$$g^{\mathrm{TM}}(z,t) = \xi^{\mathrm{TM}}\left[n_c(z,t) + n_y(z,t) - N_0\right]. \quad (2)$$

The field is considered to be propagated along the z-direction and thus only x-transition (corresponding to TE mode) and y-transition (corresponding to TE mode) are possible. Here $n_x(z,t)$ and $n_y(z,t)$ are denoted as number of holes involved in the x-transitions and y-transitions, respectively, and $n_c(z,t)$ is the electron in the conduction band. The total number of electronic states involved in the transitions is N_0. $\xi^{\mathrm{TM/TE}}$ is the gain coefficient corresponding to TM/TE mode. Actually, in this case, the valence band is considered to be divided into two subbands which are heavy hole band (HHB) and light hole band (LHB). The x-transitions taking place with heavy holes are much greater in number than the y-transitions whereas the y-transitions taking place with the light holes are much greater in number than the x-transitions. When there is no anisotropy in the SOA, then the x-transitions and the y-transitions take place symmetrically. But in the case of SOA with anisotropy, the y-transitions will be promoted over the x-transitions. Actually, in this case, the carrier distribution corresponding to x-transition (N_{TE}) and y-transition (N_{TM}) is not same. The population imbalance factor "f" signifies the ratio of the carrier distribution of TE mode and TM mode; that is, $f = N_{\mathrm{TE}}/N_{\mathrm{TM}}$.

Now the rate equations for $n_x(z,t)$ and $n_y(z,t)$ can be expressed as

$$\begin{aligned}\frac{\partial n_x(z,t)}{\partial t} &= -\frac{n_x(z,t)-\bar{n}_x}{T_1} - \frac{n_x(z,t)-fn_y(z,t)}{T_2} \\ &\quad - g^{\mathrm{TE}}(z,t)\,S^{\mathrm{TE}}(z,t), \\ \frac{\partial n_y(z,t)}{\partial t} &= -\frac{n_y(z,t)-\bar{n}_y}{T_1} - \frac{fn_y(z,t)-n_x(z,t)}{T_2} \\ &\quad - g^{\mathrm{TE}}(z,t)\,S^{\mathrm{TM}}(z,t),\end{aligned} \quad (3)$$

where T_1 and T_2 are electron-hole recombination time and interhole relaxation time, respectively. "f" is the hole population imbalance factor. $S^{\mathrm{TM/TE}}(z,t)$ is the photon number which is proportional to power of the input signal.

$\bar{n}_x$ and $\bar{n}_y$ are the respective equilibrium values and these depend on the electric biasing current (I) as follows:

$$\begin{aligned}\bar{n}_x &= \frac{f}{1+f} * \frac{I}{e}T_1, \\ \bar{n}_y &= \frac{1}{1+f} * \frac{I}{e}T_1,\end{aligned} \quad (4)$$

where "e" is the electric unit charge.

Now solving (3) under steady state condition (neglecting the ultrafast nonlinear optical processes), the carrier density for respective modes is

$$n_x = \frac{\left((I/e)\,(f/(1+f)) + \xi^{\mathrm{TE}}S^{\mathrm{TE}}N_0/\left(1+\varepsilon S^{\mathrm{TE}}\right)\right)}{\left(1/\tau_{\mathrm{sp}} + \xi^{\mathrm{TE}}S^{\mathrm{TE}}\,(2+1/f)\,/\,(1+\varepsilon S^{\mathrm{TE}})\right)}, \quad (5)$$

$$n_y = \frac{\left((I/e)\,(1/(1+f)) + \xi^{\mathrm{TM}}S^{\mathrm{TM}}N_0/\left(1+\varepsilon S^{\mathrm{TM}}\right)\right)}{\left(1/\tau_{\mathrm{sp}} + \xi^{\mathrm{TM}}S^{\mathrm{TM}}\,(2+f)\,/\,(1+\varepsilon S^{\mathrm{TM}})\right)}. \quad (6)$$

Here S^{TE} is total photon density of the pump beam and the probe beam. Now, the power of each mode is related to the photon density by the equation

$$P^{\mathrm{TE/TM}} = S^{\mathrm{TE/TM}} * \frac{\hbar\omega v_g^{\mathrm{TE/TM}}}{L}, \quad (7)$$

where "L" is the length of the SOA, $v_g^{\mathrm{TE/TM}}$ is the group velocity of the respective mode, and ω is frequency of optical signal.

So, the gain of the SOA can be expressed in terms of power of the input signal as well as electric biasing current using (5), (6), and (7). When the electric current is sufficiently high (greater than 160 mA) and the power of the probe beam

Table 1: Parameters of SOA.

Parameter	Symbol	Value	Units
Confinement factor			
For TE mode	Γ^{TE}	0.2	
For TM mode	Γ^{TM}	0.14	
Phase modulation coefficient			
For TE mode	α^{TE}	5	
For TM mode	α^{TM}	5	
Modal loss			
For TE mode	α_{int}^{TE}	0.27	ps^{-1}
For TM mode	α_{int}^{TM}	0.27	ps^{-1}
Gain coefficient			
For TE mode	ξ^{TE}	$7.0 * 10^{-9}$	ps^{-1}
For TM mode	ξ^{TM}	$6.5 * 10^{-9}$	ps^{-1}
Group velocity			
For TE mode	V_g^{TE}	100	μm/ps
For TM mode	V_g^{TM}	100	μm/ps
Hole population imbalance factor	f	0.5	
Electron-hole recombination time	T_1	500	ps
Hole-hole relaxation time	T_2	0.1	ps
Optical transition state number	N_0	10^8	
SOA length	L	800	μm

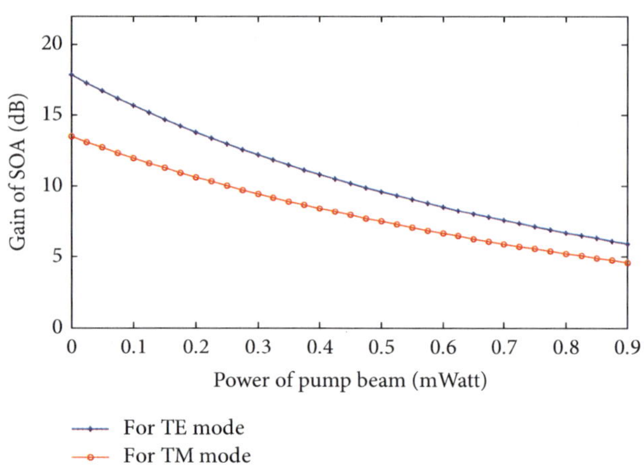

Figure 4: Variations of gain of SOA for different modes.

is very low, then the gain of the SOA depends only on the power of the pump beam. Variations of the gain of the SOA for different modes are different. In Figure 4, the variations of the gain of the SOA for different modes with respect to the power of the pump beam are shown.

Now, if $A(z,t)$ is the weakly varying time-space dependent electric field component of the beam propagating along Z direction then the equation of motion can be written in the form of

$$\left(\frac{\partial}{\partial t} + v_g^{TE/TM}\frac{\partial}{\partial z}\right) A^{TE/TM}(z,t) = \frac{1}{2}\Gamma^{TE/TM}\left(1 + i\alpha^{TE/TM}\right) g^{TE/TM}(z\cdot t)\, A^{TE/TM}(z,t) - \frac{1}{2}\alpha_{int}^{TE/TM} A^{TE/TM}(z,t). \quad (8)$$

The meaning of the symbols, used in the above equation, and their values are shown in Table 1 [14].

Each mode of $A(z,t)$ is related to the intensity $S(z,t)$ by the equation

$$A^{TE/TM}(z,t) = \sqrt{S^{TE/TM}(z,t)e^{j\Phi^{TE/TM}(z,t)}}. \quad (9)$$

From (5), the phase of each mode at the output can be expressed as

$$\Phi^{TE/TM}(L,t) = \frac{1}{2}\frac{\alpha^{TE/TM}\Gamma^{TE/TM}g^{TE/TM}L}{v_g^{TE/TM}}, \quad (10)$$

where "L" is the length of the SOA.

So the phase difference between the TE mode and TM mode at the output is given by

$$\Delta\Phi = \frac{1}{2}\left(\frac{\alpha^{TE}\Gamma^{TE}g^{TE}}{v_g^{TE}} - \frac{\alpha^{TM}\Gamma^{TM}g^{TM}}{v_g^{TM}}\right)L. \quad (11)$$

The phase difference between two modes depends on the gain of the respective modes. For the SOA with anisotropy, the gain of the different modes is not equal and also depends on the power of the input beams (using (5), (6), and (7)). If the power of the probe beam is very low and input injected current is fixed (more than 160 mA) then the phase difference between two modes depends on the power of the pump beam.

Now, the output intensity "S_o" of different modes of the probe beam can be expressed in terms of input intensity "S_i" by the equation

$$S_o^{TE/TM} = S_i^{TE/TM} e^{(\Gamma^{TE/TM}g_0^{TE/TM} - \alpha_{int}^{TE/TM})(L/v_g^{TE/TM})}. \quad (12)$$

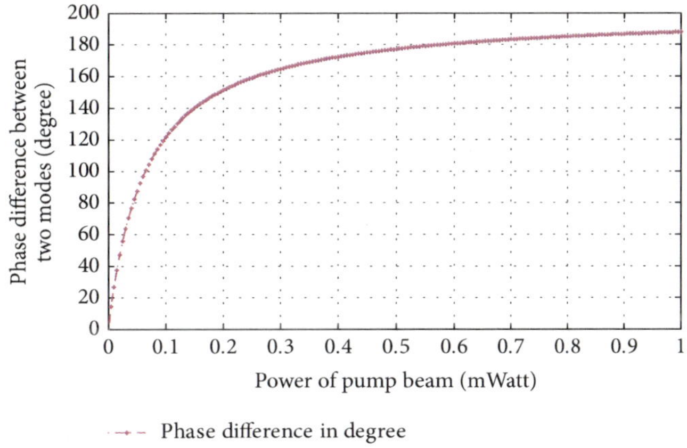

FIGURE 5: Variation of phase difference between TE mode and TM mode with respect to power of the input pump beam.

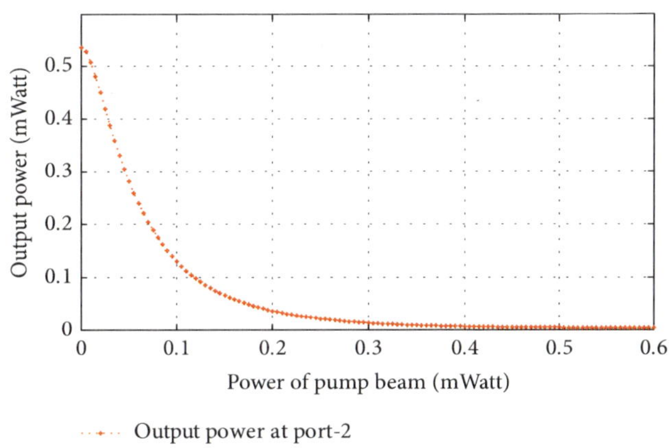

FIGURE 6: Variation of the output power at port 2 with respect to input pump power.

The resultant output can be expressed as

$$S_o = S_o^{\mathrm{TE}} + S_o^{\mathrm{TM}} + 2\sqrt{S_o^{\mathrm{TE}} \cdot S_o^{\mathrm{TM}}} \cdot \cos(\Delta\Phi). \qquad (13)$$

So the amplification in decibel unit by the SOA is

$$G_o^{\mathrm{TE/TM}} = 10\log_{10}\left(\frac{S_o^{\mathrm{TE/TM}}}{S_i^{\mathrm{TE/TM}}}\right). \qquad (14)$$

Using (1)–(7) and (11), the variation of the phase difference between TE mode and TM mode with respect to the power of the pump beam is plotted in Figure 5. It can be observed from Figure 5 that when the power of the pump beam is 0.5 mWatt, then phase difference between two modes becomes almost 180 degrees. Due to the presence of PBS, the output probe beam is transmitted to port 1 or port 2 according to the polarization direction of the output beam. The variation of the output power at port 2 with respect to pump beam is shown in Figure 6. When the power of the pump beam is 0.4 mWatt or more, the power at port 2 is almost zero. Then, the total power is developed at port 1.

In the above theory, the linearly polarized input probe signal is fed into the SOA and then this signal is decomposed into two mutually perpendicular linearly polarized components (TE mode and TM mode). These two components move through the SOA with unequal velocity due to the presence of anisotropy in SOA and different carrier dynamics. These two components recombine again at the output of the SOA with a certain phase difference between them. The direction of the polarization of the output signal depends on the phase difference between two modes and thus the polarization direction of the output signal is rotated. As the phase difference between two modes depends on the power of the pump beam for the fixed input biasing current, the direction of the polarization of the output signal of the SOA also depends on the power of the input pump beam. Now, the PBS is set in such a way that in the absence of the pump beam the PBS passes the output beam to port 2.

So, based on the above working principle, switching action of PSW can be stated as follows: in the presence of the pump beam, power is developed at port 1; that is, port-1 remains "ON" and port-2 remains "OFF," and in the absence of the pump beam power is developed at port 2; that is, port 2 is "ON" and port 1 is "OFF." Thus, the PSW can be a worthy choice as an optical switching element for designing of various types of optical processors. In this article, the optical Peres gate is designed using PSWs and the technique of the

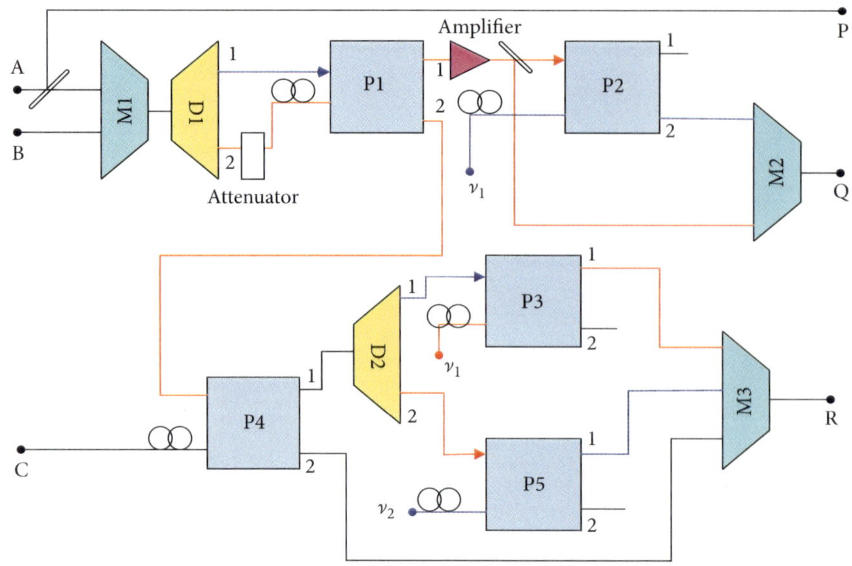

FIGURE 7: Scheme of optical Peres gate using PSWs.

designing of the optical Peres gate is discussed in the next section.

3. Scheme of Optical Peres Gate

The proposed optical Peres gate is consisting of five PSWs (P1, P2, P3, P4, and P5), three MUXs (M1, M2, and M3), and two DMUXs (D1 and D2) as shown in Figure 7. Three inputs of the Peres gate are denoted as the inputs "A," "B," and "C" and outputs of the Peres gate are denoted as outputs "P," "Q," and "R" as stated in the Introduction of the article. In this circuit, at first, the input signal "A" is split up into two equal parts with the help of a beam splitter. One part of this signal gives the output "P." Other parts of the input signal "A" and the input signal "B" both are fed into DMUX (D1) via MUX (M1). Channel 1 of D1 supplies the pump beam for PSW (P1) and channel 2 supplies the probe beam for PSW (P1). DMUX routes the optical signals in the specified channel according to the frequency of the input signals. As channel 2 carries the optical signal of frequency ν_2, PSW (P1) gets the probe beam of frequency ν_2 only when at least any one of the input signals "A" and "B" is of frequency ν_2. An attenuator is placed on channel 2 of D1. Thus the power of the probe beam of P1 is sufficiently low. As the pump beam for P1 is supplied through channel 1 of D1, when any one of two inputs "A" and "B" is of frequency ν_1 then, P1 gets the pump power. Port 1 output of P1 is amplified twice by an amplifier and then split up equally with the help of a beam splitter. One part is fed as the pump beam of PSW (P2), and another part served as one input of the MUX (M2). Port 2 of P2 is connected to the input of M2. A probe beam of frequency ν_1 is fed into P2. The output of M2 gives the output "Q" of the circuit. On the other hand, port 2 of P1 supplies the pump power for PSW (P4) where input "C" is acting as the probe beam of it. Port 1 of P4 is fed into DMUX (D2) whereas port 2 of P4 is fed into MUX (M3). Channel 1 and channel 2 of D2 supply the pump beams for PSWs P3 and P5, respectively. The two optical beams (with very low intensities) of frequencies ν_2 and ν_1, respectively, are acting as the probe beam of P3 and P5, respectively. Port 1 outputs of P3 and P5 are fed into MUX (M3). The output of M3 gives the "R" output of the circuit. Port 2 output terminals of P3 and P5 are open. Port 1 of P2 is also open. Frequency encoded truth table of Peres gate is shown in Table 2. Now some cases are discussed below.

TABLE 2: Truth table of frequency encoded Peres gate.

Frequency of the input beam			Frequency of the output beam		
A	B	C	P	Q	R
ν_1	ν_1	ν_1	ν_1	ν_1	ν_1
ν_1	ν_1	ν_2	ν_1	ν_1	ν_2
ν_1	ν_2	ν_1	ν_1	ν_2	ν_1
ν_1	ν_2	ν_2	ν_1	ν_2	ν_2
ν_2	ν_1	ν_1	ν_2	ν_2	ν_1
ν_2	ν_1	ν_2	ν_2	ν_2	ν_2
ν_2	ν_2	ν_1	ν_2	ν_1	ν_2
ν_2	ν_2	ν_2	ν_2	ν_1	ν_1

In all cases, output "P" is same as the input "A."

Case 1 (both the input signals "A" and "B" are of frequency ν_1 and the input "C" is of frequency ν_1/ν_2). Here, both the inputs "A" and "B" are fed into D1 via M1. D1 routes both the signals in channel 1. So in this case no beam is transmitted via channel 2 of D1. Thus, P1 does not get any probe beam. In the absence of the probe beam, there is no output. Thus, no power is developed at port 1 as well as port 2 of P1. So, P2 does not get any power as pump beam. In the absence of the pump beam, the power is developed at port 2 and this output beam goes to the output of M2. As the frequency of the probe beam is ν_1 for P2, the output of M2 is of frequency ν_1. Thus, the output "Q" will be of frequency ν_1.

Here, port 2 of P1 is OFF. Thus, P4 does not get any power as the pump power. In the absence of the pump beam, the input signal "C" switches to port 2. If the frequency of the input "C" is ν_1, then port 2 output is of frequency ν_1 and this beam goes to the output of M3. Thus, the output "R" will be of frequency ν_1. If the frequency of the input signal "C" is ν_2 then the output "R" will be of frequency ν_2. In case of P3 and P5, there are no pump beams. Thus, these PSWs contribute nothing to the input of M3.

Case 2 (any one of the inputs "A" and "B" is of frequency ν_1 and another is of frequency ν_2; the input "C" is of frequency ν_1/ν_2). Here, both the inputs "A" and "B" are fed into D1 via M1. D1 routes the input of frequency ν_1 to channel 1 and this beam serves as the pump beam of P1. Another input of frequency ν_2 is routed via channel 2 and this signal is treated as the probe beam of P1 after passing the attenuator. So here, P1 gets the pump beam and also gets the probe beam of frequency ν_2. In the presence of the pump beam, the probe beam switches to port 1 of it. Thus, P1 gives the optical beam of frequency ν_2 at port 1. This beam is amplified by the amplifier and split up equally into two parts with the help of the beam splitter. One part of the beam served as the pump beam of P2 and switches the probe to port 1 which is open. Another part of the output of P1 is fed into M2 and gives the output "Q" at the output of M2. Thus, the output "Q" is of frequency ν_2 for this case.

Here, the pump beam is present for P1, and thus port 2 of P1 does not supply any beam for P4. Similar to the previous case, the output "R" is same as input "C."

Case 3 (both the inputs "A" and "B" are of frequency ν_2; the input "C" is of frequency ν_1). Here, both the input beams "A" and "B" are routed through channel 2 of D1 and serve as the probe of P1. But P1 does not get any power as pump beam. The probe beam of frequency ν_2 switches to port 2. So P2 does not receive any pump beam and thus the power is developed at port 2 of P2. As the probe beam of frequency ν_1 is fed into P2, the output (at port 2) becomes of frequency ν_1 and this beam reaches the output "Q" via MUX (M2).

Here, P4 gets sufficient power as the pump beam from port 2 of P1. This pump beam switches the input signal "C" to port 1 of P4. The output beam from port 1 of P4 is fed into DMUX D2. As the frequency of the input "C" is ν_1, the output beam of P4 routed through channel 1 of D2 and this signal serves as the pump beam of P3. The probe beam for P3 is of frequency ν_2. Due to the presence of the pump beam, the output signal goes to port 1 and this output signal is of frequency ν_2. This output signal reaches the output of M3 and gives the output "R" of the Peres gate. Thus, for this condition, the output "R" is of frequency ν_2.

Here, P5 does not receive any pump beam and thus contributes nothing to the output "R."

Case 4 (all the input signals are of frequency ν_2). Similar to Case 3, here also, both the inputs "A" and "B" are of frequency ν_2 and thus the output "Q" is of frequency ν_1. In this case, also, P4 gets sufficient power as the pump beam from port 2 of P1 and so input signal "C" switches to port 1 of P4. As the frequency of the input signal "C" is ν_2 for this case, this signal is routed through channel 2 of D2 and serves as the pump beam of P5. The probe beam for P5 is of frequency ν_1. So the beam of frequency ν_1 reaches the output of the M3 via port 1 of P5. The output of M2 gives the output "R" of the Peres gate. On the other hand, P3 does not receive any pump beam and thus contributes nothing to the M2.

The simulated power spectrums of the inputs and outputs of the Peres gate are shown in Figures 8(a), 8(b), 8(c), and 8(d). The authors have represented the inputs by Gaussian pulses of beam width 0.4 nm (full width half maxima, that is, FWHM). Here optical beams of wavelength 1557 nm (corresponding frequency is 1.9255×10^{14}, i.e., ν_2) are encoded as the binary state "1," and the state "0" is encoded by the optical beam of wavelength 1552 nm and frequency 1.9317×10^{14} Hz (ν_1). Required power for input signals is calculated in Discussion of the article.

Blue colored and red colored pulses belong to the wavelengths 1552 nm (representing binary state "0") and 1557 nm (representing binary state "1"), respectively. In Figure 8(a), all the inputs of the Peres gate are in binary state "0," and thus these are represented by Gaussian pulses of wavelength 1552 nm. The output signals are also of wavelength 1552 nm for this condition. This simulation result is in accordance with the truth table of the frequency encoded Peres gate. In Figure 8(b), both the input signals "A" and "C" are of wavelength 1552 nm (representing binary state "0") and input signal "B" is of wavelength 1557 nm (representing binary state "1"). Simulation results show that the output signals "P" and "R" are of wavelength 1552 nm (representing binary state "0") and the output signal "Q" is of wavelength 1557 nm (representing binary state "1"). In Figure 8(c), both the input signals "A" and "B" are of wavelength 1557 nm (representing binary state "1") and input signal "C" is of wavelength 1552 nm (representing binary state "0"). Simulation results show that the output signals "P" and "R" are of wavelength 1552 nm (representing binary state "0") and the output signal "Q" is of wavelength 1557 nm (representing binary state "1"). In Figure 8(d), all the input signals are of wavelength 1557 nm; that is, all the inputs are at binary state "1." In this case, the output signal "P" is of wavelength 1557 nm and the output signals "Q" and "R" are of wavelength 1552 nm. All the simulation results are in accordance with the truth table of Peres gate (Table 2).

4. Scheme of Optical Full Adder Circuit Using Peres Gate

The proposed optical full adder is constructed using two Peres gates (PG1 and PG2). For each Peres gate, there are three input terminals ("1," "2," and "3") and three output terminals ("1," "2," and "3") as shown in Figure 9. Authors have denoted the input signals of the full adder circuit by the optical signals "A," "B," and "C." Among the three input signals of the full adder circuit, the inputs "A" and "B" are fed into Peres gate PG1 through the input terminals "1" and "2." One constant

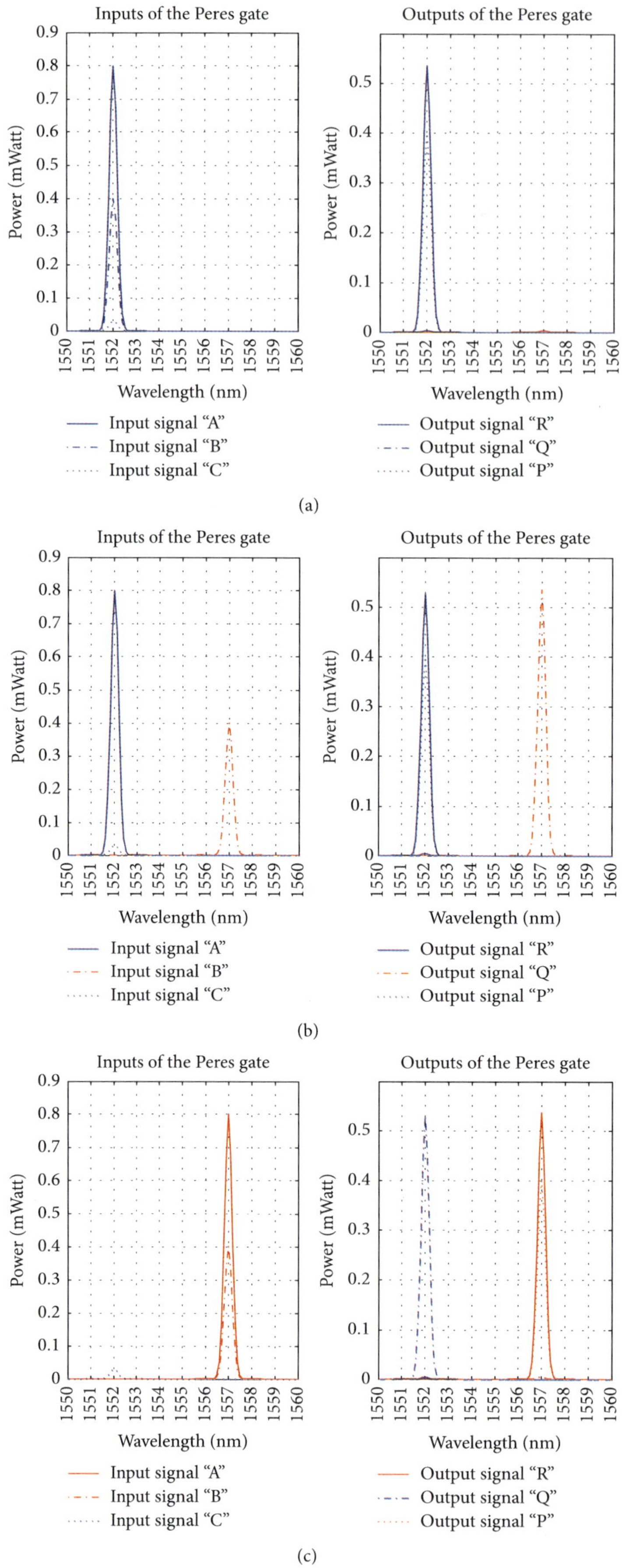

Figure 8: Continued.

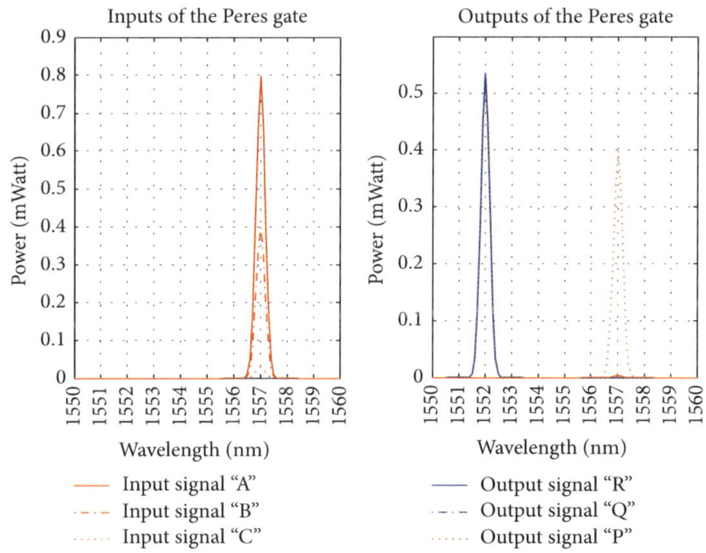

(d)

FIGURE 8: Simulated power spectrum of the optical Peres gate.

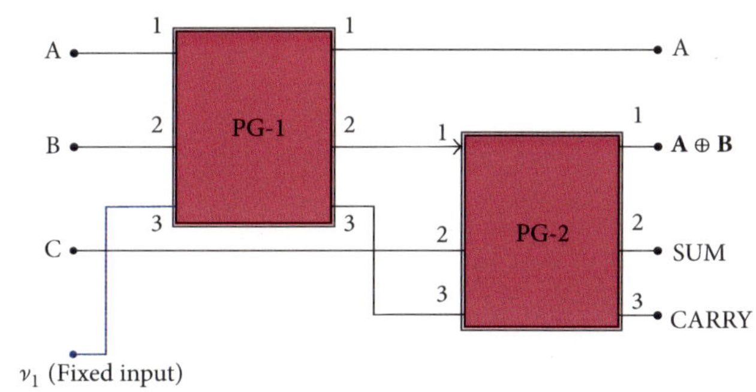

FIGURE 9: Optical full adder circuit using Peres gates.

input signal of frequency ν_1 (i.e., binary state "0") is also fed into PG1 through the input terminal "3." The first output terminal, that is, terminal 1 of PG1, gives the result which is same as input "A." Output terminal 2 of PG-1 is connected to PG2 through input terminal 1 of PG2. Output terminal 3 of PG-1 is connected to PG2 through input terminal 3 of PG2. The input signal "C" is fed into PG2 through input terminal 2.

Output terminal 2 of PG1 gives the signal which is same as $\mathbf{A} \oplus \mathbf{B}$. This is the sum of "A" and "B." This signal is fed into PG2 as the first input, whereas the input "C" of the full adder acts as the second input of PG2. Thus the first output terminal of PG2 gives $\mathbf{A} \oplus \mathbf{B}$ and output terminal 2 of PG2 gives the result $\mathbf{A} \oplus \mathbf{B} \oplus \mathbf{C}$. Output terminal 2 of PG2 gives nothing but the sum of "A," "B," and "C."

Now third input of the Peres gate PG1 is of frequency ν_1, that is, in binary state "0." Thus the logic expression of the third output of this Peres gate is $\mathbf{A}\cdot\mathbf{B}$. This is the carry of "A" and "B." This result is fed into PG2 via input terminal 3. Here output terminal 3 of PG2 gives carry of the inputs "A," "B," and "C" for all possible combinations of inputs. This result is in accordance with the truth table of the full adder, shown in Table 3. It is to be noted that as the output terminal 2 of PG-1 gives the sum of the inputs "A" and "B" and output terminal 3 of PG-1 gives the carry of the inputs "A" and "B," this Peres gate PG-1 is acting as the half adder circuit.

The simulated power spectrums of the inputs and outputs of the full adder using Peres gates are shown in Figures 10(a), 10(b), and 10(c). Similar to the optical Peres gate, here also the authors have represented the inputs and outputs by Gaussian pulses of beam width 0.4 nm (full width half maxima, that is, FWHM). Here also the optical beam of wavelength 1557 nm (corresponding frequency is 1.9255×10^{14}, that is, ν_2) is encoded as binary state "1," and state "0" is encoded by the optical beam of wavelength 1552 nm and frequency 1.9317×10^{14} Hz (ν_1).

In Figure 10(a), both the input signals "A" and "C" are in binary state "0" whereas the input "B" is in binary state "1," and these are represented by corresponding Gaussian pulses. In this condition, according to the simulation results, the sum of the input signals is in binary state "1" and thus this is represented by the Gaussian pulse of wavelength 1557 nm

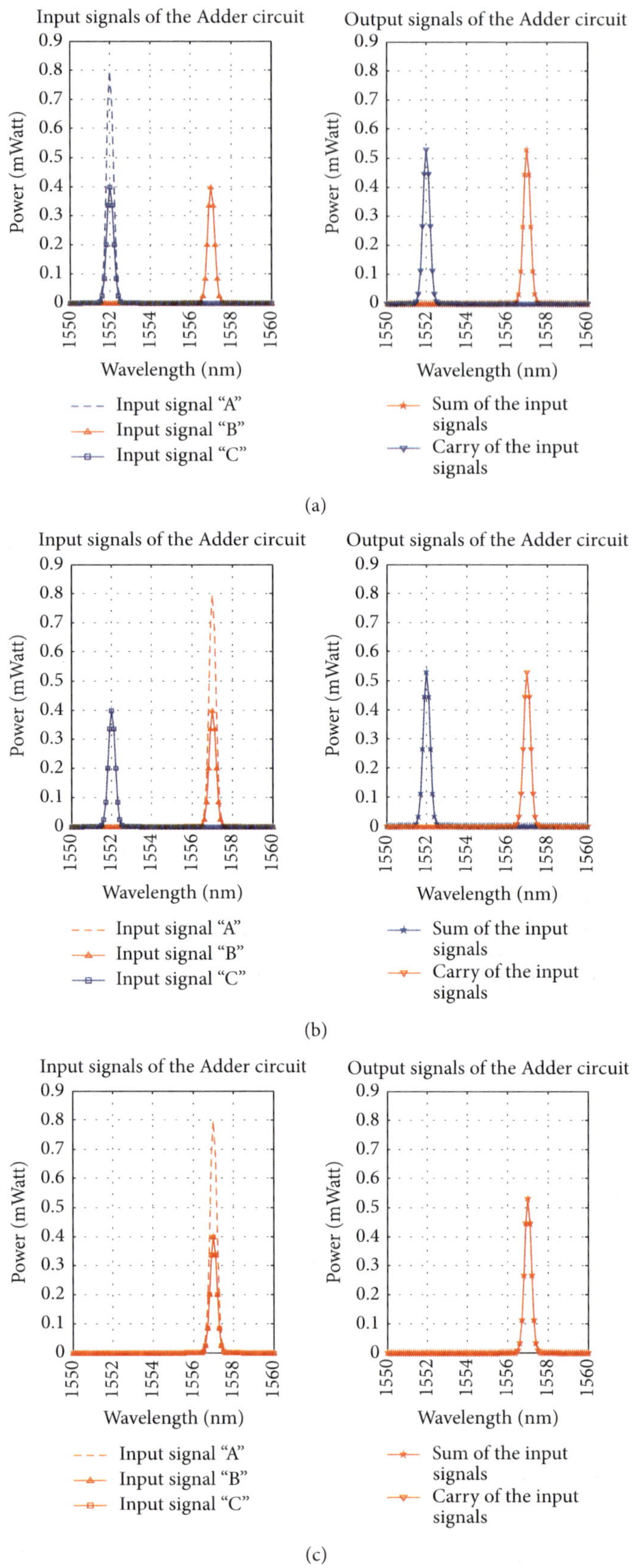

FIGURE 10: Simulated power spectrum of the optical full adder using reversible Peres gates.

TABLE 3: Truth table of frequency encoded full adder.

Frequency of the input signals			Frequency of the output signals	
A	B	C	Sum	Carry
ν_1	ν_1	ν_1	ν_1	ν_1
ν_1	ν_2	ν_1	ν_2	ν_1
ν_2	ν_1	ν_1	ν_2	ν_1
ν_2	ν_2	ν_1	ν_1	ν_2
ν_1	ν_1	ν_2	ν_2	ν_1
ν_1	ν_2	ν_2	ν_1	ν_2
ν_2	ν_1	ν_2	ν_1	ν_2
ν_2	ν_2	ν_2	ν_2	ν_2

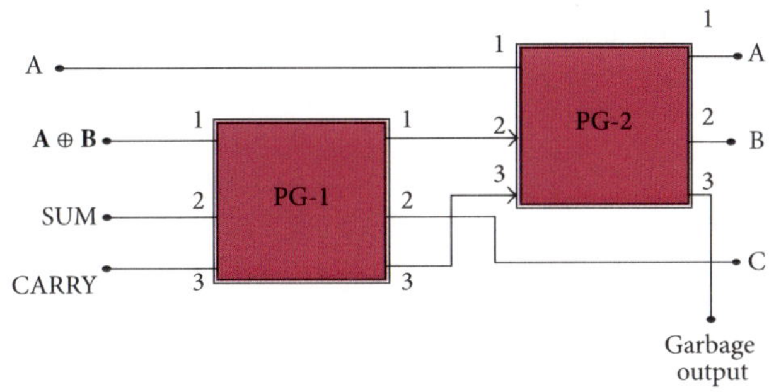

FIGURE 11: Scheme of data recovery circuit of the proposed full adder using Peres gates.

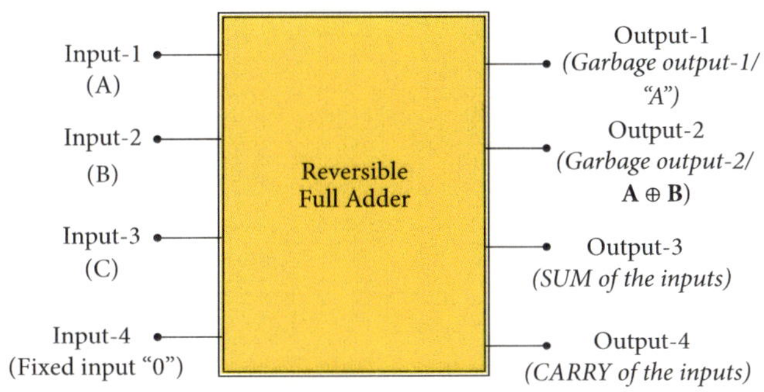

FIGURE 12: Block diagram of reversible full adder circuit.

and as the carry is in binary state "0," this is represented by the Gaussian pulse of wavelength 1552 nm. In Figure 10(b), the inputs "A" and "B" are of wavelength 1557 nm (i.e., binary state "1"), the input "C" is of wavelength 1552 nm (i.e., binary state "0"), and thus simulation results show that the sum of the inputs is of wavelength 1552 nm and the carry is of wavelength 1557 nm for this case. In Figure 10(c), all the inputs are of wavelength 1557 nm (i.e., binary state "1") and thus both the sum and the carry are of wavelength 1557 nm. Blue colored and red colored pulses belong to the wavelengths 1552 nm (representing binary state "0") and 1557 nm (representing binary state "1"), respectively.

One can easily retrieve the input information from the output information of the proposed full adder. To retrieve the input information, the authors have designed a data recovery circuit (as shown in Figure 11) which is nothing but the mirror image of the proposed full adder circuit. In Figure 11, all the outputs of the full adder circuit are fed into the data recovery circuit as the inputs of it and at the output of the data recovery circuit the input signals "A," "B," and "C" are recovered. The block diagrams of the reversible full adder and the data recovery circuit are shown in Figures 12 and 13, respectively.

5. Discussion

Commercially available tensile strained bulk JDS-uniphase SOA can be used to construct PSW for Peres gate and full adder circuit. For this type of SOA, the power of the pump beam should be at least 0.4 mWatt and the power of the probe beam should not be greater than 0.03 mWatt. In case of Peres gate, the input beam "A" is split up into two equal parts. So the power of the input "A" should be at least 0.8 mWatt and the input "B" should be of power 0.4 mWatt, and the power of input beam "C" should not be greater than 0.03 mWatt. For adder circuit, one constant input beam of frequency ν_1 is fed into the Peres gate PG1. This beam acts as the probe

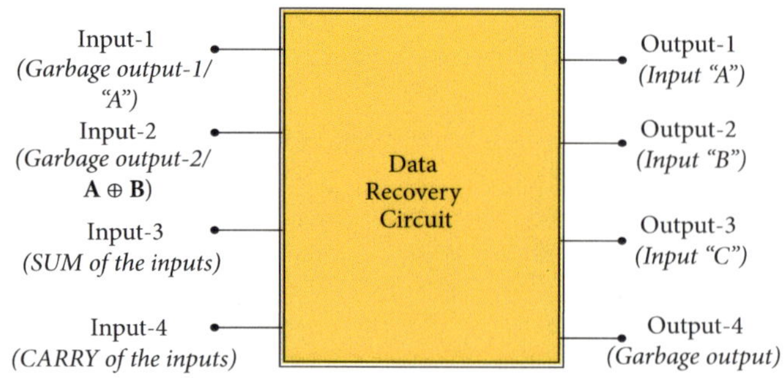

FIGURE 13: Block diagram of data recovery circuit of the proposed full adder.

beam and thus power of this beam should not be greater than 0.03 mWatt. An amplifier should be used to amplify the output signal of output terminal 2 of "PG-1" so that this signal can act as the first input signal of PG-2 and also an attenuator should be used at output terminal 3 of PG-1 so that this signal can act as the third input signal of PG-2.

When the pump beam is present, SOA shows nonlinearity and at the output both cross-gain modulation and cross-phase modulation take place. SOA shows some unwanted pattern effect due to the slow gain recovery time. This can be reduced using several techniques [18–22]. Due to spontaneous recombination of electron-hole pairs in the SOA gain medium spontaneous emission arises and this spontaneous emission is also amplified with the input signal. Amplified spontaneous emission (ASE) produces noise at the output of SOA and this is the main source of noise in SOA. This noise cannot be neglected, but, using a proper band pass filter with least bandwidth, this noise can be miniaturized [23, 24].

The state of polarization (SOP) of the light can be determined with the help of Muller matrix, Jones Matrix, Poincare's sphere, and so on [25]. If one represents the state of polarization (SOP) of the input-output signals by the Jones matrix, then each element of the matrix depends on the phase and amplitude of the signals. Thus, the change in SOP can be determined. Similarly, from the points on the surface of Poincare's sphere, the change in state of polarization can be determined.

It is also possible to measure the rotation of the state of polarization of the probe beam by replacing the PBS of the experimental setup of Figure 2 by an optical analyzer [26].

Initially, the pass axis of the polarizer and analyzer are kept parallel, and linearly polarized probe beam is made incident at an angle of 45 degrees with the active layer of the SOA. Now incident beam will decompose into TE and TM mode with equal amplitudes and propagate with equal velocities in the absence of pump beam, and at the output end of the SOA they will recombine without changing its state of polarization and finally pass through the pass axis of the analyzer with maximum intensity. In the presence of the pump beam, birefringence is developed, and therefore TE mode and TM mode propagate with unequal velocities. Consequently, a phase difference is developed between TE and TM modes depending on the power of pump beam and path length of the SOA. Finally, they recombine at the output end of the SOA with a phase difference $\Delta\Phi$. As the two modes are of equal amplitudes, resultant vibration will rotate through an angle "δ" such that "$\delta = \Delta\Phi/2$." Now if one rotates the pass axis of the analyzer through an angle δ along the direction of rotation of the state of polarization, the maximum intensity of the resultant beam is got. Thereby it is possible to measure the angle of rotation of the state of polarization of the beam using the relation "$\delta = \Delta\Phi/2$."

6. Conclusion

The proposed optical full adder circuit using reversible logic gates has actually four inputs and four outputs. In addition to three input signals ("A," "B," and "C"), one constant input signal of frequency ν_1 is applied in the circuit. At the output, we get sum and carry of the inputs of the full adder, and the circuit gives another two garbage outputs ("A" and "**A ⊕ B**"). But with the help of these two garbage outputs, one can easily recover the inputs "A" and "B." Thus the utilization of this circuit is that one which can get full information about the inputs from the knowledge of the output information. The switching speed of SOA is very high. One can expect 100 Gb/s speed from this circuit. The frequencies are chosen from the "C band" of the spectrum as SOA shows frequency independent gain in this region. The whole circuit is optical one and the data are frequency encoded; thus this circuit will be an essential component for designing optical ALU and WDM communication networks.

Conflicts of Interest

The authors declare that they have no conflicts of interest regarding the publication of this paper.

References

[1] R. Landauer, "Irreversibility and heat generation in the computing process," *International Business Machines Journal of Research and Development*, vol. 5, pp. 183–191, 1961.

[2] C. H. Bennett, "Logical reversibility of computation," *International Business Machines Journal of Research and Development*, vol. 17, pp. 525–532, 1973.

[3] Y. Van Rentergem and A. De Vos, "Optimal design of a reversible full adder," *International Journal of Unconventional Computing*, vol. 1, pp. 339–355, 2005.

[4] H. V. R. Aradhya, B. V. P. Kumar, and K. N. Muralidhara, "Design of control unit for low power AU using reversible logic," in *Proceedings of the International Conference on Communication Technology and System Design 2011, ICCTSD 2011*, pp. 631–638, India, December 2011.

[5] G. Yang, W. N. Hung, X. Song, and M. Perkowski, "Majority-based reversible logic gates," *Theoretical Computer Science*, vol. 334, no. 1-3, pp. 259–274, 2005.

[6] S. Kotiyal, H. Thapliyal, and N. Ranganathan, "Mach-Zehnder interferometer based all optical reversible NOR gates," in *Proceedings of the 2012 IEEE Computer Society Annual Symposium on VLSI, ISVLSI 2012*, pp. 207–212, Amherst, Mass, USA, August 2012.

[7] G. K. Maity, J. N. Roy, and S. P. Maity, "Mach-Zehnder interferometer based all-optical Peres gate," *Communications in Computer and Information Science*, vol. 192, no. 3, pp. 249–258, 2011.

[8] S. Kotiyal, H. Thapliyal, and N. Ranganathan, "Mach-Zehnder interferometer based design of all optical reversible binary adder," in *Proceedings of the 2012 Design, Automation & Test in Europe Conference & Exhibition (DATE 2012)*, pp. 721–726, Dresden, Germany, March 2012.

[9] D. Mandal, M. K. Mandal, and S. K. Garai, "Frequency encoded data based optical full adder using reversible Toffoli gates," *Journal of Optics (India)*, vol. 45, no. 2, pp. 197–207, 2016.

[10] K. Prudhvi and Y. Syamala, "Transistor level implementation of digital reversible circuits," *International Journal of VLSI design & Communication Systems (VLSICS)*, vol. 5, no. 6, 2014.

[11] M. Zhao, J. De Merlier, G. Morthier, and R. G. Baets, "Dynamic birefringence of the linear optical amplifier and application in optical regeneration," *IEEE Journal of Selected Topics in Quantum Electronics*, vol. 8, no. 6, pp. 1399–1404, 2002.

[12] B. S. G. Pillai, M. Premaratne, D. Abramson et al., "Analytical characterization of optical pulse propagation in polarization-sensitive semiconductor optical amplifiers," *IEEE Journal of Quantum Electronics*, vol. 42, no. 10, pp. 1062–1077, 2006.

[13] L.-Q. Guo and M. J. Connelly, "A Mueller-matrix formalism for modeling polarization azimuth and ellipticity angle in semiconductor optical amplifiers in a pump-probe scheme," *Journal of Lightwave Technology*, vol. 25, no. 1, pp. 410–420, 2007.

[14] H. J. S. Dorren, D. Lenstra, Y. Liu, M. T. Hill, and G.-D. Khoe, "Nonlinear polarization rotation in semiconductor optical amplifiers: Theory and application to all-optical flip-flop memories," *IEEE Journal of Quantum Electronics*, vol. 39, no. 1, pp. 141–148, 2003.

[15] S. Mandal, D. Mandal, and S. K. Garai, "An all-optical method of developing data communication system with error detection circuit," *Optical Fiber Technology*, vol. 20, no. 2, pp. 120–129, 2014.

[16] D. Mandal, S. Mandal, and S. K. Garai, "A new approach of developing all-optical two-bit-binary data multiplier," *Optics & Laser Technology*, vol. 64, pp. 292–301, 2014.

[17] D. Mandal and S. K. Garai, "All-optical binary logic unit (BLU) using frequency encoded data," *Optical Fiber Technology*, vol. 22, pp. 56–67, 2015.

[18] L. Zhang, I. Kang, A. Bhardwaj et al., "Reduced recovery time semiconductor optical amplifier using p-type-doped multiple quantum wells," *IEEE Photonics Technology Letters*, vol. 18, no. 22, pp. 2323–2325, 2006.

[19] F. Girardin, G. Guekos, and A. Houbavlis, "Gain recovery of bulk semiconductor optical amplifiers," *IEEE Photonics Technology Letters*, vol. 10, no. 6, pp. 784–786, 1998.

[20] R. J. Manning and D. A. O. Davies, "Three-wavelength device for all-optical signal processing," *Optics Expresss*, vol. 19, no. 12, pp. 889–891, 1994.

[21] J. Wang, A. Marculescu, J. Li et al., "Pattern effect removal technique for semiconductor-optical-amplifier-based wavelength conversion," *IEEE Photonics Technology Letters*, vol. 19, no. 24, pp. 1955–1957, 2007.

[22] J. Leuthold, D. M. Marom, S. Cabot, J. J. Jaques, R. Ryf, and C. R. Giles, "All-Optical Wavelength Conversion Using a Pulse Reformatting Optical Filter," *Journal of Lightwave Technology*, vol. 22, no. 1, pp. 186–192, 2004.

[23] A. A. El Aziz, E. K. El Nayal, M. H. Aly, A. K. Aboulseoud, and W. P. Ng, "Impact of amplified spontaneous emission on the travelling wave semiconductor optical amplifier performance," in *Proceedings of the 2013 2nd International Workshop on Optical Wireless Communications, IWOW 2013*, pp. 123–127, UK, October 2013.

[24] S. K. Garai, "All-optical quaternary logic gates - An extension of binary logic gates," *Optics & Laser Technology*, vol. 67, pp. 125–136, 2015.

[25] D. Goldstein, *Polarized Light*, Marcel Dekker, New York, NY, USA, 2nd edition, 2003.

[26] S. K. Garai, "A novel method of developing all optical frequency encoded Fredkin gates," *Optics Communications*, vol. 313, pp. 441–447, 2014.

8

Phase-Shift Keying Modulated Data Signal using SOA-MZI-Based All-Optical Logic AND Gate at 80 Gb/s

Amer Kotb[1,2] **and Yasser Mohamed**[2]

[1] *The Guo China-US Photonics Laboratory, Changchun Institute of Optics, Fine Mechanics, and Physics, Chinese Academy of Sciences, Changchun 130033, China*
[2] *Department of Physics, Faculty of Science, University of Fayoum, Fayoum 63514, Egypt*

Correspondence should be addressed to Amer Kotb; amer_22003@yahoo.com

Academic Editor: Gong-Ru Lin

All-optical logic AND operation with phase-shift keying (PSK) modulated data signals based on a semiconductor optical amplifier (SOA) assisted Mach-Zehnder interferometer is numerically analyzed and investigated. By solving the rate equations of the SOA, the dependence of the quality factor on the critical parameters of the input signal and the SOA is examined, including the impact of the amplified spontaneous emission to obtain realistic results. The obtained results confirm that the all-optical logic AND scheme with PSK modulated data signals is capable of operating at a data rate of 80 Gb/s with both logical correctness and acceptable quality compared to the system using on-off-keying signal.

1. Introduction

Modulation is the transfer process of information over a medium. Recently, advanced optical modulation formats have attracted increased attention. Some of these formats carry information through the amplitude, which is known as amplitude modulation (AM). In AM, the amplitude of the carrier changes in response to the information by keeping the frequency and the phase constants. In many applications, AM is being replaced by frequency and phase modulation technologies even though it is not as widely used as it was in previous years. However, AM is not efficient in terms of its use of power and bandwidth and is prone to high levels of noise. In contrast, phase modulation (PM) signals convey the information in the phase itself. PM schemes are attracting much interest for use in the ultrafast optical communications systems because they are less sensitive to nonlinear fiber transmission impairments, require lower optical signal-to-noise ratio, are more robust against pattern-dependent degradations, and can give higher spectral efficiency compared to conventional intensity modulation formats [1, 2]. This fact has spurred great interest in conducting all-optical operations on phase-shift keying (PSK) modulated data so as to realize exclusively by the means of the light critical network functionalities that are compatible with the specific format. All-optical modulation based on nonlinear silicon photonics waveguide has been demonstrated [3–7]. Such all-optical logic operations are an integral part of building all-optical data networks, where packet routing, data buffering, and wavelength conversion are expected to be processed in the optical plane. For this purpose, a technical option that has been widely adopted is to employ semiconductor optical amplifiers (SOAs). In fact, the SOAs' devices have technologically matured to the point that they have become the primary choice for the implementation of all-optical logic gates. SOA is the most attractive optoelectronics device due to its compact size, low power requirement, temperature stability, and ease to connect with other optoelectronics devices. In recent years, the performances of the all-optical logic AND gate based on several different schemes using SOA have been reported [8–26]. The AND is involved in the accomplishment of numerous tasks in the optical domain both in fundamental and in system-oriented level [26]. Various SOA-based schemes for the demonstration of all-optical logic gates using PSK modulated data signals have been reported [27–33]. This activity has also included the execution of the Boolean AND operation on the

PSK modulated data signals [31], which has been possible owing to the capability of the SOAs to handle the phase modulated signals as well [32]. However, these demonstrations cannot conveniently keep pace with the excessive increase in single-channel data rates due to the slow SOA's gain and phase recovery. The performances of all-optical logic XOR and OR gates for 100 Gb/s PSK modulated data signal in SOAs assisted Mach-Zehnder interferometer (MZI) have been numerically analyzed [34, 35]. In this work, we continue and generalize the relevant results of previous works [31, 35] by investigating the performance of an all-optical AND gate using PSK modulated data signals based on SOAs-MZI at a data rate of 80 Gb/s. The dependence of the gate's quality factor (QF) on the critical parameters of the input signals and the SOAs has been examined and assessed. For realistic results, the effect of the amplified spontaneous emission (ASE) on the QF has also been included. The outcome of our work complements the set of all-optical gates that have been addressed so far for PSK signals.

2. SOA-MZI Model

The schematic diagram and truth table for AND operation is shown in Figure 1, where an optical band-pass filter (OBPF) is used to reject out the spectral components other than the switched signal.

The semiconductor materials used in the SOA should be direct band-gap and match lattices for the radiative recombination processes. The semiconductor materials used in this simulation are InGaAsP/InP with wavelengths of 1300 nm or 1550 nm. The operation of the SOA-based MZI can be theoretically studied by means of the rate equation model [8–12]. This simulation has been prepared and run via Wolfram Mathematica. This simulation takes into account the nonlinear effects of the interband and intraband. More specifically, the carrier heating (CH) results from thermalization of carriers in the entire energy-band following the pulse. This is a fast process occurring in the time scale of 300 fs. The input pulse reduces the gain at the photon-energy of this pulse; that is, in the gain spectrum it burns a hole. The process is known as a spectral hole burning (SHB). By taking into consideration both the CH and SHB effects, the time-dependent gain for each SOA is given by [9, 23]

$$\begin{aligned}\frac{dh_{CD}(t)}{dt} &= \frac{h_0 - h_{CD}(t)}{\tau_C} \\ &\quad - \left(\exp\left[h_{CD}(t) + h_{CH}(t) + h_{SHB}(t)\right] - 1\right)\frac{P(t)}{E_{sat}}, \\ \frac{dh_{CH}(t)}{dt} &= -\frac{h_{CH}(t)}{\tau_{CH}} \\ &\quad - \frac{\varepsilon_{CH}}{\tau_{CH}}\left(\exp\left[h_{CD}(t) + h_{CH}(t) + h_{SHB}(t)\right] - 1\right)P(t), \\ \frac{dh_{SHB}(t)}{dt} &= -\frac{h_{SHB}(t)}{\tau_{SHB}} \\ &\quad - \frac{\varepsilon_{SHB}}{\tau_{SHB}}\left(\exp\left[h_{CD}(t) + h_{CH}(t) + h_{SHB}(t)\right] - 1\right)P(t) \\ &\quad - \frac{dh_{CD}(t)}{dt} - \frac{dh_{CH}(t)}{dt}.\end{aligned} \tag{1}$$

The total gain $G(t)$ of each SOA is given by

$$G(t) = \exp\left[h_{CD}(t) + h_{CH}(t) + h_{SHB}(t)\right], \tag{2}$$

where function "h" is the SOA's gain integrated over its length due to the carrier depletion (CD), CH, and SHB. $G_0 = \exp[h_0]$ is the unsaturated power gain. E_{sat} is the saturation energy, which is linked to the saturation power (P_{sat}) through $E_{sat} = P_{sat}\tau_c$, where τ_c is the carrier lifetime. $P(t)$ is input optical power injected into the SOAs. τ_{CH} and τ_{SHB} are the temperature relaxation rate and the carrier-carrier scattering rate, respectively. ε_{CH} and ε_{SHB} are the nonlinear gain suppression factors due to CH and SHB, respectively.

The carrier density-induced phase change of each SOA is given by [23]

$$\Phi(t) = -0.5\left[\alpha h_{CD}(t) + \alpha_{CH} h_{CH}(t)\right], \tag{3}$$

where α is the traditional linewidth enhancement factor, α_{CH} is the linewidth enhancement due to CH, and α_{SHB} is the linewidth enhancement due to SHB. The contribution α_{SHB} is almost null since SHB produces a nearly symmetrical spectral hole centered at the signal wavelength. Kramers-Kronig integrand becomes antisymmetric around the operating frequency and its integral remains small [24].

The input pulses used in this simulation are assumed to be Gaussian-shaped pulses, which have power described by [9]

$$\begin{aligned}P_{A,B}(t) &\equiv P_{in}(t) \\ &= \sum_{n=-\infty}^{n=+\infty} a_{nA,B}\frac{2\sqrt{\ln(2)}E_0}{\sqrt{\pi}\tau_{FWHM}}\exp\left(-\frac{4\ln(2)(t-nT)^2}{\tau_{FWHM}^2}\right),\end{aligned} \tag{4}$$

where $a_{nA,B}$ stands for the nth pulse, which can take the logical value of "1" or "0" with equal probability. E_0 is the input pulse energy and τ_{FWHM} is the pulse width (full-width at half-maximum). n is the length of the pseudorandom binary sequence (PRBS), 127 bit, and T is the bit period.

The modulated PSK signals injected into the SOAs have the following power function [36]:

$$P_{PSK}(t) = \exp\left(J\pi P_{A,B}(t)\right). \tag{5}$$

The split components of the PSK modulated signal B suffer the optical properties of the SOAs that have been perturbed by the respective data in the two arms of the MZI. When they interfere, the resultant AND output obeys [23]

$$\begin{aligned}P_{AND}(t) &= 0.25P_{PSKB}(t)\Big(G_1(t) + G_2(t) \\ &\quad - 2\sqrt{G_1(t)G_2(t)}\cos\left[\Phi_1(t) - \Phi_2(t)\right]\Big),\end{aligned} \tag{6}$$

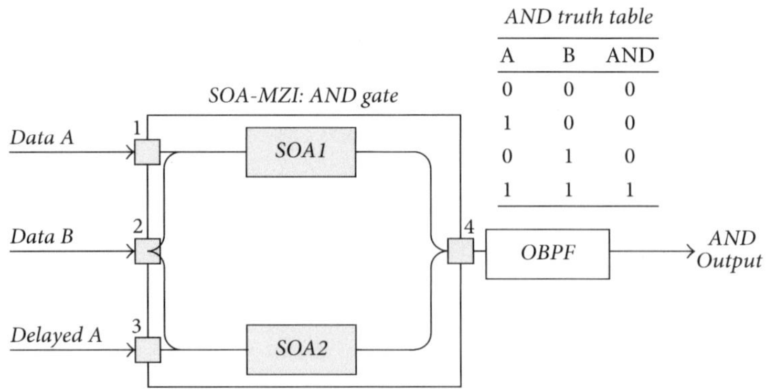

FIGURE 1: Schematic diagram and truth table for AND operation based on SOA-MZI.

where P_{PSKB} is the modulated PSK input power of signal B. $G_{1,2}(t)$ and $\Phi_{1,2}(t)$ are the time-dependent total gains and the phase changes of the signal B inside SOA1 and SOA2, respectively.

The parameters used in this simulation are $I = 100\,\text{mA}$, $G_0 = 30\,\text{dB}$, $P_0 = 0.006\,\text{pJ}$, $\tau_{FWHM} = 1.5\,\text{ps}$, $\tau_c = 200\,\text{ps}$, $\tau_{CH} = 0.3\,\text{ps}$, $\tau_{SHB} = 0.1\,\text{ps}$, $\varepsilon_{CH} = \varepsilon_{SHB} = 0.08\,\text{W}^{-1}$, $\alpha = 4$, $\alpha_{CH} = 1$, and $P_{sat} = 25\,\text{mW}$. The instantaneous optical power inside each SOA for the AND operation is given by

$$P_1(t) = P_{PSKA}(t) + 0.5P_{PSKB}(t),$$
$$P_2(t) = P_{\text{Delayed PSKA}}(t) + 0.5P_{PSKB}(t). \quad (7)$$

At zero injected current, the spontaneous emission process appeared and is amplified through the optically active region of the amplifier. The amplifiers degrade the SNR of the amplified signal primarily because of ASE, which is added to the signal in this simulation. The ASE can be used to measure the position of the gain peak, the pulse width of the gain curve, and the gain ripple. The power of the ASE has been numerically added to the output power of the AND operation using the following basic equation [23, 37]:

$$P_{ASE} = N_{SP}(G_0 - 1)K\upsilon B_0, \quad (8)$$

where K is Planck's constant, B_0 is the optical bandwidth, and υ is the optical frequency.

3. All-Optical AND Model

In order to realize the AND operation, we have used a similar scheme to that in [8, 10]. The logic AND operation is achieved between the two PSK modulated data signals by means of cross-phase modulation (XPM) in SOAs 1 and 2 located in the two arms of the MZI. The input data A is injected into the upper arm of the SOA-MZI, while its delayed copy with a time delay (τ) is injected into the lower arm. Concurrently, data B at wavelength λ_2 acts as enabling beam and is injected into the configuration of the middle port of the MZI. This signal is halved into two components of equal amplitude and then is injected into the middle arm of the MZI. The input signals A and B have two different arbitrary PRBS. Initially, the two arms of the MZI are balanced in phase and the output is zero. Then data A and its delayed replica are combined to create a phase window experienced by data B. When A = "0", this phase window does not exist, and the output is "0" for both B = "1" and B = "0". However, when A = "1," the phase window opens allowing the split copies of signal B = "1" to interfere nondestructively, which results in a nonzero output. Thus, according to this switching mode, the circuit is produced "1" at λ_2 only when both data A and B are "1", which is functionally the same as an all-optical AND gate.

4. Results

The operation of the AND gate has been analyzed using a numerical solution of the SOA's rate equations in (1). Figure 2 illustrates the simulation results and the eye diagram for the Boolean AND operation, from which it can be noticed that the eye is clear and open. The eye diagram is with the degrading effects while the pseudo-eye diagram is without any degrading effects. The QF gives information for the optical signal-to-noise ratio (SNR) in digital transmission. The performance of the AND operation has been evaluated by the measurement of the QF, which is given by $QF = (P_1 - P_0)/(\sigma_1 + \sigma_2)$, where $P_{1,0}$ are the average powers of the expected "1"s and "0"s and $\sigma_{1,2}$ are the corresponding standard deviations. The QF gives the information of the optical SNR in digital transmission. In this simulation, the achieved QF is 34.6 at 80 Gb/s, which is well over the critical limit of 6 required to keep the bit-error-rate (BER) less than 10^{-9} [23]. In order to obtain realistic results, the metric characteristics of the AND operation, such as the BER and the extinction ration (ER), have been calculated. The BER is linked to the QF through the following basic equation [23]:

$$\text{BER} = 0.5\,\text{erfc}\left[\frac{QF}{\sqrt{2}}\right], \quad (9)$$

where erfc is the error-function. The calculated BER in this simulation is 1.26×10^{-262}.

The ER is defined in dB by the following basic equation [38, 39]:

$$\text{ER (dB)} = 10\log_{10}\left[\frac{P_1^{\min}}{P_0^{\max}}\right], \quad (10)$$

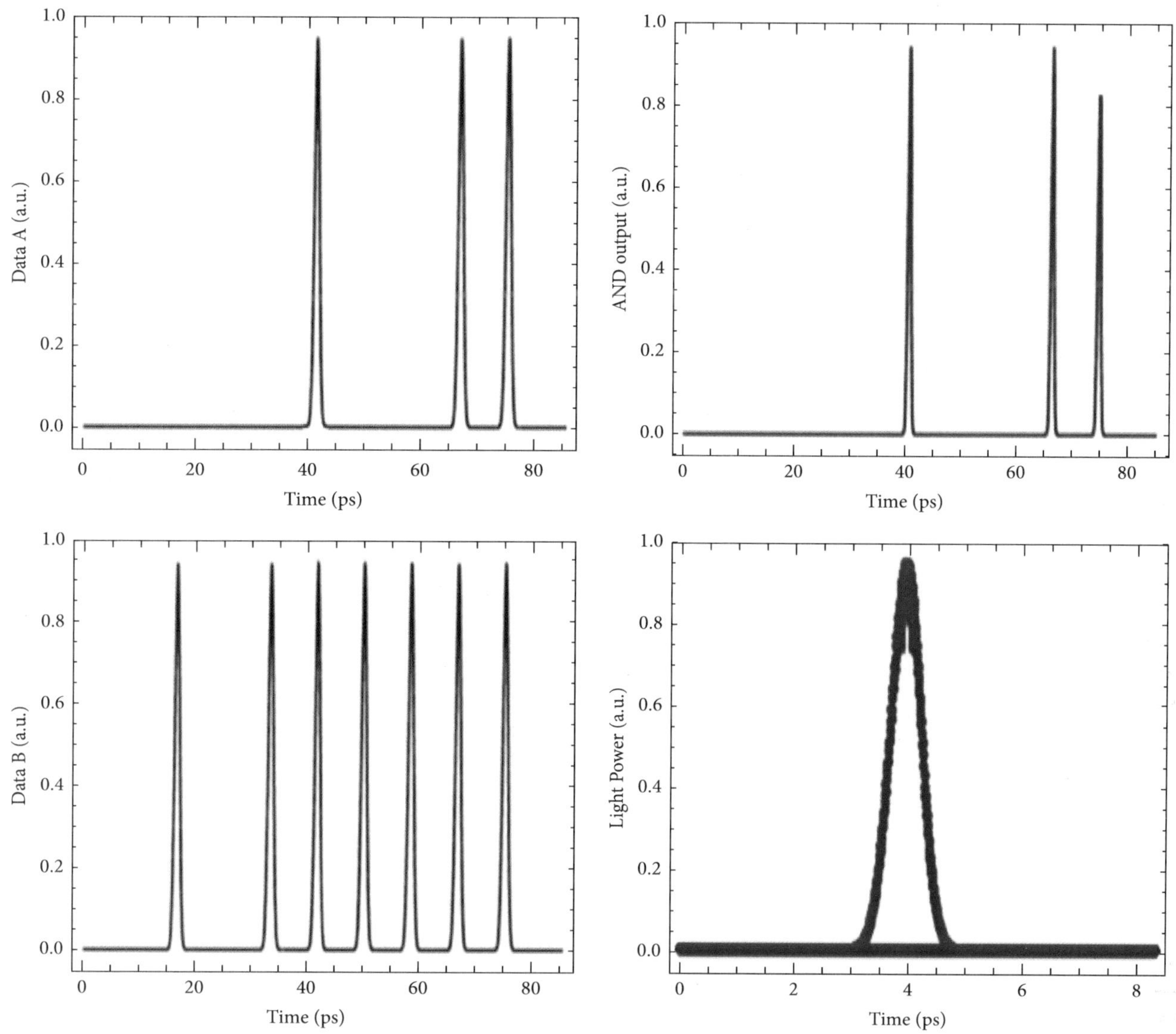

FIGURE 2: Simulation results and the corresponding eye diagram for AND operation. The achieved QF is 34.6 at a data rate of 80 Gb/s.

where $P_1^{\min}$ is the minimum value of the peak power of logic "1" and $P_0^{\max}$ is the maximum value of the peak power of logic "0". In this simulation, the calculated ER using (10) is around 40 dB.

The simulated QF depends on the pulse energy and the injected SOA injection current of SOA-MZI-based for the AND operation at $N_{sp} = 2$ and 4 is shown, respectively, in Figures 3(a) and 3(b). As shown in Figure 3(a), a high input pulse energy causes a strong saturation, which reduces the value of the QF. On the other hand, the QF is increased as more carriers are supplied via higher bias currents as shown in Figure 3(b). The injection current plays an important role in the amplification process. Moreover, the injection current required to obtain a QF of 34.6 is a few tens of mAs with the PSK modulated signals using SOA compared to hundreds of mAs with the SOA based on other modulation formats [23].

Figures 4(a) and 4(b) show the dependence of the QF on the SOA's carrier lifetime (τ_c) and the pulse width (τ_{FWHM}) at $I = 50$ mA and 100 mA, respectively. For a standard SOA, the value of the carrier lifetime varies from 100 ps to 300 ps depending on the type of the semiconductor material [23]. In our simulation case, the value of this parameter can be changed several times to calculate the corresponding QF for each value. By this way, the value of the carrier lifetime, which makes high QF, can be easily optimized. Since the carrier lifetime determines the speed of gain recovery of the SOA [23], the QF is higher for smaller values of this parameter as shown in Figure 4(a). A similar trend is observed in Figure 4(b), where the QF is decreased with increasing the pulse width. The QF is decreased with increasing the pulse width because two neighboring pulses tend to overlap for wider pulse widths.

The calculated QF depends on the length and the thickness of the SOA's active region as shown in Figures 5(a) and 5(b), respectively. The typical length and thickness of the standard SOA are 2 mm and 0.4 μm, respectively [23]. By

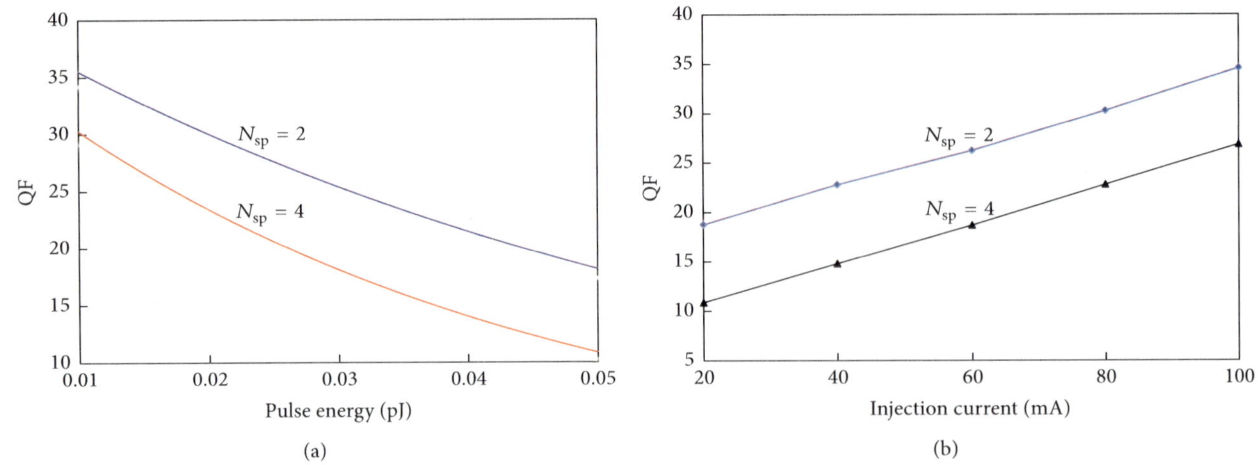

FIGURE 3: QF versus (a) pulse energy and (b) SOA's injection current for AND operation at $N_{sp} = 2$ and 4.

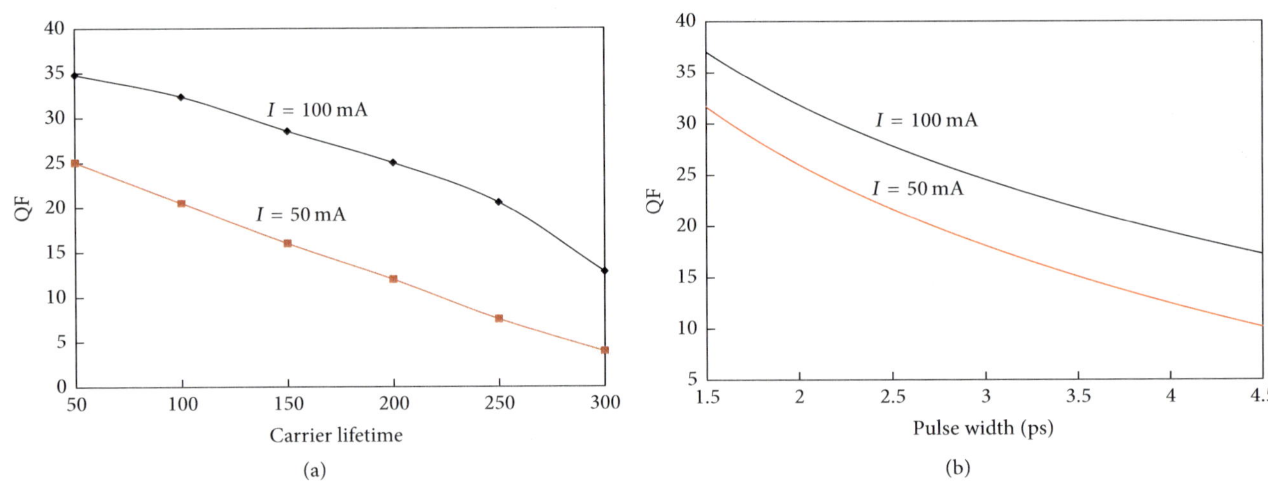

FIGURE 4: QF versus (a) SOA's carrier lifetime and (b) pulse width for AND operation.

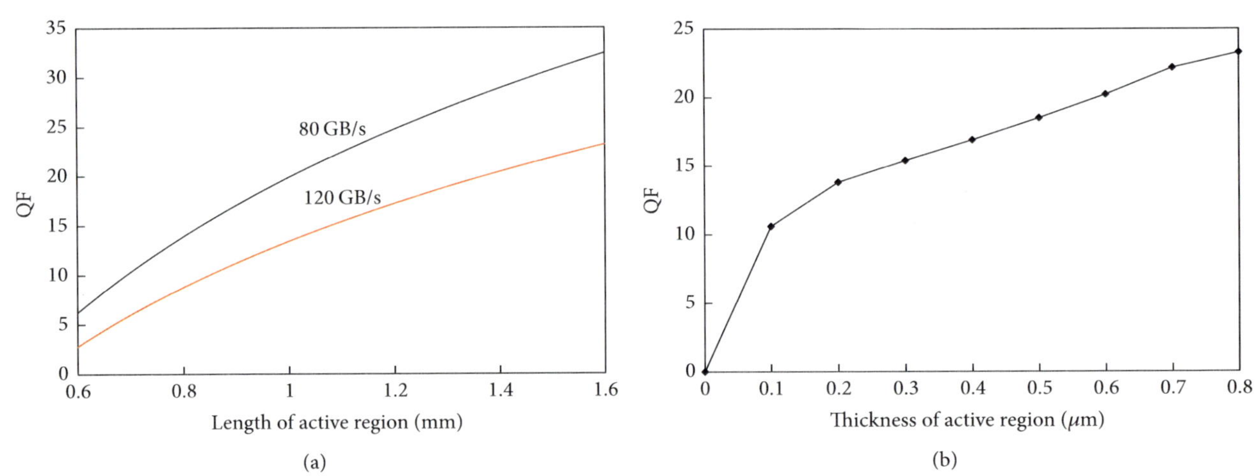

FIGURE 5: QF versus SOA's active region (a) length at 80 Gb/s & 120 Gb/s and (b) thickness for AND operation.

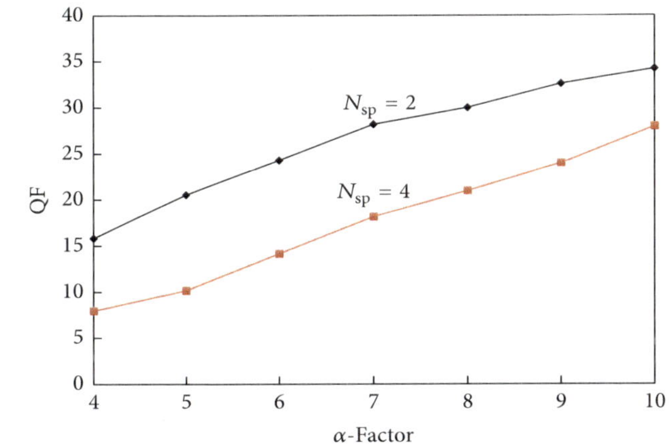

FIGURE 6: QF versus SOA's traditional linewidth enhancement factor (α-factor) for AND operation at $N_{sp} = 2$ and 4.

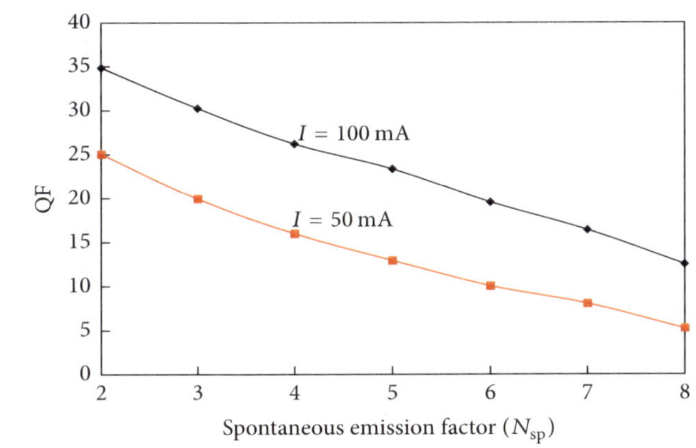

FIGURE 7: QF versus spontaneous emission factor (N_{sp}) of SOA-MZI-based AND operation at $I = 50$ mA and 100 mA.

increasing the length of the SOA, the overall gain increases exponentially and then the QF as shown in Figure 5(a), which has been measured at two different data rates. A thick active region is used for fast gain recovery and hence an acceptable performance of the Boolean functions. The optical gain region is determined by the active region thickness. The carrier density is increased with increasing the thickness of the active region and hence the QF as shown in Figure 5(b).

The dependence of the QF on the traditional linewidth enhancement factor (α-factor) at $N_{sp} = 2$ and 4 is shown in Figure 6. This figure confirms that the QF is increased with α-factor due to the effect of this parameter on the induced phase change and accordingly on the magnitude of switching [40]. Moreover, for the PSK modulated signals, the QF is higher for a small α-factor.

The QF versus N_{sp} of a SOA-MZI-based AND operation at $I = 50$ mA and 100 mA is shown in Figure 7. It is clearly seen that the QF is decreased with increasing the N_{SP}, which is considered as a noise. This simulation has been carried out at $G_0 = 30$ dB, $B_0 = 3$ nm, and a signal wavelength of 1550 nm. The effect of ASE is experimentally verified by adding a few nm wide optical unmodulated signals to the streams data with negligible saturation effects.

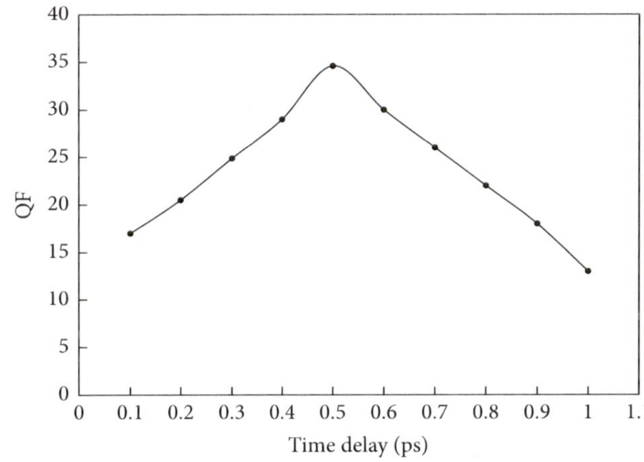

FIGURE 8: QF versus time delay for AND operation.

The time delay (τ) of delayed data A is a very important parameter for the performance of the AND gate. The optimum time delay used in this simulation that maximizes the QF to almost 34 is 0.5 ps. Figure 8 shows that a shorter τ leads to a sharp fall of the QF. At high τ, the magnitude of the phase switching window between the data A and its delayed decreases and hence the corresponding QF.

5. Conclusion

In conclusion, the performance of ultrafast all-optical logic AND operation with phase-shift keying (PSK) modulated data signals in semiconductor optical amplifier (SOA) assisted Mach-Zehnder interferometer was numerically simulated. The obtained results show that the all-optical logic AND gate based on PSK modulated data signals was capable of operating at a data rate of 80 Gb/s with higher output quality factor than that used for SOA based on other modulation formats.

Conflicts of Interest

The authors declare that they have no conflicts of interest.

References

[1] K. Mishina, S. Kitagawa, and A. Maruta, "All-optical modulation format conversion from on-off-keying to multiple-level phase-shift-keying based on nonlinearity in optical fiber," *Optics Express*, vol. 15, no. 13, pp. 8444–8453, 2007.

[2] G. Charlet, "Progress in optical modulation formats for high-bit rate WDM transmissions," *IEEE Journal of Selected Topics in Quantum Electronics*, vol. 12, no. 4, pp. 469–483, 2006.

[3] C.-L. Wu, S.-P. Su, and G.-R. Lin, "All-optical modulation based on silicon quantum dot doped SiOx: SI-QD waveguide," *Laser & Photonics Reviews*, vol. 8, no. 5, pp. 766–776, 2014.

[4] Q. Xu and M. Lipson, "All-optical logic based on silicon microring resonators," *Optics Express*, vol. 15, no. 3, pp. 924–929, 2007.

[5] R. Osgood, J. B. Driscoll, W. Astar et al., "Nonlinear silicon photonics," *Nature Photonics*, vol. 4, pp. 535–544, 2010.

[6] D. T. Spencer, J. F. Bauters, M. J. R. Heck, and J. E. Bowers, "Integrated waveguide coupled Si3N4 resonators in the ultrahigh-Q regime," *Optica*, vol. 1, no. 3, pp. 153–157, 2014.

[7] G.-R. Lin, S.-P. Su, C.-L. Wu et al., "Si-rich SiNx based Kerr switch enables optical data conversion up to 12 Gbit/s," *Scientific Reports*, vol. 5, Article ID 9611, 2015.

[8] A. Kotb, S. Ma, Z. Chen, N. K. Dutta, and G. Said, "Effect of amplified spontaneous emission on semiconductor optical amplifier based all-optical logic," *Optics Communications*, vol. 284, no. 24, pp. 5798–5803, 2011.

[9] A. Kotb, *All-Optical Logic Gates Using Semiconductor Optical Amplifiers*, Lambert Academic Publishing, Germany, 2012.

[10] A. Kotb, "AND gate based on two-photon absorption in semiconductor optical amplifier," *Optoelectronics Letters*, vol. 9, no. 3, pp. 181–184, 2013.

[11] A. Kotb and K. E. Zoiros, "On the design of all-optical gates based on quantum-dot semiconductor optical amplifier with effect of amplified spontaneous emission," *Optical and Quantum Electronics*, vol. 46, no. 8, pp. 977–989, 2014.

[12] A. Kotb, "Simulation of high quality factor all-optical logic gates based on quantum-dot semiconductor optical amplifier at 1 Tb/s," *Optik—International Journal for Light and Electron Optics* , vol. 127, no. 1, pp. 320–325, 2016.

[13] H. Dong, H. Sun, Q. Wang, N. K. Dutta, and J. Jaques, "80 Gb/s All-optical logic AND operation using Mach-Zehnder interferometer with differential scheme," *Optics Communications*, vol. 265, no. 1, pp. 79–83, 2006.

[14] H. Dong, H. Sun, Q. Wang, N. K. Dutta, and J. Jaques, "All-optical logic and operation at 80 Gb/s using semiconductor optical amplifier based on the Mach-Zehnder interferometer," *Microwave and Optical Technology Letters*, vol. 48, no. 8, pp. 1672–1675, 2006.

[15] H. Sun, Q. Wang, H. Dong, and N. K. Dutta, "All-optical logic performance of quantum-dot semiconductor amplifier-based devices," *Microwave and Optical Technology Letters*, vol. 48, no. 1, pp. 29–35, 2006.

[16] S. Ma, Z. Chen, H. Sun, and N. K. Dutta, "High speed all optical logic gates based on quantum dot semiconductor optical amplifiers," *Optics Express*, vol. 18, no. 7, pp. 6417–6422, 2010.

[17] D. Nesset, M. C. Tatham, and D. Cotter, "All-optical AND gate operating on 10 Gbit/s signals at the same wavelength using four-wave mixing in a semiconductor laser amplifier," *IEEE Electronics Letters*, vol. 31, no. 11, pp. 896-897, 1995.

[18] H. Soto, C. A. Díaz, J. Topomondzo, D. Erasme, L. Schares, and G. Guekos, "All-optical AND gate implementation using cross-polarization modulation in a semiconductor optical amplifier," *IEEE Photonics Technology Letters*, vol. 14, no. 4, pp. 498–500, 2002.

[19] X. Zhang, Y. Wang, J. Sun, D. Liu, and D. Huang, "All-optical AND gate at 10 Gbit/s based on cascaded single-port-coupled SOAs," *Optics Express*, vol. 12, no. 3, pp. 361–366, 2004.

[20] L. Q. Guo and M. J. Connelly, "All-optical and gate with improved extinction ratio using signal induced nonlinearities in a bulk semiconductor optical amplifier," *Optics Express*, vol. 14, no. 7, pp. 2938–2943, 2006.

[21] J. H. Kim, Y. I. Kim, Y. T. Byun et al., "All-optical logic gates using semiconductor optical-amplifier-based devices and their applications," *Journal of the Korean Physical Society*, vol. 45, no. 5 I, pp. 1158–1161, 2004.

[22] X. Chen, L. Huo, Z. Zhao, L. Zhuang, and C. Lou, "Study on 100-Gb/s reconfigurable all-optical logic gates using a single semiconductor optical amplifier," *Optics Express*, vol. 24, no. 26, pp. 30245–30253, 2016.

[23] N. K. Dutta and Q. Wang, *Semiconductor Optical Amplifiers*, World Scientific Publishing Company, Singapore, 2nd edition, 2013.

[24] J. H. Kim, B. C. Kim, Y. T. Byun et al., "All-optical AND gate using cross-gain modulation in semiconductor optical amplifiers," *Japanese Journal of Applied Physics*, vol. 43, no. 2, pp. 608–610, 2004.

[25] B.-K. Kang, J. H. Kim, Y. T. Byun et al., "All-optical AND gate using probe and pump signals as the multiple binary points in cross phase modulation," *Japanese Journal of Applied Physics*, vol. 41, no. 5 B, pp. L568–L570, 2002.

[26] E. Dimitriadou and K. E. Zoiros, "On the feasibility of 320 Gb/s all-optical and gate using quantum-dot semiconductor optical amplifier-based mach-zehnder interferometer," *Progress in Electromagnetics Research B*, no. 50, pp. 113–140, 2013.

[27] J. Xu, X. Zhang, Y. Zhang, J. Dong, D. Liu, and D. Huang, "Reconfigurable all-optical logic gates for multi-input differential phase-shift keying signals: design and experiments," *Journal of Lightwave Technology*, vol. 27, no. 23, pp. 5268–5275, 2009.

[28] Z. Li, Y. Dong, C. Lu et al., "Comparison of cross-gain modulation effect of Manchester-duobinary, RZ-DPSK, NRZ-DPSK, RZ, and NRZ modulation formats in SOAs," *IEEE Photonics Technology Letters*, vol. 18, no. 24, pp. 2680–2682, 2006.

[29] D. Kong, Y. Li, H. Wang et al., "All-optical XOR gates for QPSK signal based optical networks," *IEEE Electronics Letters*, vol. 49, no. 7, pp. 485-486, 2013.

[30] X. Chen, Y. Yu, and X. Zhang, "All-optical logic minterms for three-input demodulated differential phase-shift keying signals at 40 Gb/s," *IEEE Photonics Technology Letters*, vol. 23, no. 2, pp. 118–120, 2011.

[31] I. Kang, C. Dorrer, and J. Leuthold, "All-optical XOR operation of 40 Gbit/s phase-shift-keyed data using four-wave mixing in semiconductor optical amplifier," *IEEE Electronics Letters*, vol. 40, no. 8, pp. 496–498, 2004.

[32] K. Chan, C.-K. Chan, L. K. Chen, and F. Tong, "Demonstration of 20-Gb/s all-optical XOR gate by four-wave mixing in semiconductor optical amplifier with RZ-DPSK modulated inputs," *IEEE Photonics Technology Letters*, vol. 16, no. 3, pp. 897–899, 2004.

[33] N. Deng, K. Chan, C.-K. Chan, and L.-K. Chen, "An all-optical XOR logic gate for high-speed RZ-DPSK signals by FWM in semiconductor optical amplifier," *IEEE Journal of Selected Topics in Quantum Electronics*, vol. 12, no. 4, pp. 702–707, 2006.

[34] A. Kotb, "An OR gate for 100 Gb/s phase-shift keying signals in semiconductor optical amplifier-based delayed interferometer," *Australian Basic and Applied Science*, vol. 8, pp. 449–453, 2014.

[35] A. Kotb, "Analysis of all-optical logic XOR gate for 100 Gb/s phase-shift keying modulated data signals in semiconductor optical amplifier-based Mach-Zehnder interferometer," *Optical and Quantum Electronics*, vol. 47, no. 5, pp. 1063–1070, 2015.

[36] W. Li, H. Hu, X. Zhang, and N. K. Dutta, "High speed all optical logic gates using binary phase shift keyed signal based on QD-SOA," *International Journal of High Speed Electronics and Systems*, vol. 24, no. 3-4, Article ID 1550005, 2015.

[37] G. Talli and M. J. Adams, "Amplified spontaneous emission in semiconductor optical amplifiers: modelling and experiments," *Optics Communications*, vol. 218, no. 1-3, pp. 161–166, 2003.

[38] E. Dimitriadou and K. E. Zoiros, "On the design of reconfigurable ultrafast all-optical NOR and NAND gates using a single

quantum-dot semiconductor optical amplifier-based Mach-Zehnder interferometer," *Journal of Optics (United Kingdom)*, vol. 14, no. 10, Article ID 105401, 2012.

[39] T. Houbavlis, K. E. Zoiros, G. Kanellos, and C. Tsekrekos, "Performance analysis of ultrafast all-optical Boolean XOR gate using semiconductor optical amplifier-based Mach-Zehnder Interferometer," *Optics Communications*, vol. 232, no. 1-6, pp. 179–199, 2004.

[40] R. Bonk, T. Vallaitis, J. Guetlein et al., "The input power dynamic range of a semiconductor optical amplifier and its relevance for access network applications," *IEEE Photonics Journal*, vol. 3, no. 6, pp. 1039–1053, 2011.

9

Optical Spectra Properties and Continuous-Wave Laser Performance of Tm,Y:CaF_2 Single Crystals

Jingxin Ding,[1,2] Beibei Zhao,[1] Weiwei Ma,[1] Hao Yu,[1] Xiaobo Qian,[1] Lingchen Kong,[3] Jingya Wang,[1] Guoqiang Xie,[3] Anhua Wu,[1] Fanming Zeng,[2] and Liangbi Su[1,4]

[1] *Synthetic Single Crystal Research Center, Key Laboratory of Transparent and Opto-Functional Inorganic Materials, Shanghai Institute of Ceramics, Chinese Academy of Sciences, Shanghai, China*
[2] *School of Materials Science and Engineering, Changchun University of Science and Technology, Changchun 130022, China*
[3] *Key Laboratory for Laser Plasmas (Ministry of Education), Collaborative Innovation Center of IFSA (CICIFSA), Department of Physics and Astronomy, Shanghai Jiao Tong University, Shanghai 200240, China*
[4] *State Key Laboratory of High Performance Ceramics and Superfine Microstructure, Shanghai Institute of Ceramics, Chinese Academy of Sciences, Shanghai, China*

Correspondence should be addressed to Liangbi Su; suliangbi@mail.sic.ac.cn

Academic Editor: Wonho Jhe

3 at.% Tm, x at.% Y:CaF_2 crystals ($x = 0, 0.5, 1, 2,$ and 3) were grown by the vertical Bridgman method and investigated. Codoping Y^{3+} ions can manipulate the local structure of Tm^{3+} ions in the CaF_2 crystal and then improve the spectroscopic properties. Compared with 3 at.% Tm:CaF_2, 3 at.% Tm, 3 at.% Y:CaF_2 crystal has several advantages. Firstly, the absorption cross section is improved from 0.35×10^{-20} cm^{-2} to 0.45×10^{-20} cm^{-2} at 767 nm, and the fluorescence intensity had elevated 3.4 times. Secondly, the linewidth of the fluorescence spectrum and lifetime also increased from 164 nm to 191 nm and from 6.16 ms to 8.15 ms at room temperature, respectively. Furthermore, quantum efficiency improved from 58.2% to 80.3%. The maximum laser output power of 583 mW and slope efficiency of 25.3% were achieved in 3 at.% Tm, 3 at.% Y:CaF_2 crystal under 790 nm diode pumping.

1. Introduction

Calcium fluorides, as laser substrates, possess various advantages of large size, high thermal conductivity, well controlled crystal growth processes, and low nonlinear refractive coefficient. Trivalent rare-earth ions, like Tm, Nd, Pr, doped CaF_2 crystals behave broad, smooth absorption and emission spectra due to heterovalent substitution of Ca^{2+} within the structure without loss of structural integrity [1]. Various RE^{3+} optical centers could be formed in this fluoride by substituting divalent cation ions, and the excessive charge of rare-earth ions is compensated by interstitial fluorine. The Nd^{3+} doped CaF_2 crystal as a laser-pumped-amplifier medium has been abandoned due to a very serious concentration quenching effect which results from the clustering of the neodymium ions and kinds of cross-relaxation type energy transfer processes, which weaken their emission quantum efficiency [2]. However, the [Nd^{3+}-Nd^{3+}] quenching pairs in clusters can be easily dissociated by codoping buffer ions such as Y^{3+} ions [3–8], La^{3+} ions [9], and Sc^{3+} ions [10]. For example, Nd,Y:CaF_2 crystal, Y^{3+} ions were codoped in Nd:CaF_2 crystal which substitute for Ca^{2+} forming complicated local structure that performs an effect on spectroscopic properties [8].

As the solid-state lasers medium, the doped Tm^{3+} calcium fluoride crystal proves the potential to achieve efficient compact diode-pumped lasers with an oscillation wavelength near 2 μm which could be directly pumped around 790 nm ($^3H_6 \rightarrow {}^3H_4$ absorption transition) due to much lower non-radiative losses caused by multiphonon relaxation [11]. Tm ions act both as a sensitizer and activator in a single-doped sample, meaning that a higher concentration is necessary for effective absorption of 800 nm laser excitation. Compared to oxide crystals, pure CaF_2 crystal has a thermal conductivity as high as 10 W/cm·K [12] and low phonon energy

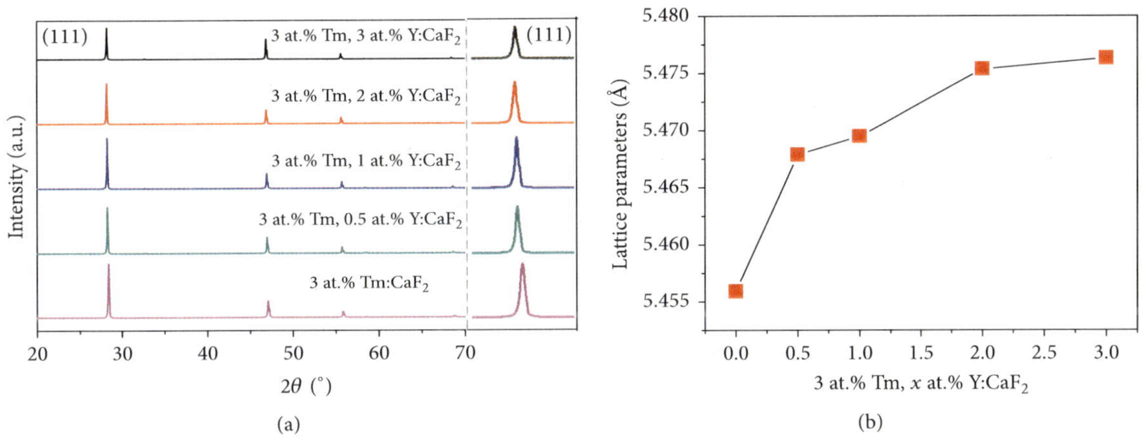

FIGURE 1: The XRD patterns (a) and lattice parameters (b) of 3 at.% Tm, x at.% Y:CaF_2 crystals.

(maximum value of 495 cm^{-1}) [13], making it very suitable for a laser host crystal and becoming one of the first laser hosts in the early 1960 nm. The predecessors have done some researches on Tm^{3+} doped CaF_2 single crystals and found that Tm:CaF_2 crystals possessed broadband absorption and emission properties [11, 14, 15]. The spectroscopic investigation on Tm^{3+} doped crystals indicates that Tm^{3+} ions interactions occur (forming the cross relaxation) at relatively low dopant concentrations (nearly 1%), that is, two excited ions in the $^3F_4$ upper ground level 7F6 for one absorbed pump photon. The distinction between isolated and clustered ions is observed in the emission spectra due to different doped concentration. Besides, nonradiative processes related to $^3H_5$-$^3F_4$ induced by [Tm-Tm] clusters in the crystals could result in strong heat generation and distortions reducing quantum efficiency [16]. To achieve higher power laser at 2 μm, one of the efficient ways is to prepare Tm:CaF_2 with higher Tm deponent concentrations avoiding clustering. However, a spectral region around 1.45 μm associated with a $^3H_4 \rightarrow {}^3F_4$ optical transition could be vanished when the Tm^{3+} dopant concentration is beyond 1.34% due to [Tm-Tm] clusters [14], and the energy may be absorbed by the ground stated $^3H_6$. However, when the concentration of Tm becomes higher, some other possible ways of energy transfer would occur, such as $^1G_4$ + $^3H_4 \rightarrow {}^3F_4$ + $^1D_2$ [17], $^3H_5 \rightarrow {}^3H_6$ [18], and $^3H_5$ + $\text{photo}_{800\,nm} \rightarrow {}^1G_4$. Therefore, we take the advantage of Y^{3+} as buffer ions reported in Nd,Y:CaF_2 [3, 8] crystals to prepare the Tm,Y:CaF_2 crystals to break the [Tm-Tm] clusters and increase the Tm^{3+} emission intensity as well as efficiency in higher concentration, and the laser performance and wide emission spectrum have been expected. Importantly, the incorporation of Y^{3+} ions to the influences of spectroscopic properties in Tm^{3+} doped CaF_2 single crystals had not been investigated systematically.

In this paper, to improve high pump absorption efficiency and high gain per unit length, a series of 3 at.% Tm and x at.% Y:CaF_2 crystals have been grown by the vertical Bridgman method, and spectroscopic properties were studied systematically. We have carried out laser experiments and obtained continuous-wave laser output.

2. Experimental

The single crystal samples, namely, 3 at.% Tm, x at.% Y:CaF_2 (x = 0, 0.5, 1, 2, 3) crystals (at.%, atom percent), were grown by the traditional vertical Bridgman method. High purity fluorides crystalline powders (4 N), CaF_2, TmF_3, YF_3, were used as starting materials, and 1 wt% PbF_2 was selected as an oxygen scavenger avoiding oxidation and volatilization additionally. These materials were completely mixed by molar ratios and filled into an assembled platinum crucible. The growth parameters are as follows: the temperature of the melt around 130°C, the pulling rate 0.8 mm/h, the cooling rate 20°C/h. The samples (the same thickness of about 2 mm) were handled with cutting and double-face optically polishing for spectral measurement.

By recording absorption and emission spectra, we investigated the spectroscopic properties of the crystals. The absorption spectra were measured by using a Jasco V-570 UV/VIS/NIR spectrophotometer. The fluorescence spectra and lifetime were obtained with a FLS980 time-resolved fluorimeter with grating blazed at 1820 nm and detected using a Hamamatsu InSb. Measuring of fluorescence spectra was performed under pumping at 808 nm with a CW laser operation. All the measurements were conducted at room temperature.

3. Results and Discussion

3.1. Phase Identification and Crystal Structure. 3 at.% Tm, x at.% Y:CaF_2 crystals (x = 0, 0.5, 1, 2, 3) have been analyzed by powder XRD and behave the purity phase CaF_2 without any impure peaks as shown in Figure 1(a). The XRD patterns of the crystals have matched well with the JCPDS standard card of CaF_2 ICSD 00-075-0363 indicating that the fluorite cubic structure (Fm-3 m) has not been changed by the increasing concentration of yttrium. Tm^{3+} and Y^{3+} ions substitute for Ca^{2+} ions in the CaF_2 lattice, and smaller F^- ions have taken placed in the interstitial positions of the empty cubes to compensate the charge and maintain electrical neutrality leading to the smaller Bragg's angels and larger lattice parameters

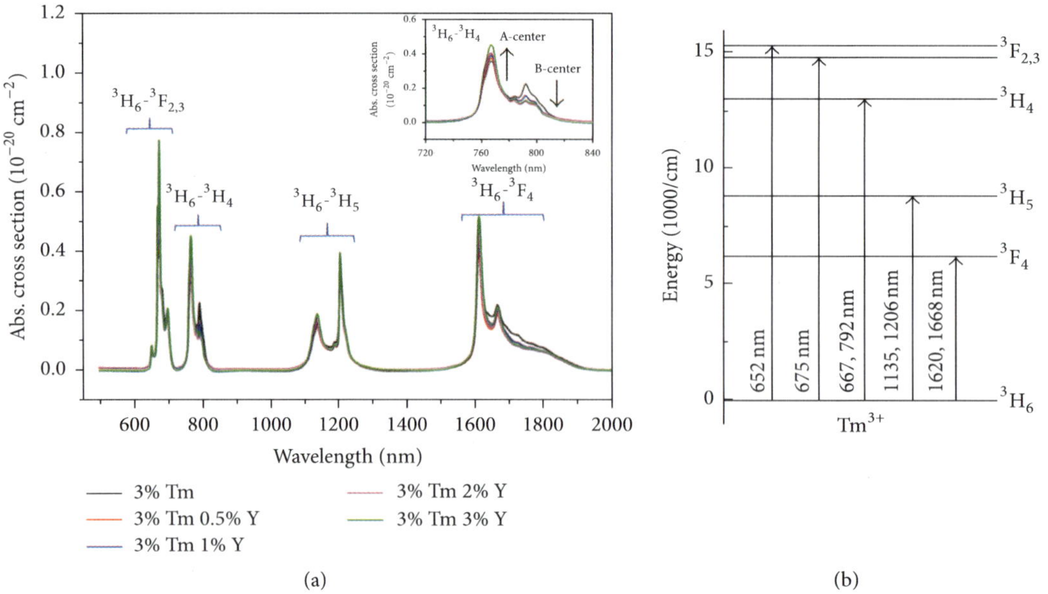

FIGURE 2: Absorption spectra (a) and energy level diagrams (b) of 3 at.% Tm, x at.% Y:CaF_2 crystals at 300 K.

which could be distinguished clearly on the enlarged view of (111) in Figure 1(a). The lattice parameters of series of 3 at.% Tm, x at.% Y:CaF_2 (x = 0.5, 1, 2, 3) single crystals are 5.46785 Å, 5.46947 Å, 5.47535 Å, 5.4763 Å, respectively, much larger than 5.4559 Å of Tm:CaF_2, and increase with the rising codoping Y^{3+} ion concentration as shown in Figure 1(b). These observations confirm that Tm^{3+} and Y^{3+} ions had been effectively doped into the host lattice of CaF_2.

3.2. Absorption and Emission Properties. The absorption spectra from 500 nm to 2000 nm at room temperature of 3 at.% Tm, x at.% Y are shown in Figure 2(a). Due to various splitting energy levels of $^3H_6$ and $^3F_4$, the absorption bands have been divided into several peaks. Several main absorption bands, $^3H_6$-$^3F_2$ (652 nm), $^3H_6$-$^3F_3$ (675 nm), $^3H_6$-$^3H_4$ (667 nm, 792 nm), $^3H_6$-$^3H_5$ (1135 nm, 1206 nm), $^3H_6$-$^3F_4$ (1620 nm, 1668 nm), have been marked in the spectra. Clearly, all the absorption cross sections are increased with the Y ions. Additionally, the absorption $^3H_6$-$^3H_4$, which is usually used for diode pumping, inset in the picture has been analyzed in detail at 767 nm and 792 nm in the same bands caused by different emission centers, defined as A-center and B-center.

The largest absorption cross section at 767 nm of 3 at.% Tm, 3 at.% Y:CaF_2 improves to 0.45×10^{-20} cm^{-2}, much larger than that reported in [14]. On the other side, the absorption cross section at 792 nm decreases gradually from 0.22×10^{-20} cm^{-2} to 0.12×10^{-20} cm^{-2} with the increasing Y ions. Y ions play an important role in modulating spectral performance. The absorption spectra of Tm^{3+} ions can be significantly altered by codoping with Y^{3+} ions. The changeable absorption cross sections indicate that the local structure and symmetry of the calcium fluoride crystal have been modified by changing the amount of deponent codoping Y^{3+} ions. The increasing phenomenon could be attributed to the stronger crystal field caused by interstitial F^- ions in the lattice induced by the codoped Y^{3+} separating the [Tm-Tm] clusters, caused by a considerable high doping concentration 3 at.% much larger than 1.34% [11, 19], to an appropriate distance and forming A-centers instead of B-centers. By doping Y^{3+} ions, B-center has been broken, forming more A-centers as a result. Anyway, it is clear that codoping Tm:CaF_2 with Y^{3+} ions slightly broadens the absorption bands $^3H_6$-$^3H_4$, which should be profitable for LD pumping. The broad wavelength tunability indicated an efficient ground-state stark splitting with the introduction of Y^{3+} ions in the as-grown Tm,Y:CaF_2 crystal.

The fluorescence spectra of 3 at.% Tm, x at.% Y:CaF_2 crystals, corresponding to $^3F_4$-$^3H_6$ emission transition of Tm^{3+} around 1.8 μm, excited by 767 nm are reported in Figure 3. The emission spectra of the five crystals consist of four bands, peaking at 1611 nm (6207 cm^{-1}), 1666 nm (6002 cm^{-1}), 1820 nm (5494 cm^{-1}), 1856 nm (5387 cm^{-1}), respectively. The carves of Tm,Y:CaF_2 demonstrate several intense separate local maxima compared to the one of the Tm:CaF_2, indicating that the Y^{3+} ions codoping modulate the emission spectral structure of Tm^{3+} ions in CaF_2 hosts. The emission intensity of 2873 a.u. at 1820 nm of the 3 at.% Tm, 3 at.% Y:CaF_2 crystals is 3.4 times higher than that of the 3 at.% Tm:CaF_2 crystal (842 a.u.), whose value is the largest above all the samples. As has been discussed in the absorption cross sections, Y^{3+} codoping breaks the Tm^{3+} ion clusters and increases the fluorescence quantum efficiency which could also be proved in Table 1. Additionally, the luminescence intensity of the 1.8 μm band is improved by doping Y^{3+}.

Figure 4 shows the logarithm of the emission intensity at 1820 nm of Tm^{3+} as a function of the decay time in 3

TABLE 1: The detailed values of σ_{abs}, σ_{em}, τ_{rad}, τ_{em}, η of 3 at.% Tm, x at.% $Y:CaF_2$ at 1820 nm.

Crystals	σ_{abs}/cm^2	σ_{em}/cm^2	τ_{rad}/ms	τ_{em}/ms	$\eta/\%$
3 at.% $Tm:CaF_2$	0.063	$1.206 * 10^{-20}$	10.58	6.16	58.2
3 at.% Tm, 0.5 at.% $Y:CaF_2$	0.054	$1.026 * 10^{-20}$	11.43	7.25	63.4
3 at.% Tm, 1 at.% $Y:CaF_2$	0.057	$1.090 * 10^{-20}$	10.24	6.53	63.7
3 at.% Tm, 2 at.% $Y:CaF_2$	0.054	$1.026 * 10^{-20}$	10.82	7.55	69.8
3 at.% Tm, 3 at.% $Y:CaF_2$	0.056	$1.088 * 10^{-20}$	10.14	8.15	80.3

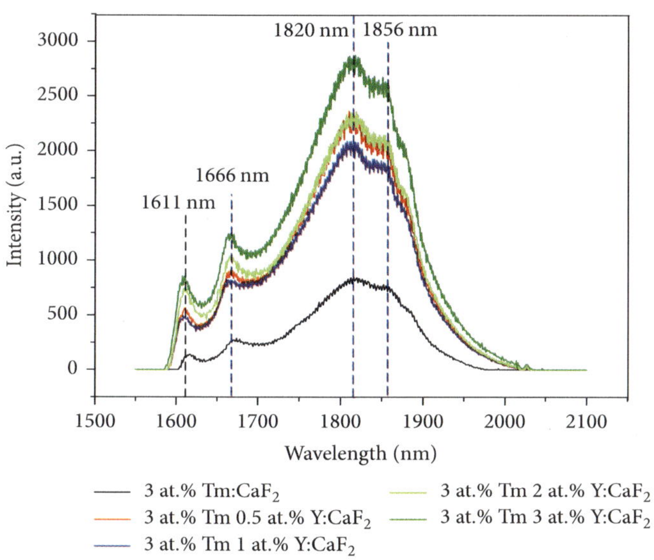

FIGURE 3: Fluorescence spectra for $^3F_4$-$^3H_6$ transition of 3 at.% Tm, x at.% Y CaF_2 crystals at 300 K.

at.% Tm, x at.% $Y:CaF_2$ crystals excited by 767 nm at room temperature. The straight lines indicated that the decay was consistent with a first-order exponential and the emission lifetimes were labelled as arrows. The emission lifetimes were fitted to be 6.16 ms, 7.25 ms, 6.53 ms, 7.55 ms, 8.15 ms for 3 at.% Tm, x at.% $Y:CaF_2(x = 0, 0.5, 1, 2, 3)$, respectively, which is in the order of 5 ms [11] much shorter than the longer lifetime 15 ms [20]. Due to the higher concentration of Tm^{3+}, the emission lifetimes caused by the fluorescence of clustered thulium centers and the tetragonal optical centers are responsible for the longer lifetime 15 ms [11, 20]. It also could be discussed that the emission centers with higher symmetry could extend the emission lifetime of the energy level $^3F_4$ which could benefit the pump efficiency. The shorter emission lifetime means that in the 3 at.% Tm, x at.% $Y:CaF_2$ crystals, [Tm-Tm] clusters take a dominant station affecting the lifetime of $^3F_4$ compared to these tetragonal optical centers. The emission lifetimes of $Tm,Y:CaF_2$ crystals were longer than that of the $Tm:CaF_2$ crystal, indicating clearly that

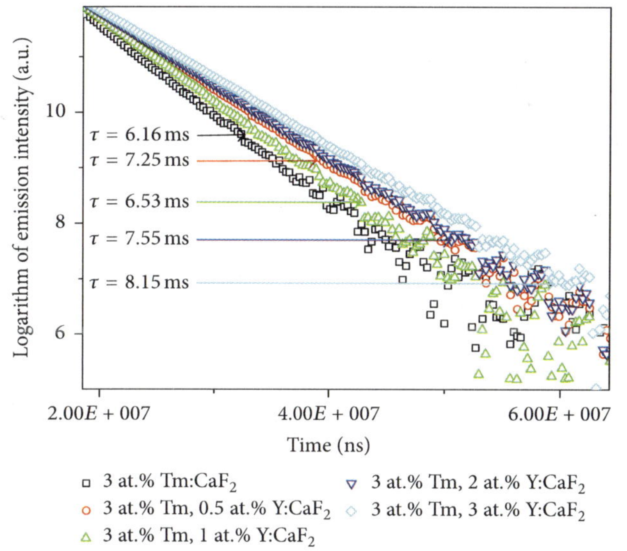

FIGURE 4: Fluorescence lifetimes of energy level $^3F_4$ of 3 at.% Tm, x at.% $Y:CaF_2$ crystals.

codoping Y^{3+} ions as buffer ions increase the fluorescence lifetime of Tm ions.

3.3. Calculations for Spectral Parameters. In this session, some spectral parameters including the emission cross section σ_{em}, the radiation lifetimes τ_{rad}, the quantum efficiency η, the quality factor $\sigma_{em} * \tau_{em}$, the effective linewidth $\Delta\lambda$ have been calculated to measure the quality of these crystals. The emission cross section σ_{em} and the radiation lifetimes τ_{rad} have been calculated by the reciprocity method and the Fuchtbauer-Ladenburg (FL) equation, respectively, and the absorption cross sections σ_{abs} could be obtained from the absorption spectra in Figure 2(a).

$$\sigma_{em}(\lambda) = \sigma_{abs}(\lambda) \frac{Z_l}{Z_u} \exp\left[\frac{hc}{k_B T}\left(\frac{1}{\lambda_{ZL}} - \frac{1}{\lambda}\right)\right], \quad (1)$$

where λ_{ZL} will be referred to as the "zero line" wavelength ($\lambda_{ZL} = 1666$ nm wavelength associated with the transition between the lowest stark components of each multiplet $^3H_6$ and $^3F_4$) and Z_l/Z_u represents the ratio of the partition functions of the lower and upper states and the value is 1.512 [14].

$$\sigma_{em} = \frac{\lambda_{peak}^4}{8\pi n^2 \Delta\lambda \cdot \tau_{rad}}, \quad (2)$$

where λ_{peak} is the wavelength of the maximum emission intensity (here is 1820 nm) and n stands for the refractive index (the refractive index of calcium fluoride is 1.442 at 1820 nm). We can take advantage of (2) for the value of τ_{rad}. In theory, the product of σ_{em} and τ_{rad} is inversely proportional to $\Delta\lambda$. It indicates that the result of the experiment is nearly

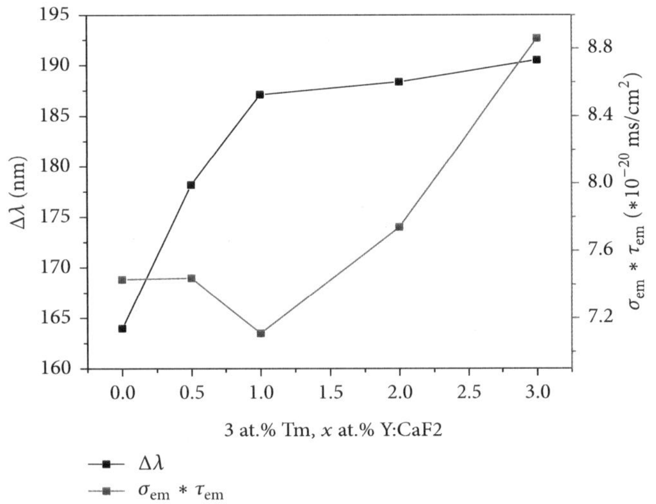

FIGURE 5: $\tau_{rad} * \sigma_{em}$ and $\Delta\lambda$ obtained from 3 at.% Tm, x at.% Y:CaF_2 crystals.

in agreement with that of the theory. Equation (2) can also be expressed as

$$\sigma_{em} \cdot \tau_{em} = \frac{\lambda^4_{peak}}{8\pi c n^2} \cdot \frac{1}{\Delta\lambda} \cdot \frac{\tau_{em}}{\tau_{rad}} = \frac{\lambda^4_{peak}}{8\pi c n^2} \cdot \frac{1}{\Delta\lambda} \cdot \eta, \quad (3)$$

where $\lambda^4_{peak}/8\pi c n^2$ can be regarded as a constant and η is the quantum efficiency. The calculated results have been shown in Table 1 and Figure 5.

The emission cross section σ_{em} varies from $1.026 * 10^{-20}/cm^2$ to $1.088 * 10^{-20}/cm^2$. The quantum efficiency of emission at 1820 nm is 58.2%, 63.4%, 63.7%, 69.8%, 80.3% for 3 at.% Tm, x at.% Y ($x = 0, 0.5, 1, 2, 3$), respectively. As discussed above, cooping Y ions as buffer ions actually benefit the quantum efficiency increasing the quantum efficiency effectively, indicating that the efficiency of the fluoresce is very sensitive to the cationic coordination [21], and the highest quantum efficiency has been increased to 80.3%.

We can see the difference of $\Delta\lambda$ (where $\Delta\lambda$ is the effective linewidth which can be obtained by measurement) from Figure 4. The change trend of the effective linewidth at 1820 nm is depicted in Figure 5, which is almost increasing with codoping Y^{3+} ion concentration. $\Delta\lambda$ at 1820 nm is 163.97 nm, 178.16 nm, 187.12 nm, 188.37 nm, 190.52 nm for 3 at.% Tm, x at.% Y:CaF_2($x = 0, 0.5, 1, 2, 3$), respectively, behaving the superiority for LD pumping. Compared with single-doped one, the effective linewidth of the codoping crystals increased rapidly, and the trend turns out to be saturated when the concentration of Y^{3+} grows higher. It indicates that the effect of Y^{3+} concentration on the effective linewidth causes saturation which is in favor of femtosecond laser output.

3.4. Laser Performance. Taking both emission intensity and lifetime into consideration, two samples, 3 at.% Tm:CaF_2 and 3 at.% Tm, 3 at.% Y:CaF_2, were applied in laser experiments as laser-pumped-amplifier mediums. The size of the crystal is 4 mm × 4 mm × 6 mm and the end faces were optically polished flat and parallel without being coated. The continuous-wave (CW) experiment with a fiber-coupled AlGaAs diode laser as the pump source emitting at 790 nm was carried out at room temperature, and the setup for testing was shown in Figure 6.

In this experiment, the output mirror transmission is 2%, and the folded cavity consisted of three mirrors: M1, M2, and M3, having the same radium of curvature of 10 cm. The pump light was focused into the crystal through a 1:1 optical imaging module. The pump source was provided by a laser diode around 790 nm. The laser output powers of two samples were depicted in Figure 7, we obtained two different laser curves. To the sample, 3 at.% Tm, 3 at.% Y:CaF_2 crystal, its laser slope efficiency and maximum output power of the crystal are 25.3% and 583 mW with the 2% transmission output coupler, while a lower slope efficiency of 15.9% and maximum output power 159 mW were obtained with 3 at.% Tm:CaF_2 crystal. The maximum pump power was limited by the absorption capacity of the 3 at.% Tm:CaF_2 crystal. We found its excited state absorption tended to be saturated when the absorbed pump power was over 1.6 W. To avoid damage, the absorbed pump power of 3 at.% Tm:CaF_2 was set below 1.6 W corresponding to the incident pump power of 3 W, and the absorbed pump power of 3 at.% Tm, 3 at.% Y:CaF_2 crystal was set below 2.5 W corresponding to the incident pump power of 4.5 W.

4. Conclusions

The codoping Y^{3+} ions Tm:CaF_2 crystals were successfully grown by vertical Bridgman method, and the properties of the series crystals were analyzed systematically. Absorption of $^3H_6$-$^3H_4$ is caused by A-center at 767 nm and B-center at 792 nm, and the absorption cross section of A-center is increased while the absorption cross section of B-center is decrease by codoping Y^{3+}. Emission intensity and effective linewidth of emission at 1820 nm are greatly improved when the concentration increased to 3 at.%. The quantum efficiency is enhanced to 80.3% by codoping Y^{3+} ions compared to the undoped crystals. It demonstrates codoping Y^{3+} ions have a positive effect on spectroscopic properties. In laser experiment, we finally obtained a maximum laser output power of 583 mW and slope efficiency of 25.3% in codoped sample.

Conflicts of Interest

The authors declare that they have no conflicts of interest.

Acknowledgments

This work was financially supported by the National Natural Science Foundation of China (Grants nos. 61422511, 61635012, and 61475089) and The National Key Research and Development Program of China (Grant no. 2016YFB0402101).

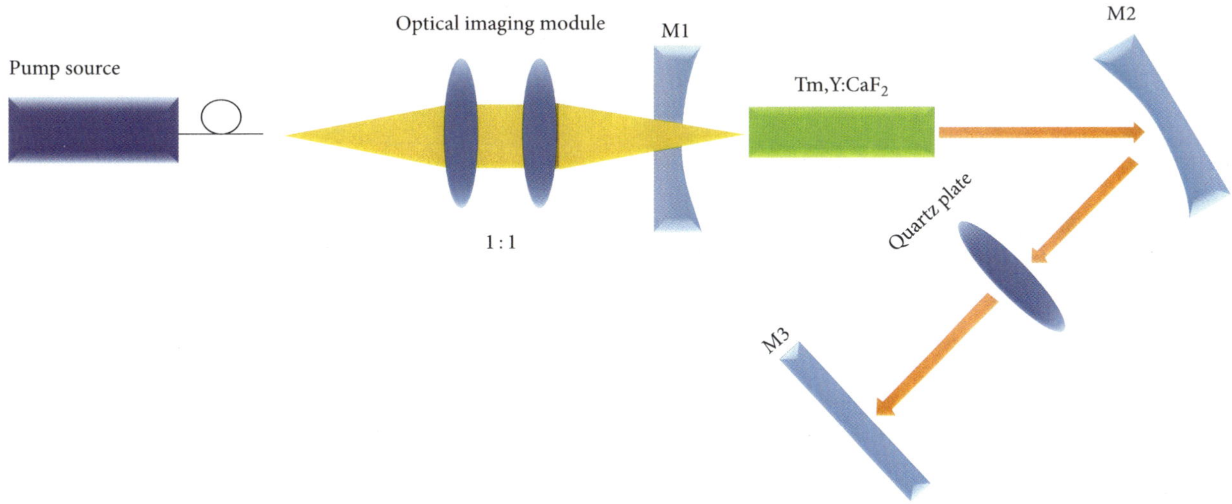

FIGURE 6: Schematic of the experimental setup for 1.8 μm laser operation.

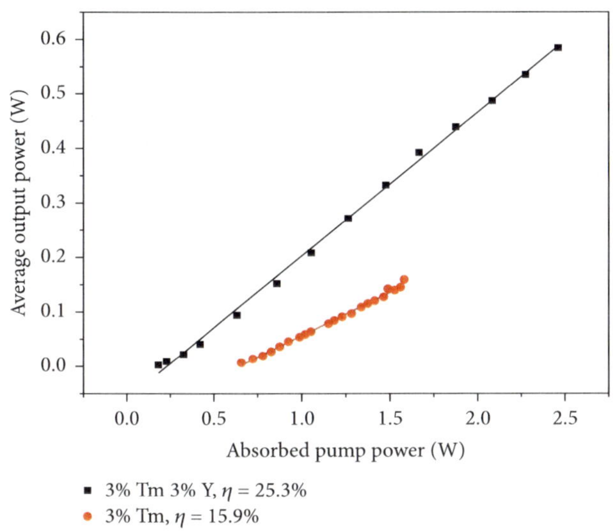

FIGURE 7: Laser output power versus absorbed pump power curve for 3 at.% Tm:CaF_2 and 3 at.% Tm, 3 at.% Y:CaF_2 crystals.

References

[1] C. R. A. Catlow, A. V. Chadwick, G. N. Greaves, and L. M. Moroney, "Direct observations of the dopant environment in fluorites using EXAFS," *Nature*, vol. 312, no. 5995, pp. 601–604, 1984.

[2] S. A. Payne, J. A. Caird, L. L. Chase, L. K. Smith, N. D. Nielsen, and W. F. Krupke, "Spectroscopy and gain measurements of Nd^{3+} in SrF_2 and other fluorite-structure hosts," *Journal of the Optical Society of America B: Optical Physics*, vol. 8, no. 4, pp. 726–740, 1991.

[3] T. P. J. Han, G. D. Jones, and R. W. G. Syme, "Site-selective spectroscopy of Nd^{3+} centers in CaF_2:Nd^{3+} and SrF_2:Nd^{3+}," *Physical Review B: Condensed Matter and Materials Physics*, vol. 47, no. 22, pp. 14706–14723, 1993.

[4] N. E. Kask and L. S. Kornienko, "EPR of Nd^{3+} ions in fluorite," *Journal of Experimental and Theoretical Physics*, vol. 26, pp. 331–335, 1968.

[5] A. A. Kaminskii, V. V. Osiko, A. M. Prochoro, and Y. K. Voronko, "Spectral investigation of stimulated radiation of Nd^{3+} in CaF_2-YF_3," *Physics Letters*, vol. 22, pp. 419–421, 1966.

[6] K. S. Bagdasarov, Y. K. Voronko, A. A. Kaminskii, L. V. Krotova, and V. V. Osiko, "Modification of the optical properties of CaF_2-TR^{3+} crystals by yttrium impurities," *Physica Status Solidi*, vol. 12, no. 2, pp. 905–912, 1965.

[7] T. T. Basiev, Y. K. Voronko, A. Y. Karasik, V. V. Osiko, and I. A. Shcherbakov, "Spectral migration of electronic excitation along Nd^{3+} ions in CaF_2-YF_3 crystal on selective laser excitation," *Journal of Experimental and Theoretical Physics*, vol. 75, pp. 66–74, 1978.

[8] D. Jiang, Y. Zhan, Q. Zhang et al., "Nd,Y:CaF_2 laser crystals: novel spectral properties and laser performance from a controlled local structure," *CrystEngComm*, vol. 17, pp. 7398–7405, 2015.

[9] H. Wang, J. Zhu, Z. Gao et al., "Femtosecond mode-locked Nd,La:CaF_2 disordered crystal laser," *Optical Materials Express*, vol. 6, no. 7, pp. 2184–2189, 2016.

[10] J. Zhang, J. Zhu, J. Wang, Z. Wei, L. Su, and J. Xu, "CW laser performance of Nd-La·CaF_2 and Nd-Sc·CaF_2 disordered crystals pumped by a laser diode," *Acta Photonica Sinica*, vol. 45, pp. 114001-1–114001-5, 2016.

[11] M. E. Doroshenko, O. K. Alimov, A. G. Papashvili et al., "Spectroscopic and laser properties of Tm^{3+} optical centers in CaF_2 crystal under 795 nm diode laser excitation," *Laser Physics Letters*, vol. 12, no. 12, Article ID 125701, 2015.

[12] P. J. D. Lindan and M. J. Gillan, "A molecular dynamics study of the thermal conductivity of CaF_2 and UO_2," *Journal of Physics: Condensed Matter*, vol. 3, no. 22, pp. 3929–3939, 1991.

[13] V. Petit, L. Doualan, P. Camy, V. Ménard, and R. Moncorgé, "CW and tunable laser operation of Yb^{3+} doped CaF_2," *Applied Physics B*, vol. 78, p. 681, 2004.

[14] P. Camy, J. L. Doualan, S. Renard, A. Braud, V. Ménard, and R. Moncorgé, "Tm^{3+}:CaF_2 for 1.9 μm laser operation," *Optics Communications*, vol. 236, no. 4–6, pp. 395–402, 2004.

[15] N. M. Strickland and G. D. Jones, "Site-selective spectroscopy of Tm^{3+} centers in CaF_2:Tm^{3+}," *Physical Review B: Condensed Matter and Materials Physics*, vol. 56, no. 17, pp. 10916–10929, 1997.

[16] J. Ganem and S. R. Bowman, "Use of thulium-sensitized rare earth-doped low phonon energy crystalline hosts for IR sources," *Nanoscale Research Letters*, vol. 8, p. 455, 2013.

[17] H. Zhang, Y. Li, Y. Lin, Y. Huang, and X. Duan, "Composition tuning the upconversion emission in $NaYF_4$:Yb/Tm hexaplate nanocrystals," *Nanoscale*, vol. 3, no. 3, pp. 963–966, 2011.

[18] X. Bai, D. Li, Q. Liu, B. Dong, S. Xu, and H. Song, "Concentration-controlled emission in LaF_3:Yb^{3+}/ Tm^{3+} nanocrystals: switching from UV to NIR regions," *Journal of Materials Chemistry*, vol. 22, no. 47, pp. 24698–24704, 2012.

[19] P. J. Bendall, C. R. A. Catlow, J. Corish, and P. W. M. Jacobs, "Defect aggregation in anion-excess fluorites II. Clusters containing more than two impurity atoms," *Journal of Solid State Chemistry*, vol. 51, no. 2, pp. 159–169, 1984.

[20] M. E. Doroshenko, A. G. Papashvili, O. K. Alimov et al., "Specific spectroscopic and laser properties of Tm^{3+} ions in hot-formed CaF_2 laser ceramics," *Laser Physics Letters*, vol. 13, no. 1, Article ID 015701, 2016.

[21] G. Lakshminarayana, R. Yang, M. Mao, J. Qiu, and I. V. Kityk, "Photoluminescence of Sm^{3+}, Dy^{3+}, and Tm^{3+}-doped transparent glass ceramics containing CaF_2 nanocrystals," *Journal of Non-Crystalline Solids*, vol. 355, no. 52–54, pp. 2668–2673, 2009.

10

Continuous-Wave and Passively Q-Switched $Er:Y_2O_3$ Ceramic Laser at 2.7 μm

Zhipeng Qin,[1] Guoqiang Xie,[1] Jian Zhang,[2] Jingui Ma,[1] Peng Yuan,[1] and Liejia Qian[1]

[1] *Key Laboratory for Laser Plasmas (Ministry of Education), Collaborative Innovation Center of IFSA (CICIFSA), School of Physics and Astronomy, Shanghai Jiao Tong University, Shanghai 200240, China*
[2] *Key Laboratory of Transparent and Opto-Functional Advanced Inorganic Materials, Shanghai Institute of Ceramics, Chinese Academy of Sciences, Shanghai 201899, China*

Correspondence should be addressed to Guoqiang Xie; xiegq@sjtu.edu.cn

Academic Editor: Wonho Jhe

We report on a continuous-wave (CW) and passively Q-switched $Er:Y_2O_3$ ceramic laser in mid-infrared spectral region. In the CW regime, a maximum output power of 2.07 W is achieved at 2717.3 nm with a slope efficiency of 13.5%. Stable passive Q-switching of the $Er:Y_2O_3$ ceramic laser is demonstrated based on semiconductor saturable absorber mirror. Under an absorbed pump power of 12.4 W, a maximum average output power of 223 mW is generated with a pulse energy of 1.7 μJ and a pulse width of 350 ns at 2709.3 nm.

1. Introduction

Mid-infrared coherent resources in the 2~5 μm spectral range have attracted more and more attention due to the potential applications in spectroscopy, medical treatment, security and defense, free-space communication, and advanced sensing technologies for trace gas detection and greenhouse gas monitoring [1–5]. Rare-earth-doped laser is a general way to generate high-power and high-beam-quality mid-infrared coherent resource [6–8].

Er-doped solid-state lasers can be emitted in the spectral range of 2.7~3.0 μm by electronic transition ($^4I_{11/2} \rightarrow {}^4I_{13/2}$). So far, Er:YAG has been widely used for commercial 2.94 μm laser. For mid-infrared Er:YAG lasers, high Er-doping concentration (50%) is required to overcome self-terminating phenomenon which is caused by a long lifetime of lower laser level [9, 10]. Although high doping concentration can reduce the population in lower laser level by enhancing the energy transfer upconversion process ($^4I_{13/2} + {}^4I_{13/2} \rightarrow {}^4I_{9/2} + {}^4I_{15/2}$), it will bring serious thermal effects due to heat deposited in a small region, which will limit the available output power and beam quality. $Er:Y_2O_3$ is an excellent alternative for mid-infrared solid-state laser materials. For $Er:Y_2O_3$ gain medium, a much lower doping concentration (e.g., 2%~7%) is feasible for laser emission [11–14], which significantly alleviates the thermal effects. Besides, $Er:Y_2O_3$ has a larger product of emission cross section (~1×10^{-19} cm^2) and upper laser level lifetime (2.4 ms) in comparison with Er:YAG [14], which is favorable to achieve a high laser gain. Due to the high melting point, the growth of $Er:Y_2O_3$ single crystal is challenging. $Er:Y_2O_3$ ceramic has similar optical performances to the $Er:Y_2O_3$ single crystal, and the ceramic possesses the advantages of short fabrication period, large size, and low cost. So far, CW operation of $Er:Y_2O_3$ ceramic laser at room temperature has been reported with an output power of 320 mW and a slope efficiency of 6.5% [12]. At cryogenic temperature of 77 K, a CW output power of 14 W was generated [13]. For passive Q-switching operation, only an average power of 6 mW was generated in $Er:Y_2O_3$ ceramic laser [11]. Besides, passively Q-switched lasers in $Er:CaF_2$, $Er:SrF_2$, $Er:Lu_2O_3$, and Er:Pr:GGG crystals were also reported (Table 1) [15–19].

Here we demonstrate a passively Q-switched $Er:Y_2O_3$ ceramic laser at the wavelength of 2.7 μm. In the CW regime, an output power of 2.07 W is obtained with a slope efficiency of 13.5%. The Q-switched operation is realized based on

TABLE 1: Passively Q-switched Er-doped bulk lasers near 3 μm spectral region.

Gain medium	SA	P_{ave} [mW]	λ_0 [μm]	τ [ns]	E [μJ]	P_{peak} [W]	Ref.
$Er:Y_2O_3$	BP	6	2.72	4470	0.48	0.1	[11]
$Er:CaF_2$	Graphene	172	2.80	1324	2.74	2.1	[15]
$Er:SrF_2$	BP	180	2.79	702	2.34	3.3	[16]
$Er:Lu_2O_3$	MoS_2	1030	2.84	335	8.5	25.4	[17]
$Er:Lu_2O_3$	C_3N_4	1090	2.840	351	11.1	31.6	[18]
Er:Pr:GGG	Graphene	186	2.71	360	1.54	4.3	[19]
$Er:Y_2O_3$	SESAM	223	2.71	350	1.71	4.9	This work

SA, saturable absorber; P_{ave}, average power; P_{peak}, peak power.

SESAM. Under an absorbed pump power of 12.4 W, the Q-switched $Er:Y_2O_3$ ceramic laser emits an average power of 223 mW with a pulse width of 350 ns at repetition rate of 130.6 kHz, resulting in a pulse energy of 1.7 μJ.

2. Experimental Setup

The experimental setup schematics of CW and passively Q-switched $Er:Y_2O_3$ ceramic laser are shown in Figure 1. A commercial 970 nm laser diode (LD) was adopted as the pump source, delivering a maximum power of 30 W from a multimode fiber with a core diameter of 105 μm and a numerical aperture of 0.22. After collimating and focusing by a pair of lenses F_1 and F_2 ($f_1 = 18$ mm and $f_2 = 75$ mm), the pump light was imaged into the gain medium. The gain medium of $Er:Y_2O_3$ ceramic had a length of 6.6 mm, a cross section of $2 \times 3\ \text{mm}^2$, and an Er-doping concentration of 7%. In order to remove the generated heat while pumping, the $Er:Y_2O_3$ ceramic was wrapped in indium foil and tightly fixed in a copper holder which was cooled by a thermoelectric cooler. The gain medium was antireflectively coated for both pump and laser wavelengths ($R < 0.5\%$). For CW operation, a linear cavity structure was adopted (Figure 1(a)), which consisted of a plane-concave input mirror M_1 (ROC = 50 mm) and a plane-plane output coupler (OC). Figure 1(b) shows the experimental setup schematic of SESAM Q-switched laser.

$Er:Y_2O_3$ ceramic laser had a total cavity length of 25 mm. The employed SESAM (SAM-2400-1-10ps-e, BATOP GmbH) has total absorbance of 5% and 18% at 2700 nm and 2717 nm, respectively.

3. Results and Discussion

3.1. CW Laser Performance. We used different OCs ($T = 1\%$, 3%, and 5%, resp.) to test the CW output powers. Figure 2(a) shows the CW output power versus absorbed pump power with an optimal OC ($T = 5\%$). The laser threshold was ~1.6 W. The output laser power increased linearly with absorbed pump power with a slope efficiency of 13.5%. A maximum CW power of 2.07 W was generated. The output power rollover was not observed even at the maximum pump power, indicating that CW output power could be further increased with higher pump power. At the CW output power of 1 W, we measured the output spectrum, as shown in Figure 2(b). The spectrum peak is at 2717.3 nm with a full width at half maximum (FWHM) of 4.6 nm. With the knife-edge method, we measured the beam quality of CW output laser. The measured M^2 factor of the beam was 2.84 (Figure 2(c)). We also captured the output beam profile with a thermosensitive card (VRC6S, Thorlabs). The inset of Figure 2(c) shows that the laser beam has a round beam profile. In order to test the power stability of the $Er:Y_2O_3$ ceramic laser, we monitored the output power fluctuation for one hour, as shown in Figure 2(d). The output power fluctuation (Δ_{RMS}, the ratio of root-mean-square error to average value) is ~0.65%, indicating that the output power of $Er:Y_2O_3$ ceramic laser is stable.

3.2. SESAM Q-Switched $Er:Y_2O_3$ Laser. In the Q-switched $Er:Y_2O_3$ ceramic laser, we employed a plane-concave OC with transmission of 1%. After introducing the SESAM in the cavity, laser threshold was significantly increased due to the absorption loss of SESAM. The laser threshold was 4.8 W of absorbed pump power. As the absorbed pump power was increased to 9.5 W, Q-switching operation was initiated. Q-switched pulse train was captured by a mid-infrared photodetector (VIGO system, PCI-9) and displayed on a digital oscillator, as shown in Figure 3(a). In order to achieve short Q-switched pulses, we built up a short laser cavity of 25 mm, which emitted the shortest pulse of 350 ns (Figure 3(b)).

Figure 4(a) shows the output power, pulse energy, and peak power evolution with the absorbed pump power. They all increase with the absorbed pump power. At the absorbed pump power of 12.4 W, we obtained a maximum average output power of 223 mW and a maximum pulse energy of 1.7 μJ, corresponding to the peak power of 4.9 W. Beyond the absorbed pump power of 12.4 W, the laser quenching appeared. The laser quenching phenomenon can be attributed to serious thermal lens effect in the gain medium, which results in formation of instable cavity. The focal length of thermal lens was directly measured to be 35 mm at the absorbed pump power of 12.4 W. We added the thermal lens to ABCD propagation matrix and the formation of instable cavity was verified by the numerical calculation. The Q-switched pulse could be recovered by slightly decreasing the incident pump power. The evolution of repetition rate and pulse width with absorbed pump power is shown

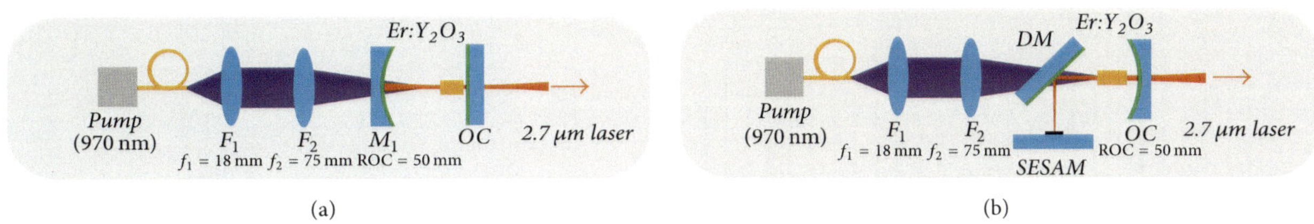

FIGURE 1: Experimental setup schematic of $Er:Y_2O_3$ ceramic laser in CW operation (a) and Q-switching operation (b). ROC: radius of curvature; OC: output coupler; DM: dichroic mirror; SESAM: semiconductor saturable absorber mirror.

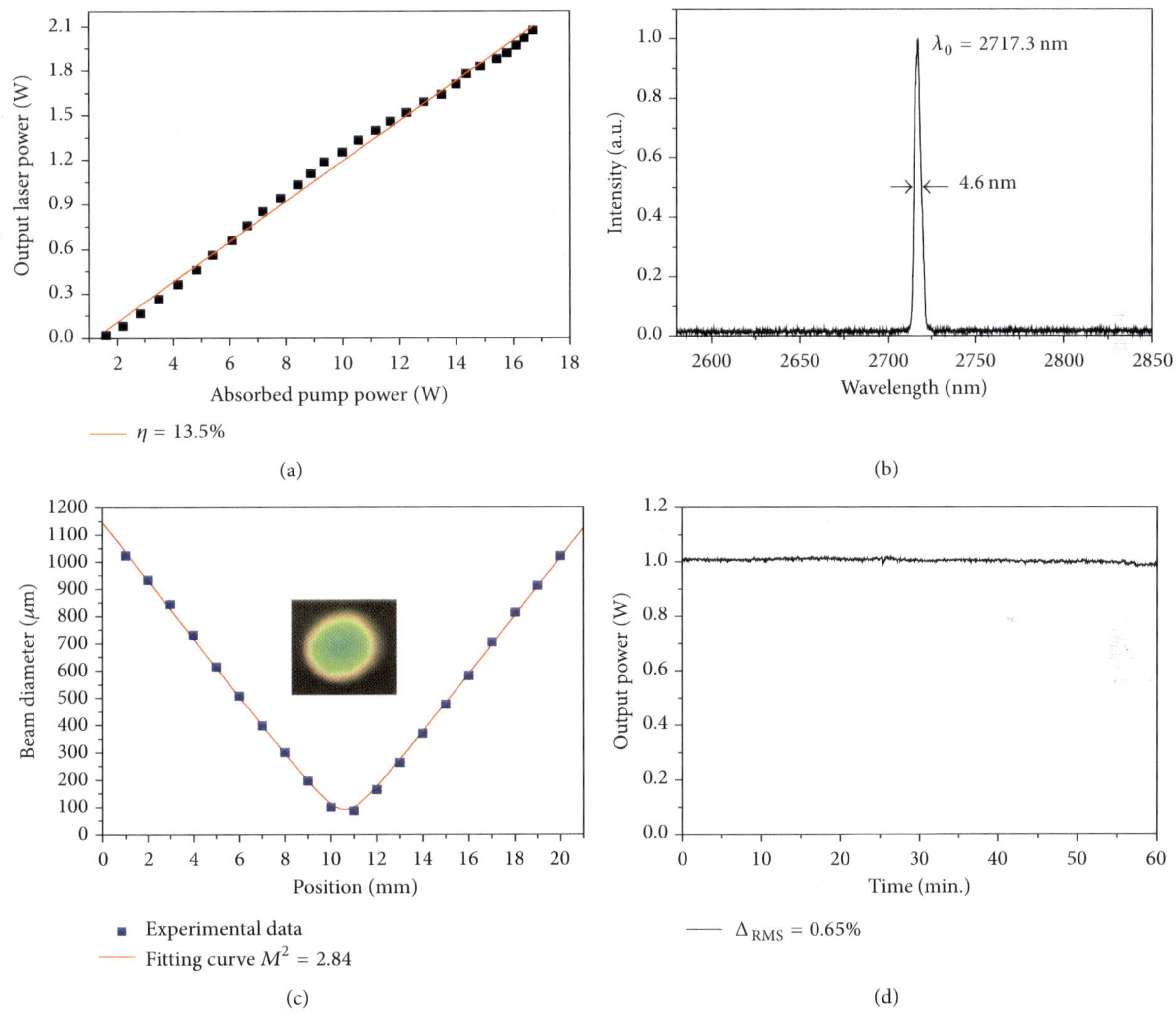

FIGURE 2: (a) CW output power versus absorbed pump power. (b) Spectrum of CW laser. (c) Measurement of the beam quality. Inset: beam profile. (d) CW output power stability of $Er:Y_2O_3$ ceramic laser.

in Figure 4(b). The pulse repetition rate increased from 105.3 kHz to 130.6 kHz in the absorbed pump power range of 9.5~12.4 W. However, the pulse duration decreased from 434 ns to 350 ns and it tended to saturate beyond the absorbed pump power of 12 W.

Figure 5 shows the Q-switched pulse spectrum. The spectral peak is at 2709.3 nm with a FWHM of 2.4 nm. Compared with the CW spectrum, there is a blue shift in Q-switching pulse spectrum, which could be attributed to the larger loss (18%) of the SESAM, measured at the wavelength of 2717 nm. Consequently, the Q-switched $Er:Y_2O_3$ ceramic laser was forced to oscillate at shorter wavelength of 2709 nm. At this wavelength, the Q-switched $Er:Y_2O_3$ ceramic laser suffered from an reabsorption loss under high pump power [12], which dramatically reduced the average output power.

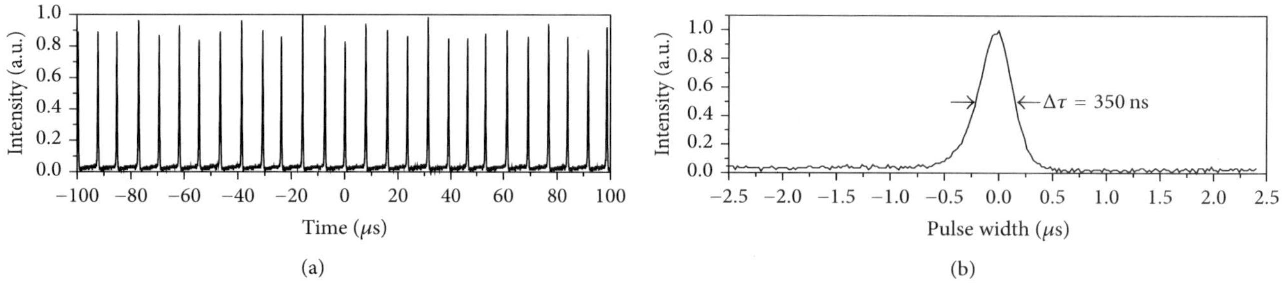

FIGURE 3: (a) Typical pulse train of passively Q-switched Er:Y_2O_3 ceramic laser. (b) Single pulse profile.

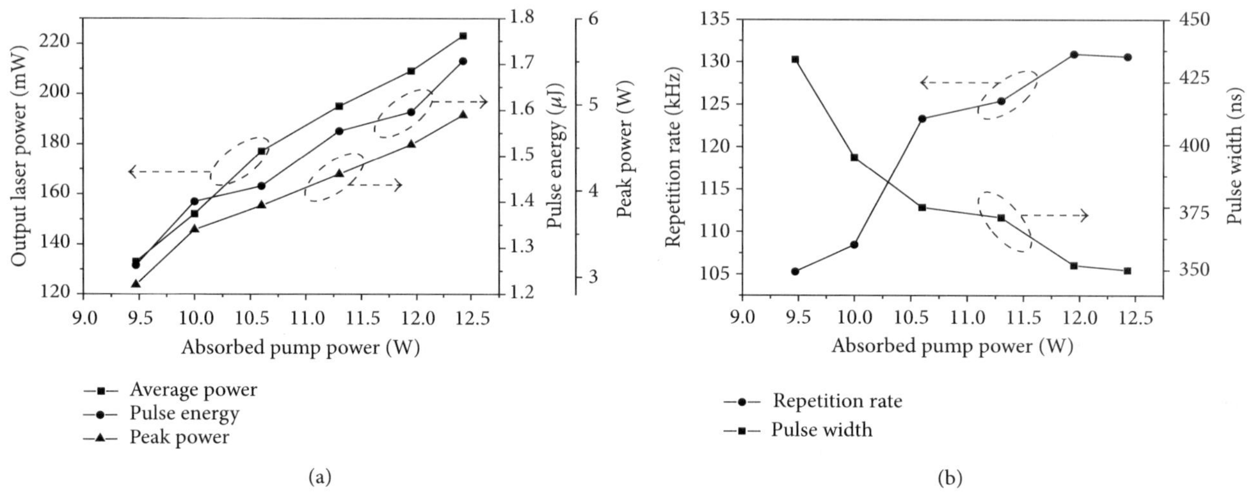

FIGURE 4: (a) Average output power, pulse energy, and peak power evolution with absorbed pump power. (b) Pulse repetition rate and pulse width evolution with absorbed pump power.

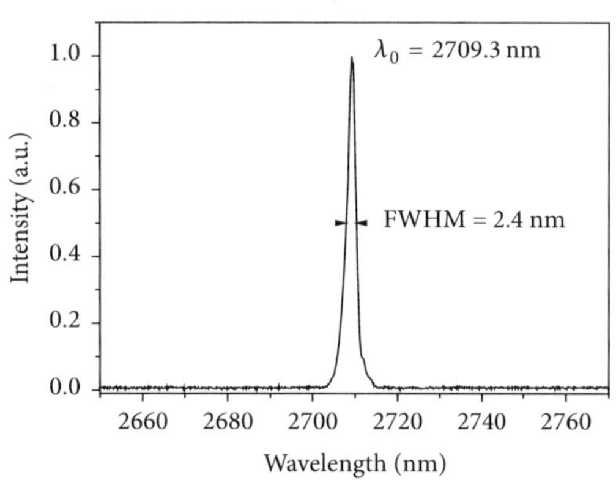

FIGURE 5: Optical spectrum of the Q-switched pulse.

4. Conclusions

In conclusion, we experimentally demonstrated a CW and passively Q-switched Er:Y_2O_3 ceramic laser at 2.7 μm wavelength. A maximum CW output power of 2.07 W was achieved with a slope efficiency of 13.5%. In Q-switching operation, the laser emitted stable Q-switched pulses with an average output power of 223 mW, a pulse energy of 1.7 μJ, a pulse width of 350 ns, and a repetition rate of 130.6 kHz at 2709.3 nm. Our research results show that Er:Y_2O_3 ceramic is an excellent gain material for CW and Q-switched laser at 2.7 μm wavelength.

Conflicts of Interest

The authors declare that they have no conflicts of interest.

Acknowledgments

This work is partially supported by the National Basic Research Program of China (Grant no. 2013CBA01505), the Shanghai Excellent Academic Leader Project (Grant no. 15XD1502100), the National Natural Science Foundation of China (Grant no. 61675130), the National Postdoctoral Program for Innovative Talents (no. 190375), and the project funded by China Postdoctoral Science Foundation (no. 190375).

References

[1] B. G. Lee, M. A. Belkin, R. Audet et al., “Widely tunable single-mode quantum cascade laser source for mid-infrared spectroscopy,” *Applied Physics Letters*, vol. 91, no. 23, Article ID 231101, 2007.

[2] R. Kaufmann and R. Hibst, "Pulsed 2-94-μm erbium–YAG laser skin ablation—experimental results and first clinical application," *Clinical and Experimental Dermatology*, vol. 15, no. 5, pp. 389–393, 1990.

[3] V. W. S. Chan, "Optical space communications," *IEEE Journal of Selected Topics in Quantum Electronics*, vol. 6, no. 6, pp. 959–975, 2000.

[4] D. Halmer, S. Thelen, P. Hering, and M. Mürtz, "Online monitoring of ethane traces in exhaled breath with a difference frequency generation spectrometer," *Applied Physics B: Lasers and Optics*, vol. 85, no. 2-3, pp. 437–443, 2006.

[5] J. J. Scherer, J. B. Paul, H. J. Jost, and M. L. Fischer, "Mid-IR difference frequency laser-based sensors for ambient CH4, CO, and N2O monitoring," *Applied Physics B: Lasers and Optics*, vol. 110, no. 2, pp. 271–277, 2013.

[6] Z. Qin, G. Xie, C. Zhao, S. Wen, P. Yuan, and L. Qian, "Mid-infrared mode-locked pulse generation with multilayer black phosphorus as saturable absorber," *Optics Expresss*, vol. 41, no. 1, pp. 56–59, 2016.

[7] Z. Qin, G. Xie, H. Zhang et al., "Black phosphorus as saturable absorber for the Q-switched Er:ZBLAN fiber laser at 2.8 μm," *Optics Express*, vol. 23, no. 19, pp. 24713–24718, 2015.

[8] S. B. Mirov, V. V. Fedorov, D. Martyshkin, I. S. Moskalev, M. Mirov, and S. Vasilyev, "Progress in Mid-IR Lasers Based on Cr and Fe-Doped II-VI Chalcogenides," *IEEE Journal of Selected Topics in Quantum Electronics*, vol. 21, no. 1, 2015.

[9] E. Arbabzadah, S. Chard, H. Amrania, C. Phillips, and M. Damzen, "Comparison of a diode pumped Er:YSGG and Er:YAG laser in the bounce geometry at the 3 μm transition," *Optics Express*, vol. 19, no. 27, pp. 25860–25865, 2011.

[10] C. Ziolek, H. Ernst, G. F. Will, H. Lubatschowski, H. Welling, and W. Ertmer, "High-repetition-rate, high-average-power, diode-pumped 2.94-μm Er:YAG laser," *Optics Expresss*, vol. 26, no. 9, pp. 599–601, 2001.

[11] L. Kong, Z. Qin, G. Xie et al., "Black phosphorus as broadband saturable absorber for pulsed lasers from 1 μm to 2.7 μm wavelength," *Laser Physics Letters*, vol. 13, no. 4, Article ID 045801, 2016.

[12] L. Wang, H. Huang, D. Shen et al., "Room temperature continuous-wave laser performance of LD pumped Er:Lu 2O3 and Er:Y2O3 ceramic at 2.7 μm," *Optics Express*, vol. 22, no. 16, pp. 19495–19503, 2014.

[13] T. Sanamyan, M. Kanskar, Y. Xiao, D. Kedlaya, and M. Dubinskii, "High power diode-pumped 2.7–μm Er3+:Y2O3 laser with nearly quantum defect-limited efficiency," *Optics Express*, vol. 19, no. 105, pp. A1082–A1087, 2011.

[14] T. Sanamyan, J. Simmons, and M. Dubinskii, "Er3+-doped Y2O3 ceramic laser at ~2.7μm with direct diode pumping of the upper laser level," *Laser Physics Letters*, vol. 7, no. 3, pp. 206–209, 2010.

[15] C. Li, J. Liu, S. Jiang et al., "2.8 μm passively Q-switched Er:CaF2 diodepumped laser," *Optical Materials Express* , vol. 6, no. 5, pp. 1570–1575, 2016.

[16] J. Liu, J. Liu, Z. Guo et al., "Dual-wavelength Q-switched Er:SrF2 laser with a black phosphorus absorber in the midinfrared region," *Optics Express*, vol. 24, no. 26, pp. 30289–30295, 2016.

[17] M. Fan, T. Li, S. Zhao et al., "Watt-level passively Q-switched Er:Lu2O3 laser at 2.84 μm using MoS2," *Optics Expresss*, vol. 41, no. 3, pp. 540–543, 2016.

[18] M. Fan, T. Li, G. Li et al., "Graphitic C3N4 as a new saturable absorber for the mid-infrared spectral range," *Optics Expresss*, vol. 42, no. 2, pp. 286–289, 2017.

[19] Z. Y. You, Y. Wang, Y. J. Sun et al., "CW and Q-switched GGG/Er:Pr:GGG/GGG composite crystal laser at 2.7 μm," *Laser Physics Letters*, vol. 14, no. 4, Article ID 045810, 2017.

11

Performance Comparison of Wavefront-Sensorless Adaptive Optics Systems by using of the Focal Plane

Huizhen Yang,[1] Zhen Zhang,[2] and Jian Wu[2]

[1]*School of Electronic Engineering, Huaihai Institute of Technology, Lianyungang 222005, China*
[2]*School of Information and Electrical, China University of Mining and Technology, Xuzhou 221008, China*

Correspondence should be addressed to Huizhen Yang; yanghz526@126.com

Academic Editor: Nicusor Iftimia

The correction capability and the convergence speed of the wavefront-sensorless adaptive optics (AO) system are compared based on two different system control algorithms, which both use the information of focal plane. The first algorithm is designed through the linear relationship between the second moment of the aberration gradients and the masked far-field intensity distribution and the second is stochastic parallel gradient descent (SPGD), which is the most commonly used algorithm in wavefront-sensorless AO systems. A wavefront-sensorless AO model is established with a 61-element deformable mirror (DM) and a CCD. Performance of the two control algorithms is investigated and compared through correcting different wavefront aberrations. Results show that the correction ability of AO system based on the proposed control algorithm is obviously better than that of AO system based on SPGD algorithm when the wavefront aberrations increase. The time needed by the proposed control algorithm is much less than that of SPGD when the AO system achieves similar correction results. Additionally, the convergence speed of the proposed control algorithm is independent of the turbulence strength while the number of intensity measurements needed by SPGD increases as the turbulence strength magnifies.

1. Introduction

Compared with the conventional adaptive optics (AO) system [1], the complexity of wavefront-sensorless AO system [2] is reduced greatly because of no wavefront sensor in system structure. The most significant potential advantage of this kind of AO system is related to the fact that the far-field intensity distribution can be used as the feedback signal. The wavefront-sensorless AO system can be divided into two categories according to control algorithms, one is model-free AO systems and the other is model-based AO systems. The former directly uses various blind optimization algorithms as the control algorithm of the system, such as the stochastic parallel gradient descent algorithm (SPGD—stochastic parallel gradient descent) [3, 4], Alopex algorithm [5], simulated annealing [6], and genetic algorithm [7]. Model-based wavefront-sensorless AO technology often makes use of some kind of model to design the control algorithm for system, such as modal method [8, 9], the nonlinear optimization method [10], or geometric optical principles [11].

In this paper, we design a control algorithm for the wavefront-sensorless AO system through the linear relationship between the second moment of the aberration gradients and the masked far-field intensity distribution. An AO system model is established with a 61-element deformable mirror (DM) and a CCD. The correction capability and the convergence speed of the AO system are investigated and compared with SPGD control algorithm, which is the most commonly used control algorithm for wavefront-sensorless AO systems and verified to be relatively fast control method [12].

2. Control Algorithm Based on the Linear Relationship between the Second Moment of the Aberration Gradients and the Masked Far-Field Intensity Distribution

2.1. Theory Basis of Control Algorithm. The centroid of the far-field intensity distribution in geometric optics is related to the aberration of the input wavefront. A linear relationship between the second moment of the aberration gradients and the masked far-field intensity distribution can be achieved through a certain mathematical transformation [11]:

$$\mathrm{SM} \approx c_0(1 - \mathrm{MDS}), \tag{1}$$

where SM is the second moment of the aberration gradients and c_0 is a constant, the slope of the trend line, which is determined by the detector radius R. MDS is defined as follows:

$$\mathrm{MDS} = \frac{\iint I(x', y')\left[1 - r^2/R^2\right] dx'dy'}{\iint I(x', y')dx'dy'}, \tag{2}$$

where $I(x', y')$ is the far-field intensity distribution and $1 - r^2/R^2$ $(r = \sqrt{x'^2 + y'^2})$ is the mask. In practice, the mask could be implemented by the weighting of pixels within the software when the photodetector is replaced by a CCD camera.

In order to find out the control parameters of deformable mirror, N orthogonal modes are taken as predetermined bias functions and are added by the DM sequentially with coefficient α to the wavefront aberration to be corrected. Then the detector measurements are recorded and the N MDS_i $(i = 1, \ldots, N)$ are calculated according to (2). Control parameters V can be estimated by

$$V \approx \frac{S^{-1}\left(c_0 M - \alpha^2 S_m\right)}{2\alpha}, \tag{3}$$

where

$$M = -\begin{Bmatrix} \mathrm{MDS}_1 - \mathrm{MDS}_0 \\ \mathrm{MDS}_2 - \mathrm{MDS}_0 \\ \vdots \\ \mathrm{MDS}_N - \mathrm{MDS}_0 \end{Bmatrix}, \tag{4}$$

and MDS_0 is the corresponding MDS of the wavefront aberration to be corrected. The matrix S is the second moment of the wavefront gradients, which can be calculated by the bias function. The vector S_m is the diagonal vector of matrix S. S can be calculated according to the following equation, where $F(x, y)$ is the predefined bias function:

$$S(i, j) = D^{-1} \int_D \left[\frac{\partial}{\partial x} F_i(x, y) \frac{\partial}{\partial x} F_j(x, y)\right] + \left[\frac{\partial}{\partial y} F_i(x, y) \frac{\partial}{\partial y} F_j(x, y)\right] dx\, dy. \tag{5}$$

2.2. Control Algorithm Design. A set of bias functions are defined to represent wavefront aberrations. We use Zernike polynomial as the bias function. The wavefront to be corrected $\phi(x, y)$ can be written as a linear combination of the Zernike polynomial $Z(x, y)$:

$$\phi(x, y) = \sum_{i=1}^{N} \beta_i Z_i(x, y), \tag{6}$$

where N is the Zernike order number, β_i is the Zernike coefficient of the ith-order Zernike mode, and $Z_i(x, y)$ is the ith-order Zernike mode. The phase compensation $u(x, y)$, introduced by DM, can be decomposed into the superposition of the response functions of actuators linearly:

$$u(x, y) = \sum_{j=1}^{K} v_j V_j(x, y), \tag{7}$$

where v_j is the control signal, $V_j(x, y)$ is the response function of the jth actuator, and K is the number of DM actuator. Using DM to fit the wavefront aberration described by Zernike polynomials, we can get

$$\phi(x, y) = u(x, y) + e(x, y), \tag{8}$$

where $e(x, y)$ is the residual wavefront of the DM fitting. Equation (8) can be changed into the following equation after $\phi(x, y)$ and $u(x, y)$ are replaced by (6) and (7), respectively:

$$\sum_{i=1}^{N} \beta_i Z_i(x, y) = \sum_{j=1}^{K} v_j V_j(x, y) + e(x, y). \tag{9}$$

Both sides of (9) are multiplied by one influence function $V_l(x, y)$ at the same time and the integral in the aperture D is calculated:

$$D^{-1} \int_D \sum_{i=1}^{N} \beta_i Z_i(x, y) V_l(x, y) dx\, dy = D^{-1} \int_D \left[\sum_{j=1}^{K} v_j V_j(x, y) + e(x, y)\right] V_l(x, y) dx\, dy. \tag{10}$$

Equation (10) can get a matrix equation:

$$C_{vz}\beta = C_v v + \varepsilon, \tag{11}$$

where C_{vz} is the cross-correlation matrix with size $K \times N$ between the Zernike mode $Z(x, y)$ and the influence function $V(x, y)$, C_v is the coupling matrix between different influence function $V(x, y)$ with size $K \times K$, β is the Zernike coefficient with size N, and v is the control voltage vector with size K. $C_{vz}(i, j)$ and $C_v(i, j)$ have the following form, respectively:

$$C_{vz}(i, j) = D^{-1} \iint_D V_i(x, y) Z_j(x, y)\, dx\, dy \quad i = 1, \ldots, K, \quad j = 1, \ldots, N,$$
$$C_v(i, j) = D^{-1} \iint_D V_i(x, y) V_j(x, y) dx\, dy \quad i, j = 1, \ldots, K. \tag{12}$$

We can obtain the minimum variance solution of (10):

$$v^* = C_v^{-1} C_{zv} \beta. \tag{13}$$

Equation (13) provides a minimum variance solution to the control voltage vector v.

The control algorithm is divided into two parts: the preprocessing step and the iteration step. The detailed descriptions are as follows.

Preprocessing Step

(1) Calculate the second moment of the wavefront gradients S, the diagonal vector S_m, and its inverse matrix S^{-1} according to the predefined bias function $Z(x, y)$ by (5).

(2) Measure the influence functions $V(x, y)$ of the DM and compute the cross-correlation matrix C_{zv}, the coupling matrix C_v, and its inverse C_v^{-1} according to (12).

Notably, the vector S_m and the matrices S, S^{-1}, C_v, C_v^{-1}, and C_{zv} in the preprocessing step can be obtained in advance and these matrixes have nothing to do with the wavefront to be corrected. The above data can be regarded as known after DM and the bias functions are determined and are imported directly during correcting the wavefront aberrations.

Iterative Step

(1) Design the mask through the programming software.

(2) Gather the corresponding far-field intensity of the wavefront to be corrected $\phi(r)$ from CCD, calculate MDS according to (2), and call it MDS_0.

(3) Calculate the corresponding voltage vector of each Zernike mode according to (5) by substituting $a = \{0, \ldots, a_i, 0, \ldots\}$ and apply the voltage vector to the actuators of DM. The wavefront shape introduced by the deformable mirror is superimposed to the wavefront to be corrected $\phi(r)$.

(4) Get the corresponding far-field intensity of the wavefront aberration superimposed from CCD and calculate MDS according to (2). Repeating (3) and (4), we can get the obtained $\text{MDS}_1, \text{MDS}_2, \ldots, \text{MDS}_N$ for N Zernike mode, respectively. The total Zernike coefficient is $a = \{\ldots, a_i, \ldots, a_j, \ldots\}$.

(5) Compute M according to (4).

(6) Obtain the corresponding voltage vector V of the wavefront to be corrected $\phi(r)$ according to (3). The voltage vector V is applied to the actuators of DM through the high voltage amplifier.

(7) Regard the residual wavefront as the wavefront to be corrected $\phi(r)$; repeat (2)–(7) in the iterative step until the algorithm satisfies the termination conditions in advance, such as a certain number of iterations.

The flow chart of the control algorithm is shown in Figure 1, where an inner cycle of control algorithm is considered as one algorithm iteration and we make use of the number of iterations as the end condition of algorithm.

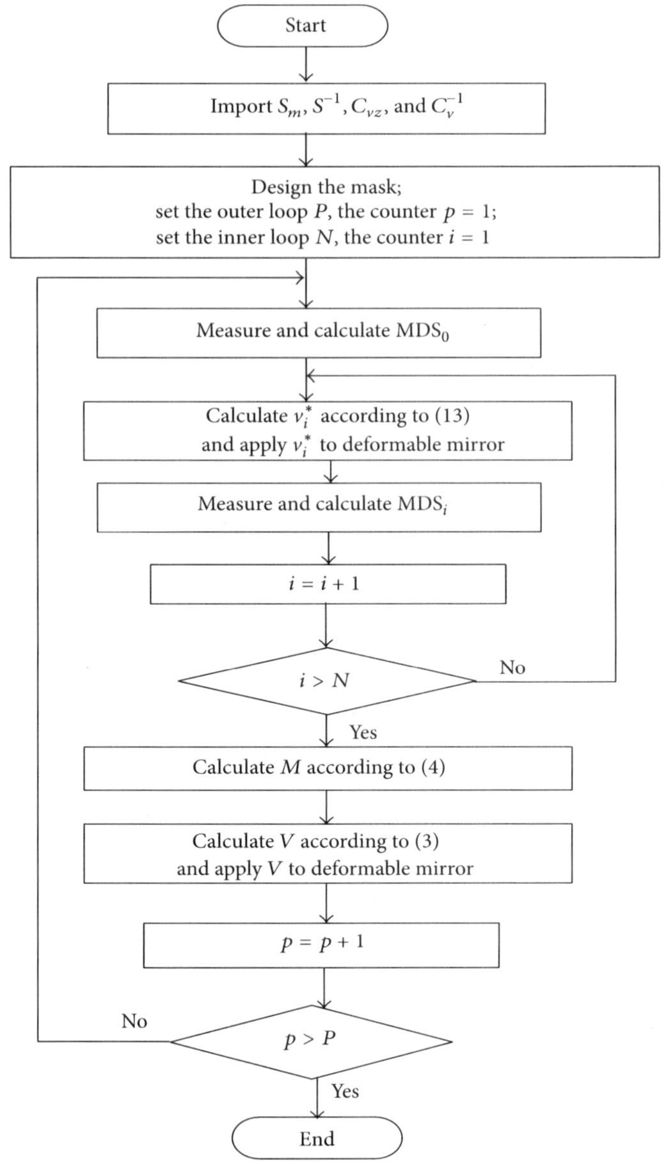

FIGURE 1: The flow chart of control algorithm for wavefront-sensorless adaptive optics system.

3. Simulation Platform of Wavefront-Sensorless AO System

As shown in Figure 2, the wavefront-sensorless AO system consists of a 61-element DM, an imaging lens, a CCD, and the control algorithm. The wavefront to be corrected is reflected to the imaging lens and the imaging lens focuses it into the CCD. The intensity is detected by CCD and provided to the control algorithm. The control algorithm generates a correction voltage vector. The voltage vector is applied to the actuators of 61-element deformable mirror.

On the basis of real measurements, we know that the response function of 61-element deformable mirror actuators is Gaussian distribution approximately [13]. Figure 3 gives the location distribution of 61-element DM actuators. The layout of all actuators is triangular.

The method proposed by Roddier, which makes use of a Zernike expansion of randomly weighted Karhunen-Loeve functions, is used to simulate atmospherically distorted wavefronts [14]. Considering that the low-order aberrations (tilts, defocus, astigmatism, etc.) have the most significant impact

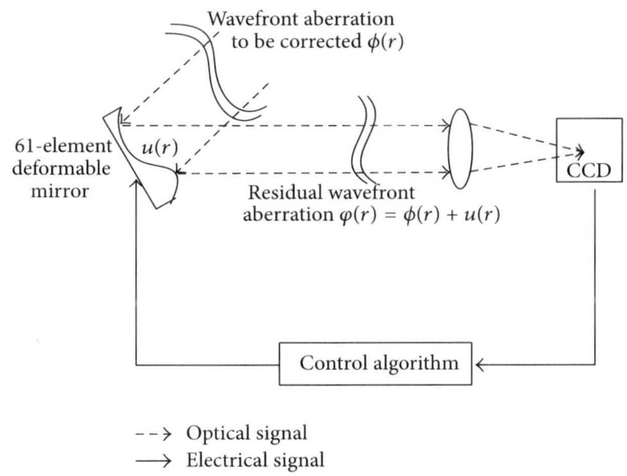

FIGURE 2: Block diagram of wavefront-sensorless adaptive optics system.

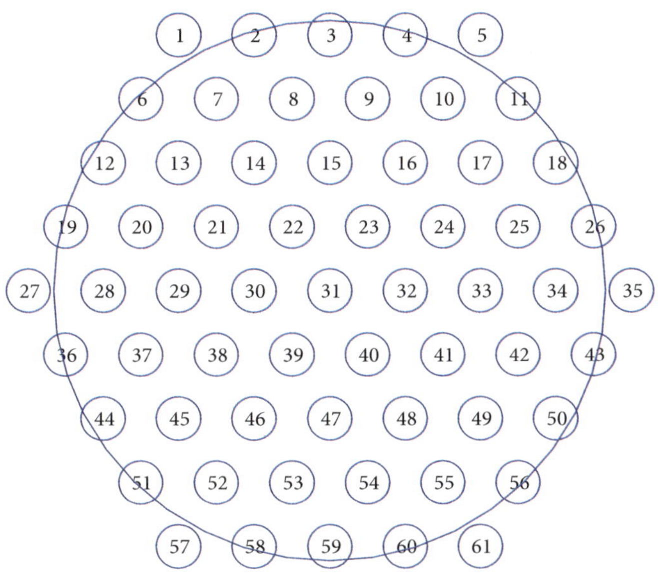

FIGURE 3: Actuators location of 61-element deformable mirror.

on image quality, we use the first 104 Zernike polynomials. The phase screens do not include the tip/tilt aberrations because the tip/tilt aberrations are usually controlled by another control loop and are considered as being removed completely in our simulation. Atmospheric turbulence strength for a receiver system with aperture size D can be characterized by the ratio D/r_0, where r_0 is the turbulence coherence length. Phase screens of different atmospheric turbulence strength can be obtained through changing r_0 in the simulation program.

4. Results and Analysis

To investigate the convergence speed and the correction capability of the wavefront-sensorless AO system with a 61-element deformable mirror based on the proposed control algorithm, we perform the adaptation process over random phase screens under five different turbulence strengths. The five different turbulence strengths are corresponding to $D/r_0 = 5, 10, 15, 20$, and 25, respectively. The averaged evolution curves over 500 frames per turbulence strength and the standard deviation evolution curves of the metric are the recorded simulation results. MR (mean radius) of the far-field intensity distribution and SR (Strehl ratio) are used to measure the correction capability of the AO system, respectively. The calculation formula of MR is as follows:

$$\mathrm{MR} = \frac{\iint_D \sqrt{(x-x_0)^2 + (y-y_0)^2}\, I(x,y)\,dx\,dy}{\int I(x,y)\,dx\,dy}, \tag{14}$$

where $(x_0, y_0) = \int (x,y) I(r)\,dx\,dy / \int I(x,y)\,dx\,dy$ is the centroid of far-field intensity distribution. The smaller the MR value, the more focused the far-field energy and the smaller wavefront aberration. MR can be easily measured either by a single photodetector with a special mask attached to it or by postprocessing a matrix detector output in the actual system. This measure appears to be the most attractive one as it gives straightforward mathematical meaning to the idea of energy spread. The Strehl ratio (SR) is defined as

$$\mathrm{SR} = \frac{F[I(x,y)]}{F[I_0(x,y)]}, \tag{15}$$

where $F[\cdot]$ is an operation, which calculates the peak intensity. $I_0(x, y)$ is the intensity distribution of ideal plane wavefront.

4.1. Adaptive Process of the Wavefront-Sensorless AO System. The number of inner cycles (iterative steps (3) and (4)) of control algorithm is 102 times because the wavefront aberration is composed of the first 104 Zernike polynomials (excluding tip and tilt) in this paper and an inner cycle of control algorithm is considered as one algorithm iteration. Figures 4(a) and 4(c) show the adaptive process of MR and SR by the control algorithm. Standard deviation (SD) curves of MR over 500 different phase realizations are presented in Figure 4(b). The SD is calculated as follows:

$$\mathrm{SD} = \frac{\left\langle (\mathrm{MR} - \langle \mathrm{MR} \rangle)^2 \right\rangle^{1/2}}{\langle \mathrm{MR} \rangle}. \tag{16}$$

From Figures 4(a) and 4(c), we can conclude that averaged curves of different metrics have converged after completing iterations and one iteration can make the AO system converge approximately. The values of SD of MR are very small after the system converges in Figure 4(b). This shows that the discrete level of MR from the averaged value is very tiny and the control algorithm has strong adaptability to 500 different phase screens.

To compare the correction capability and the convergence speed of AO system, the adaptive process of MR and SR by SPGD algorithm is given in Figure 5. Figure 5 shows that the AO system converges completely after about 1000 iterations under different turbulence strength.

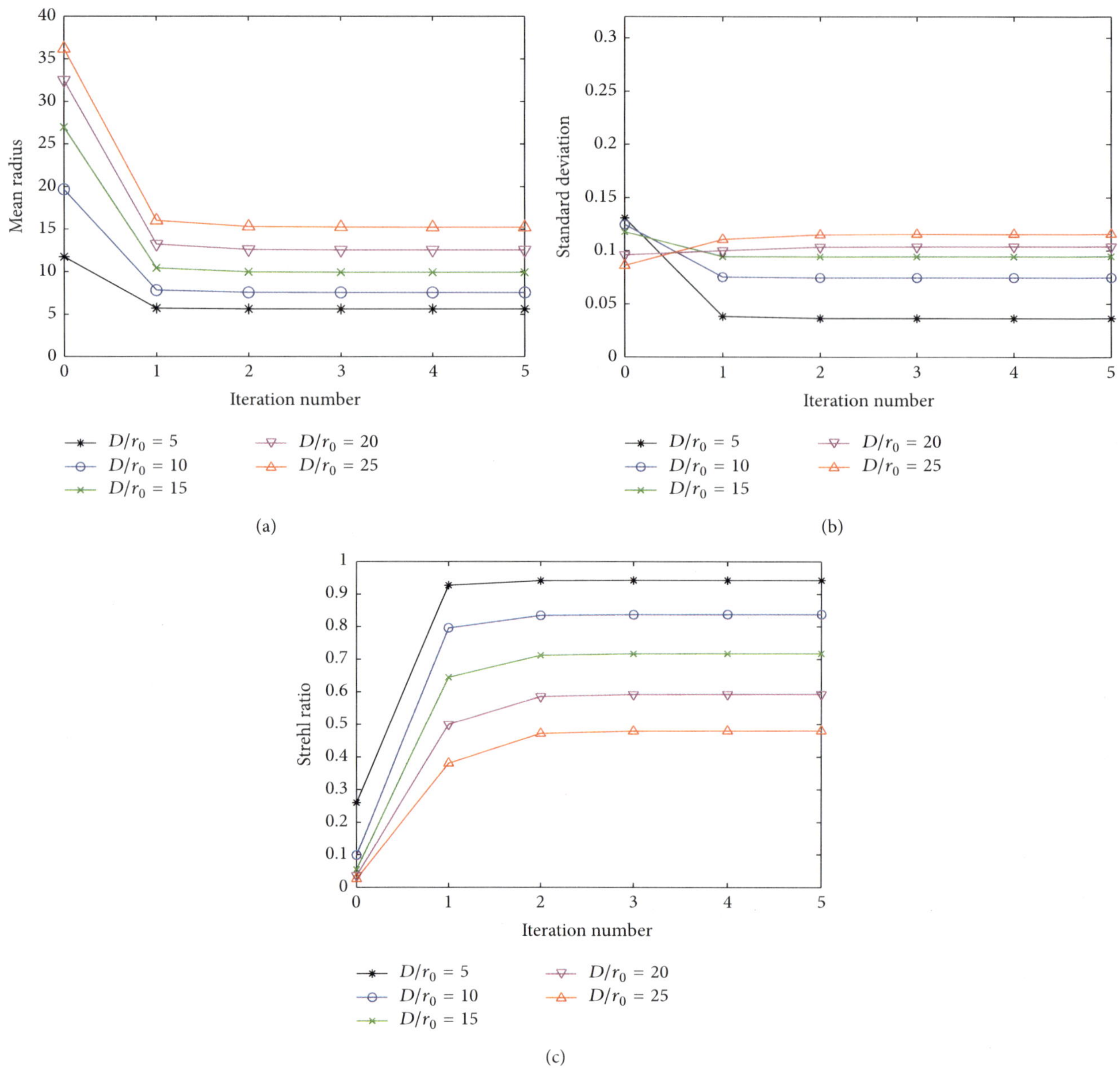

FIGURE 4: Averaged adaptation process over 500 different phase screens under different turbulence strengths. (a) MR curves, (b) standard deviation curves of MR, and (c) SR curves.

4.2. Correction Capability Analysis of the Wavefront-Sensorless AO System. For investigating the correction capability of the wavefront-sensorless AO system based on the proposed control algorithm, we also calculate the averaged MR of the initial far-field intensity distribution and the averaged ideal correction ability of the 61-element deformable mirror over 500 different phase screens. The ideal correction ability can be obtained by the minimum variance solution (4). The corresponding data are given in Figure 6.

The AO system based on two kinds of algorithms has strong correction ability on wavefront aberrations under different turbulence from Figure 6(a), in which MR values are very close to ideal correction of deformable mirror offered by the minimum variance solution. But it can be seen clearly from Figure 6(b) that the correction ability AO system based on SPGD algorithm is worse than that of AO system based on the proposed control algorithm when D/r_0 is greater than 15.

4.3. Convergence Speed Analysis of the Wavefront-Sensorless AO System. The convergence speed is an important criterion on which the control algorithm can be applied to real-time adaptive optics system. We use the number of intensity measurements to estimate the averaged convergence speed of wavefront-sensorless AO system based on two kinds of control algorithms. The number of intensity measurements can be analyzed from the control algorithm process. The total number of intensity measurements is computed through the times of inner cycle multiplied by that of the outer cycle for

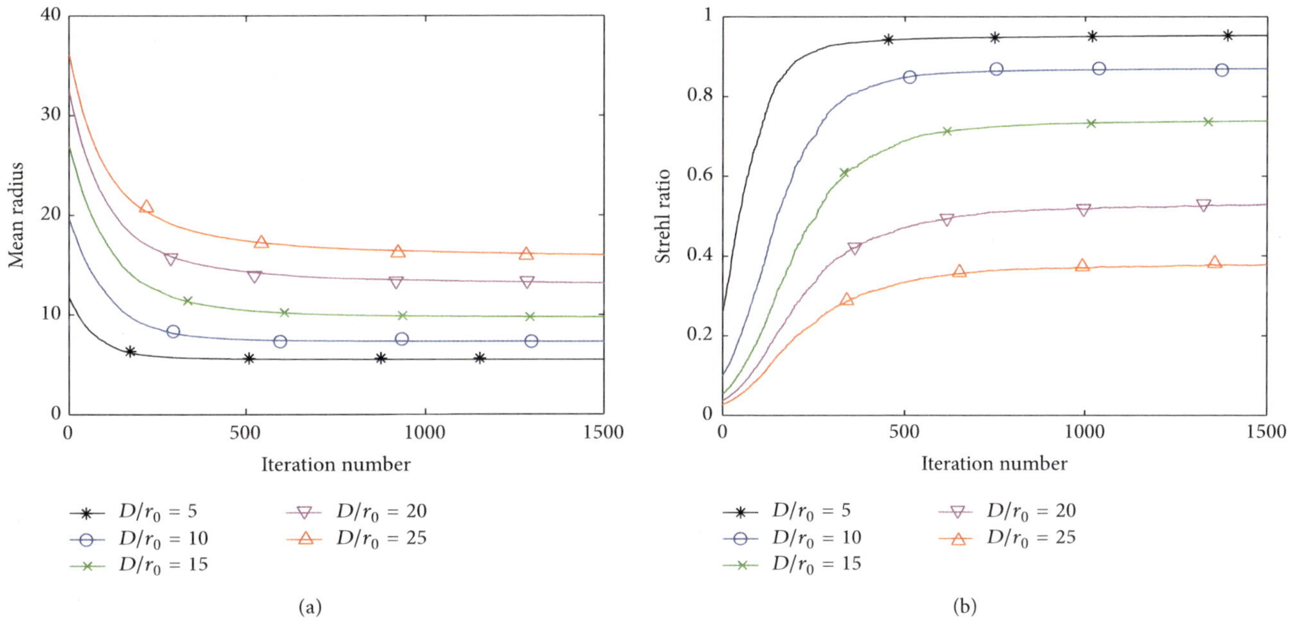

FIGURE 5: Averaged adaptation process over 500 different phase screens under different turbulence strengths by SPGD algorithm: (a) MR and (b) SR.

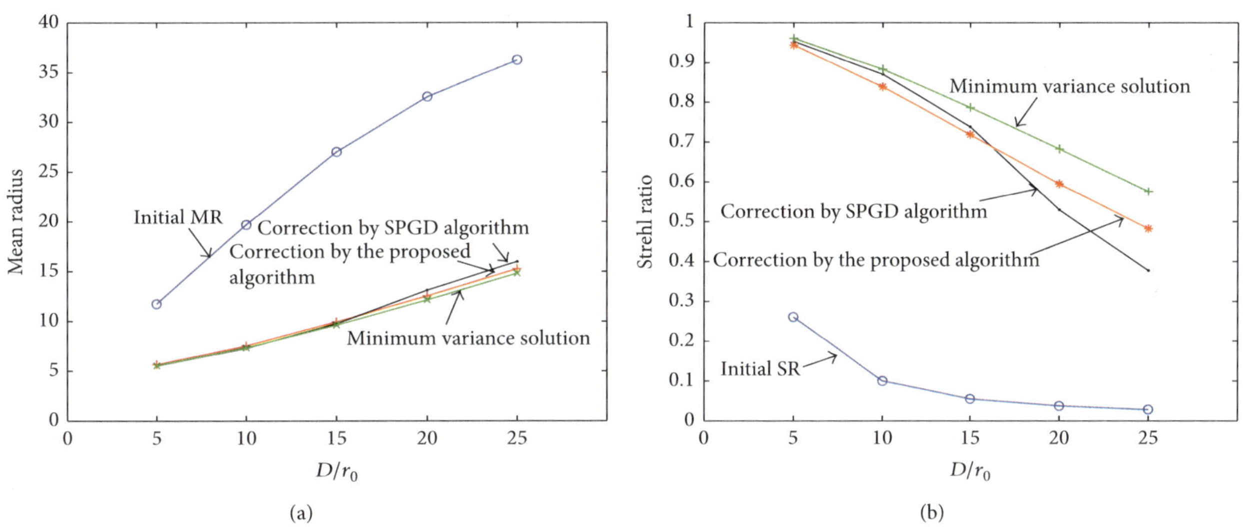

FIGURE 6: Comparisons of MR of initial, correction by SPGD algorithm, correction by proposed algorithm, and the minimum variance solution (a) and comparisons of SR (b) under different turbulences.

the wavefront-sensorless AO system based on the proposed algorithm. The number of measurements of AO system based on SPGD algorithm can be obtained through the times of algorithm iteration multiplied by two because two intensity measurements are needed by one algorithm iteration at least.

From Figure 4(a), we can see that the wavefront-sensorless AO system based on the proposed algorithm only needs one iteration for algorithm converging, which means the system needs $N+1$ photodetector measurements when N-order aberrations are corrected. The system needs 103 intensity measurements in this paper because the wavefront aberration is composed of the first 104 Zernike polynomials (excluding tip and tilt). Figure 4(a) also shows that the number of intensity measurements does not increase as the turbulence strength magnifies.

We can get the times of iteration needed by the wavefront-sensorless AO system based on SPGD from Figure 5 when MR decreases to the value where the proposed method arrives after one iteration and further achieve the number of intensity measurements through multiplying the times of iteration by two. Those data are given in Table 1, where the sign ">" indicates that the value of MR has not been reached at 12.5 and

TABLE 1: Comparison of the number of intensity measurements of SPGD with that of the proposed method.

D/r_0	5	10	15	20	25
SPGD	826	946	1458	>3000	>3000
Proposed method	103	103	103	103	103

15.2 on $D/r_0 = 20$, and 25, respectively, after the AO system completes 1500 iterations.

From data in Table 1, we can see that the number of intensity measurements needed by AO system based on SPGD is about 8 times, 9 times, and 14 times that of the wavefront-sensorless AO system based on the proposed algorithm when D/r_0 is 5, 10, and 15, respectively. More intensity measurements are needed by SPGD when D/r_0 is 20 and 25. In other words, the time needed by the wavefront-sensorless AO system based on SPGD is far more than that of the wavefront-sensorless AO system based on the proposed algorithm when the AO system achieves similar correction results. Additionally, the number of intensity measurements needed by SPGD-based AO system increases as the turbulence strength magnifies from the data in Table 1.

5. Conclusions

The centroid of the far-field intensity distribution in geometric optics is related to the aberration of the input wavefront. A linear relationship between the second moment of the aberration gradients and the masked far-field intensity distribution can be achieved through certain mathematical transformations. A closed-loop control algorithm is designed by using the linear relationship. An adaptive optics (AO) system simulation platform is established with a 61-element deformable mirror and a CCD. The convergence speed and the correction capability are investigated and compared with stochastic parallel gradient descent (SPGD) control algorithm, the most commonly used control algorithm for wavefront-sensorless adaptive optics systems through correcting wavefront aberrations under different turbulence strength.

Research results show that the control algorithm proposed has strong adaptability to different phase screens. The AO system based on two kinds of algorithms has strong correction ability on wavefront aberrations under different turbulence when the turbulence strength is relatively small. However, the correction ability of AO system based on SPGD algorithm is obviously worse than that of AO system based on the proposed control algorithm when the wavefront aberrations increase. The AO system based on the proposed control algorithm only requires $N + 1$ measurements of the far-field spot when N orders aberrations are corrected. The time needed by SPGD-based AO system is far more than that of the wavefront-sensorless AO system based on the proposed control algorithm when the AO system achieves similar correction results. Additionally, the convergence speed of the wavefront-sensorless AO system based on the proposed control algorithm is independent of the turbulence strength while the number of intensity measurements needed by SPGD-based AO system increases as the turbulence strength magnifies. Therefore, the control algorithm provided in the paper can greatly improve the convergence speed of the wavefront-sensorless AO systems, which provides the theoretical basis for applications of adaptive optics technology in real-time wavefront correction system.

Conflict of Interests

The authors declare that there is no conflict of interests regarding the publication of this paper.

Acknowledgments

This work is supported by Natural Science Foundation of Jiangsu Province of China (Grant no. BK2012653) through Huaihai Institute of Technology and Key Lab Foundation on Adaptive Optics of Chinese Academy of Sciences (Grant no. LAOF 201302).

References

[1] R. K. Tyson, *Principle of Adaptive Optics*, CRC Press, New York, NY, USA, 3rd edition, 2010.

[2] R. A. Muller and A. Buffington, "Real-time correction of atmospherically degraded telescope images through image sharpening," *Journal of the Optical Society of America A*, vol. 64, no. 9, pp. 1200–1210, 1974.

[3] M. A. Vorontsov, G. W. Carhart, M. Cohen, and G. Cauwenberghs, "Adaptive optics based on analog parallel stochastic optimization: Analysis and experimental demonstration," *Journal of the Optical Society of America A: Optics and Image Science, and Vision*, vol. 17, no. 8, pp. 1440–1453, 2000.

[4] C. Geng, W. Luo, Y. Tan, H. Liu, J. Mu, and X. Li, "Experimental demonstration of using divergence cost-function in SPGD algorithm for coherent beam combining with tip/tilt control," *Optics Express*, vol. 21, no. 21, pp. 25045–25055, 2013.

[5] M. S. Zakynthinaki and Y. G. Saridakis, "Stochastic optimization for a tip-tilt adaptive correcting system," *Computer Physics Communications*, vol. 150, no. 3, pp. 274–292, 2003.

[6] S. Zommer, E. N. Ribak, S. G. Lipson, and J. Adler, "Simulated annealing in ocular adaptive optics," *Optics Letters*, vol. 31, no. 7, pp. 939–941, 2006.

[7] P. Yang, M. Ao, Y. Liu, B. Xu, and W. Jiang, "Intracavity transverse modes controlled by a genetic algorithm based on Zernike mode coefficients," *Optics Express*, vol. 15, no. 25, pp. 17051–17062, 2007.

[8] A. Facomprez, E. Beaurepaire, and D. Débarre, "Accuracy of correction in modal sensor-less adaptive optics," *Optics Express*, vol. 20, no. 3, pp. 2598–2612, 2012.

[9] M. J. Booth, "Wavefront sensorless adaptive optics for large aberrations," *Optics Letters*, vol. 32, no. 1, pp. 5–7, 2007.

[10] H. Song, R. Fraanje, G. Schitter, H. Kroese, G. Vdovin, and M. Verhaegen, "Model-based aberration correction in a closed-loop wavefront-sensor-less adaptive optics system," *Optics Express*, vol. 18, no. 23, pp. 24070–24084, 2010.

[11] H. Linhai and C. Rao, "Wavefront sensorless adaptive optics: a general model-based approach," *Optics Express*, vol. 19, no. 1, pp. 371–379, 2011.

Broadband Degenerate Femtosecond OPO around 2060 nm

Wenlong Tian,[1,2] **Xianghao Meng**,[2] **Ninghua Zhang**,[1] **Jiangfeng Zhu**,[1] **Zhaohua Wang**,[2] **and Zhiyi Wei**[1,2]

[1]*School of Physics and Optoelectronic Engineering, Xidian University, Xi'an 710071, China*
[2]*Beijing National Laboratory for Condensed Matter Physics, Institute of Physics, Chinese Academy of Sciences, Beijing 100190, China*

Correspondence should be addressed to Zhaohua Wang; zhwang@iphy.ac.cn and Zhiyi Wei; zywei@iphy.ac.cn

Academic Editor: Guoqiang Xie

We demonstrate a broadband degenerate femtosecond optical parametric oscillator (OPO) around 2060 nm. Synchronically pumped by a commercially available 75.6 MHz Yb:KGW femtosecond laser, it is able to generate self-phase-locked 2 μm pulses with 82 mW output power and 3 dB bandwidth of 105 nm. The degenerate operation is naturally stable in free-running condition, with only 0.8% rms fluctuation in 3.5 hours. We also obtain broadband spectrum extending from 1600 nm to 2500 nm with the 3 dB bandwidth of 452 nm by increasing the intracavity power density.

1. Introduction

Femtosecond sources in the range of 2 μm, which is close to the strong water absorber lines, are of great interests for numerous applications, such as "eye-safer" remote sensing [1], high harmonic generation [2], and driven particle acceleration [3] as well as optical frequency combs [4]. For the optical frequency comb in particular, extending it to wavelength around or beyond 2 μm is attracted for a variety of applications including coherent molecular spectroscopy [5] and trace gas detection in the fingerprint region [6], however, which still faces technical challenging. Currently, there are several methods to generate ultrafast 2 μm pulses including mode-locked solid-state lasers based on Tm or Ho doped crystal [7, 8], Tm-doped fiber lasers [9], and nonlinear parametric process [10, 11]. Among them, degenerate femtosecond optical parametric oscillator (OPO) can clearly transpose near-IR frequency combs to the mid-IR, for example, transfer the properties of a frequency comb based on commercial Yb laser to the 2 μm wavelength range. In addition, a synchronically pumped OPO operated at degeneracy not only provides phase and frequency locking of the mid-IR comb lines to the pump but also results in enhanced bandwidth. Hence, degenerate femtosecond OPO is a simple approach to generate coherent and broadband mid-IR frequency combs and has been demonstrated at several different pump wavelengths ranging from 775 nm to 4 μm [12–16]. In 2012, Rudy et al. report generation of 48 fs pulses around 2070 nm from a degenerate OPO; however the average output power is only ~10 mW [17].

In this letter, we report the demonstration of a broadband degenerate femtosecond OPO synchronically pumped by a commercially available Yb:KGW femtosecond laser. With a 3 mm long magnesium oxide-doped, periodically poled, lithium niobate crystal (MgO:PPLN), self-phase-locked pulses with average output power of 82 mW are obtained, whose spectrum has a 3 dB bandwidth of 105 nm. By further improving the intracavity power density, as broad as extending from 1600 nm to 2500 nm spectrum with 3 dB bandwidth of 452 nm is also achieved.

2. Experimental Setup

Figure 1 is an illustrative view of the experimental setup of the degenerate OPO. The pump source is a commercial KLM Yb:KGW laser (Light Conversion, FLINT6.0), which can deliver 90 fs pulses at 1030 nm with up to 7 W average power. The repetition rate of the Yb:KGW laser is 75.57 MHz, corresponding to the linear cavity length of 1985 mm. We used a 3 mm-long 5 mol% MgO doped PPLN with single grating period of 31 μm as the nonlinear crystal. It is antireflective-coated at 1035 nm ($R < 0.5\%$) and 2070 nm ($R < 0.5\%$) on

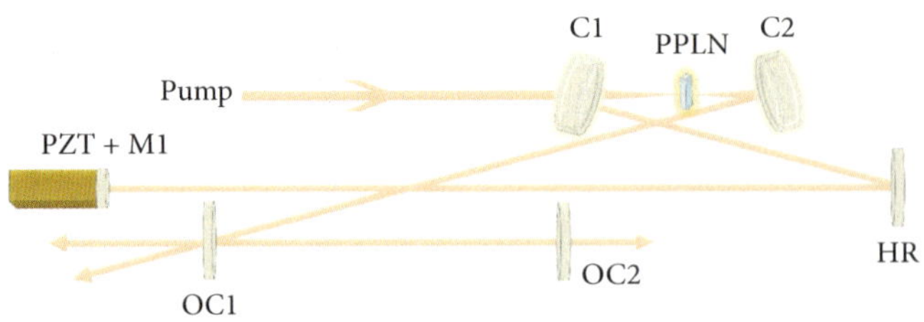

FIGURE 1: The sketch for the experimental setup of the degenerate OPO. C1 and C2: concave mirrors with ROC of 150 mm; HR: flat mirror with high reflectivity; M1: sliver mirror; PZT: piezo translator; OC1 and OC2: output couplers.

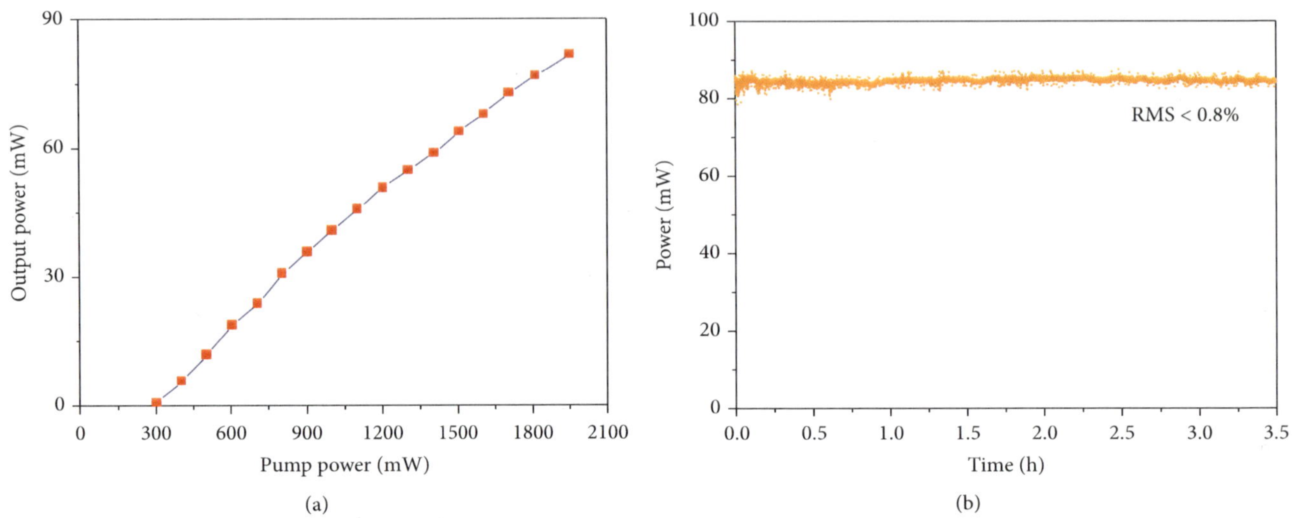

FIGURE 2: (a) Power scalable of the OPO output at degeneracy. (b) Long-term power stability of the maximum output power of the degenerate OPO.

both surfaces. Because that the phase matching of degenerate operation for such a PPLN with 31 μm grating period is realized at the temperature of 94°C, we used an oven to heat the MgO:PPLN, which can also limit the amount of photorefractive damage caused by the strong 1 μm pump. To eliminate the harmful influence on the Yb:KGW laser from reflected pump and signal, the pump laser was firstly through an isolator comprising an half-wave-plate (HWP), a polarization beam splitter (PBS), and an Faraday rotator (FR). The combination of the HWP and PBS also acted as a power attenuator, keeping the pump power focused on the PPLN less than 2 W to avoid damage on the PPLN. We used the second HWP after the FR to adjust the pump polarization relative to the crystal orientation. C1 and C2 are two concave mirrors with 150 mm radium of curvature (ROC), coated with high reflectivity over 1900–2100 nm ($R >$ 99.9%) and high transmittance around 800–1200 nm ($R <$ 10%). HR is a flat mirror with high reflectivity in the region of 1850–2200 nm ($R >$ 99.9). M1 is a sliver mirror, on which a piezo translator (PZT) was mounted to finely tune the cavity length. Two output couplers (OC) were used in the cavity, leading to output in three channels, which is convenient for monitoring or further locking. OC1 has the transmittance of 20% around 1700–2700 nm and OC2 is a 0.5 mm long broadband pass filter (BBP-1660–2250 nm) with 80% transmittance from 1660–2250 nm. The linear cavity was about 1.98 m long in length, which was matched to the pump repetition rate.

The OPO cavity could be easily aligned with the visible second harmonic generation (SHG) of the pump. After the initial alignment, we achieved successful operation of the nondegenerate OPO in free-running condition with the cavity length being finely tuned by using the PZT stage to meet the synchronous pumping condition. With further fine adjustment the cavity length or changing the temperature of the oven, OPO operation was gradually transferred to a degenerate and self-phase-locked oscillation. The output power from OC2 was dropped from 160 mW in nondegenerate operation to 82 mW in degeneracy with 2 W pump power. We measured the output power at different pump power as described in Figure 2(a). As evident, the output power linear increases with the pump power, and the threshold for degenerate operation was about 300 mW. As shown in Figure 2(b), we also measured the long-term stability of the output power at degeneracy, which was with a fluctuation of 0.8% rms in 3.5 hours, indicating that the degenerate OPO was naturally stable so that a feedback servo loop is not necessary. The fluctuation was mainly derived from the ambient temperature fluctuation, airflow, and mechanical vibration.

We performed the spectral characterization of the OPO output in near-degenerate and degenerate operation with a

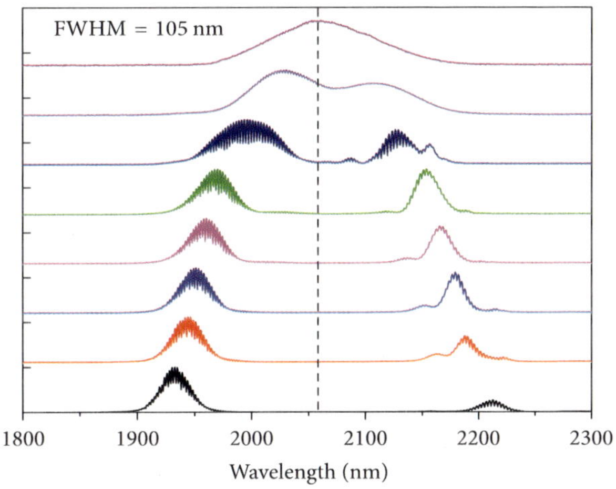

FIGURE 3: Spectra evolution from the near-degenerate operation to the degeneracy of the OPO. Dash line: the degenerate point.

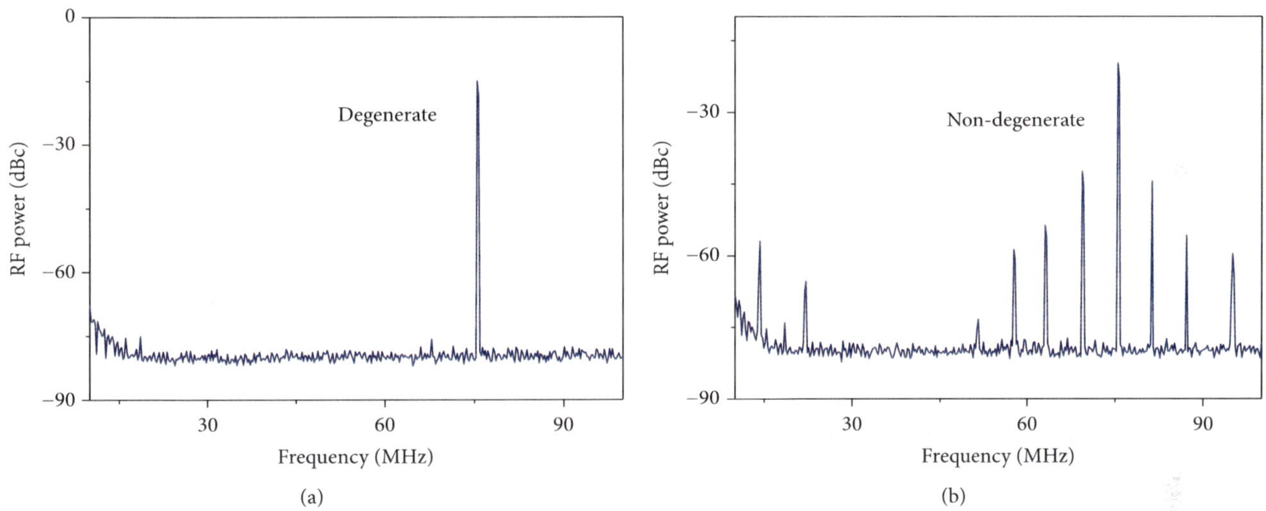

FIGURE 4: RF spectra of the degenerate and near-degenerate OPO with resolution bandwidth of 10 kHz, respectively.

commercial wavelength-meter (WaveScan, A. P. E GmbH), as presented in Figure 3. It is obvious that the OPO output spectra undergo a gradual transition from near-degenerate operation to an exactly degenerate state when increasing the cavity length. In near-degenerate operation, there are clearly two distinct signal and idler spectra, which exhibit some spiking and modulation but progressively merge into a single broadband and smooth spectrum centered at 2060 nm in degenerate operation. The full width at half maximum (FWHM) of the spectrum at degeneracy was about 105 nm, corresponding to a Fourier transform limited pulse duration of 42 fs if a sech^2 shape was assumed. However due to the lack of suitable autocorrelator, the character of the temporal profile of the degenerate pulses was not conducted.

To claim the status of the degenerate operation, the radio frequency (RF) spectra of both degenerate and near-degenerate operations were also measured via a photo detector (PD) with a 3 dB bandwidth of 1 GHz and a commercial RF spectrum analyzer (R&S FSVA40) as shown in Figure 4. At degeneracy, there is only one peak in Figure 4(a), which indicates that the signal and idler are self-phase-locked well. In the near-degenerate state, there is no stable injection locking despite the partially overlapping signal and idler spectra. This is proofed by the additional peaks in the RF spectrum (Figure 4(b)), which are contributed to the frequency beat between the overlapping tails of the signal and idler waves with different f_{ceo}.

We further investigated the relationship between the spectrum width at degeneracy and the intracavity power density, by utilizing different output couplers while maintaining the pump power. Figures 5(a)–5(d) depict the spectra characterization of the OPO output at degeneracy with different combination of OC1 and OC2. The total transmittance of OC1 and OC2 from Figures 5(a) to 5(b) tent to decrease, corresponding to the increasing of intracavity power density. Hence, it is clear to see that the spectrum of the OPO output at degeneracy is broaden as the growing of the intracavity power density, which could be attributed to the self-phase modulation. For Figure 5(d) case, the spectrum covered from 1600 nm to 2500 nm, with the FWHM of 452 nm.

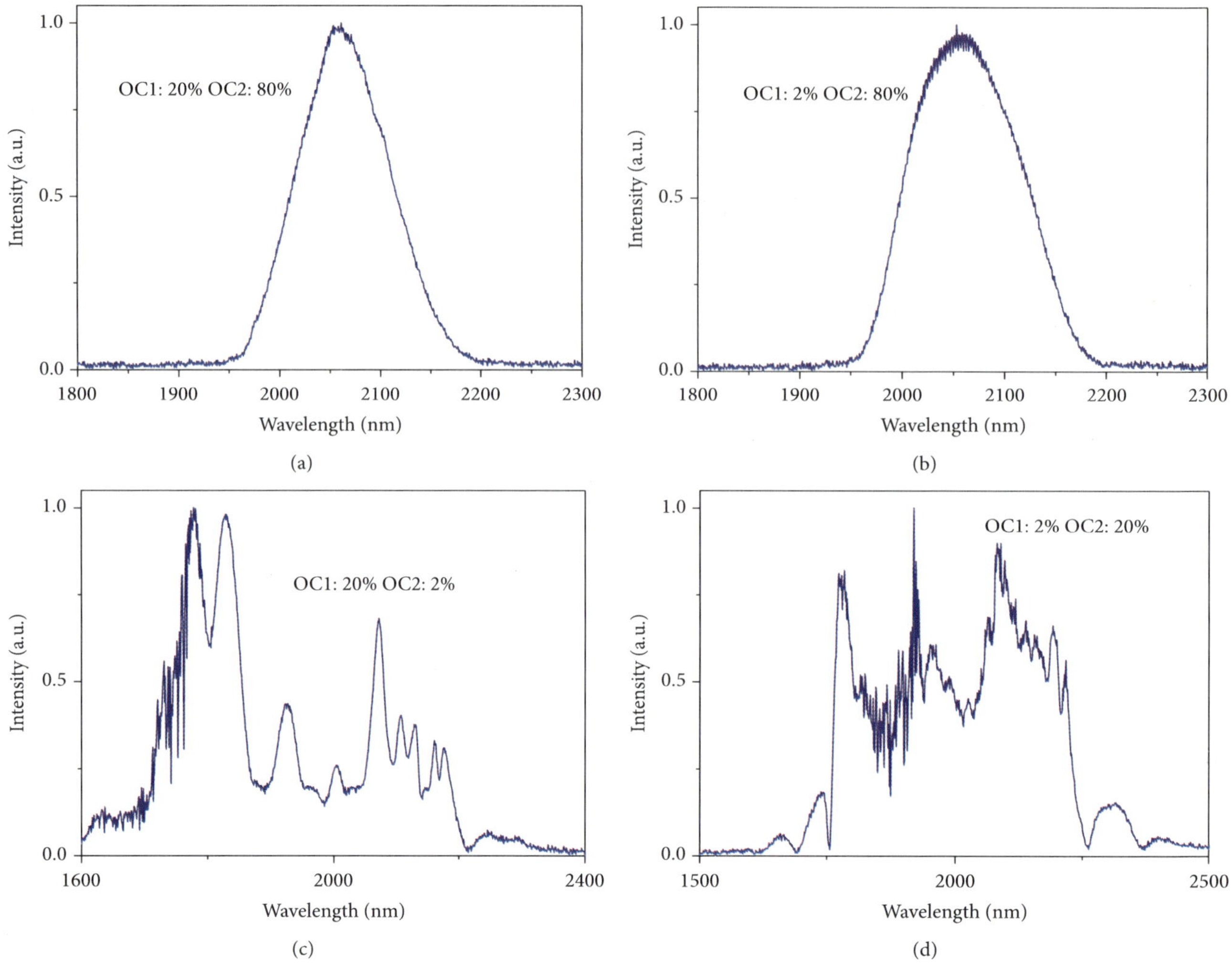

FIGURE 5: Spectra of the degenerate OPO with different output couplers.

3. Conclusion

In conclusion, we experimentally demonstrated a degenerate synchronically pumped OPO based on MgO:PPLN, generating 2 μm pulses with 82 mW average power. The spectrum of 2 μm pulses has a FWHM of 105 nm, which could support as short as 42 fs pulse duration. Without active stabilization, the degenerate OPO can be naturally stable with only 0.8% rms fluctuation in 3.5 hours. By further improving the intracavity power density, degenerate pulses with as broad as covering the region of 1600–2500 nm spectrum were successfully obtained. Since the mid-IR pulses are phase-locked to the pump laser, such broadband femtosecond source is very suitable for midinfrared frequency comb and well found lots of applications in various fields.

Conflicts of Interest

The authors declare that they have no conflicts of interest.

Acknowledgments

This work is supported by the National Natural Science Foundation of China (61705174); National Key Basic Research Program of China (2013CB922402); the Chinese Academy of Sciences Key Deployment Project (KJZD-EW-L11-03); the Fundamental Research Funds for the Central Universities (Grant no. JBX170511); the 111 Project (B17035).

References

[1] S. W. Henderson, C. P. Hale, J. R. Magee, M. J. Kavaya, and A. V. Huffaker, "Eye-safe coherent laser radar system at 2.1 μm using Tm,Ho:YAG lasers," *Optics Letters*, vol. 16, no. 10, pp. 773–775, 1991.

[2] J. Moses et al., "CTuEE2," in *Proceedings of the in Conference on Lasers and Electro-Optics/Quantum Electronics and Laser Science Conference and Photonic Applications Systems Technologies*, Optical Society of America.

[3] B. Carlsten, E. Colby, E. Esarey et al., "New source technologies and their impact on future light sources," *Nuclear Instruments and Methods in Physics Research Section A: Accelerators, Spectrometers, Detectors and Associated Equipment*, vol. 622, no. 3, pp. 657–668, 2010.

[4] A. Schliesser, N. Picqué, and T. W. Hänsch, "Mid-infrared frequency combs," *Nature Photonics*, vol. 6, no. 7, pp. 440–449, 2012.

[5] F. Keilmann, C. Gohle, and R. Holzwarth, "Time-domain mid-infrared frequency-comb spectrometer," *Optics Letters*, vol. 29, no. 13, pp. 1542–1544, 2004.

[6] F. Adler, P. Maslowski, A. Foltynowicz et al., "Mid-infrared Fourier transform spectroscopy with a broadband frequency comb," *Optics Express*, vol. 18, no. 21, pp. 21861–21872, 2010.

[7] A. Schmidt, P. Koopmann, G. Huber et al., "175 fs Tm:Lu2O3 laser at 2.07 μm mode-locked using single-walled carbon nanotubes," *Optics Express*, vol. 20, no. 5, pp. 5313–5318, 2012.

[8] V. Aleksandrov, A. Gluth, V. Petrov et al., "Tm,Ho:KLu(WO4)2 laser mode-locked near 2 μm by single-walled carbon nanotubes," *Optics Express*, vol. 22, no. 22, pp. 26872–26877, 2014.

[9] J. Bethge, J. Jiang, C. Mohr, M. Fermann, and I. Hartl, "AT5A," in *Proceedings of the Advanced Solid-State Photonics*, vol. 3.

[10] T. Petersen, J. D. Zuegel, J. Bromage, and High-average-power., "2 μm femtosecond optical parametric oscillator synchronously pumped by a thin-disk, mode-locked laser," *Optics Express*, vol. 25, no. 8, pp. 8840–8844, 2017.

[11] S. C. Kumar and M. Ebrahim-Zadeh, "Yb-fiber-based, high-average-power, high-repetition-rate, picosecond source at 2.1 μm," in *Laser & Photonics Reviews*, vol. 10, pp. 970–977, 6 edition, 2016.

[12] A. Marandi, K. A. Ingold, M. Jankowski, and R. L. Byer, "Cascaded half-harmonic generation of femtosecond frequency combs in the mid-infrared," *Optica*, vol. 3, no. 3, pp. 324–327, 2016.

[13] V. O. Smolski, S. Vasilyev, P. G. Schunemann, S. B. Mirov, K. L, and K. L. Cr. Vodopyanov, "Cr:ZnS laser-pumped subharmonic GaAs optical parametric oscillator with the spectrum spanning 3.6–5.6 μm," *Optics Letters*, vol. 40, no. 12, pp. 2906–2908, 2015.

[14] V. Ramaiah-Badarla, A. Esteban-Martin, and M. Ebrahim-Zadeh, "Self-phase-locked degenerate femtosecond optical parametric oscillator based on BiB3O6," *Laser & Photonics Reviews*, vol. 7, no. 5, pp. L55–L60, 2013.

[15] M. W. Haakestad, A. Marandi, N. Leindecker, and K. L. Vodopyanov, "Five-cycle pulses near $\lambda = 3$ μm produced in a subharmonic optical parametric oscillator via fine dispersion management," *Laser & Photonics Reviews*, vol. 7, no. 6, pp. L93–L97, 2013.

[16] N. Leindecker, A. Marandi, R. L. Byer et al., "Octave-spanning ultrafast OPO with 2.6-6.1μm instantaneous bandwidth pumped by femtosecond Tm-fiber laser," *Optics Express*, vol. 20, no. 7, pp. 7046–7053, 2012.

[17] C. W. Rudy, A. Marandi, K. A. Ingold et al., "Sub-50 fs pulses around 2070 nm from a synchronously-pumped, degenerate OPO," *Optics Express*, vol. 20, no. 25, pp. 27589–27595, 2012.

13

Design and Implementation Guidelines for a Modular Spectral-Domain Optical Coherence Tomography Scanner

Farid Atry,[1] Israel Jacob De La Rosa,[2] Kevin R. Rarick,[3] and Ramin Pashaie[2]

[1]*Department of Biomedical Engineering, University of Wisconsin–Madison, Madison, WI, USA*
[2]*Electrical Engineering Department, University of Wisconsin–Milwaukee, Milwaukee, WI, USA*
[3]*Department of Pediatrics, Medical College of Wisconsin, Milwaukee, WI, USA*

Correspondence should be addressed to Ramin Pashaie; pashaie@uwm.edu

Academic Editor: Chenggen Quan

In the past decades, spectral-domain optical coherence tomography (SD-OCT) has transformed into a widely popular imaging technology which is used in many research and clinical applications. Despite such fast growth in the field, the technology has not been readily accessible to many research laboratories either due to the cost or inflexibility of the commercially available systems or due to the lack of essential knowledge in the field of optics to develop custom-made scanners that suit specific applications. This paper aims to provide a detailed discussion on the design and development process of a typical SD-OCT scanner. The effects of multiple design parameters, for the main optical and optomechanical components, on the overall performance of the imaging system are analyzed and discussions are provided to serve as a guideline for the development of a custom SD-OCT system. While this article can be generalized for different applications, we will demonstrate the design of a SD-OCT system and representative results for *in vivo* brain imaging. We explain procedures to measure the axial and transversal resolutions and field of view of the system and to understand the discrepancies between the experimental and theoretical values. The specific aim of this piece is to facilitate the process of constructing custom-made SD-OCT scanners for research groups with minimum understanding of concepts in optical design and medical imaging.

1. Introduction

Spectral-domain optical coherence tomography (SD-OCT) [1] is a versatile imaging technique with the ability to acquire 3D images at a resolution as low as 2 μm up to several hundred micrometers deep in a scattering tissue. Optical coherence tomography can visualize the detailed structure of different biological tissue samples [2] such as brain cytoarchitecture [3], large cellular structures (e.g., [4, 5]), or retinal layers in the eye ([6–8] to name a few) or functional retinal images (e.g., [9, 10]). In addition, it can provide functional information by capturing the vascular morphology through optical coherence angiography (e.g., [11–14]) or by recording vascular blood velocity (e.g., [15–18]). Spectroscopic OCT [19] can provide additional functional information by analysis of light absorption within the bandwidth of the light source [20, 21]. Polarization-sensitive OCT (PS-OCT) [22–24] analyzes the effect of tissue on the light polarization to provide a contrast mechanism other than light intensity. Different implementation of PS-OCT has been proposed for SD-OCT structure in the past years [25–28]. Optical coherence elastography (OCE) is another application of OCT imaging, which aims to study the mechanical properties of the tissue. OCE, which was originally based on ultrasonic elastography, uses optical coherence imaging to analyze tissue displacement when it is vibrated, stretched, or compressed [29, 30]. This concept was also adapted and used for SD-OCT imaging (e.g., [31, 32]). For most OCT applications, the imaging tasks can be performed without the use of exogenous contrast agents, and in many cases it can be considered as a noninvasive imaging technique. However, OCE requires some external source of force to create vibration or mechanical stress in the tissue. In some applications, spectroscopic OCT is performed when external agents such as gold nanoparticles are introduced to the tissue to improve imaging signal-to-noise ratio (SNR). Except for polarization-sensitive OCT, the other abovementioned OCT applications are possible with a typical SD-OCT platform similar to what is discussed in this paper. For PS-OCT, some modifications to the system are necessary to precisely control the polarization of incident light and to detect the polarization state of the backscattered light. However, the authors only used the suggested design for Doppler OCT and optical coherence angiography imaging. We refer interested readers to [33, 34] for a more detailed discussion on each of these techniques. Nonetheless, a broader range of research questions can be addressed by combining SD-OCT imaging techniques with other modalities such as fluorescence microscopy or electrophysiology. These beneficial features have promoted SD-OCT applications and as a result an increasing number of commercial SD-OCT scanners have become available for basic science and clinical research applications. It is important to note other OCT configurations (swept-source OCT and time-domain OCT) that are also widely used and are capable of performing these tasks. However, in this article, the focus is on developing a guideline for a SD-OCT device.

Despite commercial developments of OCT scanners and the wide range of applications, this technology is not readily accessible to many research groups either due to the cost considerations or due to the difficulty of modifying commercial systems to meet evolving research goals. The necessary modifications are sometimes impossible to achieve or require complex optical designs or custom-machined parts which make the implementation process slow and increase the ultimate cost of the SD-OCT setup. Design and development of a custom-made OCT scanner is potentially a viable solution in such cases; however, it can also be a time-consuming process when users lack experience or knowledge in the field of optical engineering. A discussion on the optical design and practical implementation of a custom SD-OCT system can be beneficial to research groups with different backgrounds and with no or minimal knowledge of the science of optics to allow them to construct their in-house OCT system that can be tailored to their specific research needs. This can greatly accelerate research throughput of many laboratories and is the specific aim of the current manuscript.

While the theory and applications of SD-OCT are vastly presented in many great literatures (e.g., [2, 35–44]), the lack of a detailed discussion on optical design and practical implementation guidelines of this device precludes the utilization of SD-OCT scanners by many researchers. In this paper, we review the fundamental concepts of SD-OCT image formation and propound practical considerations for the design and implementation of the device, along with some guidelines for choosing proper components to obtain the desired image quality. We offer a procedure to build a cost-effective SD-OCT imaging system versatile enough to be integrated with other optical technologies. The proposed system is portable and easy to disassemble/reassemble, which facilitates its transportation. Except for a few simple custom-made optical mounts, it mainly utilizes off-the-shelf optical components, which makes the implementation more cost and time efficient.

2. Materials and Methods: Theory

OCT was proposed as an extension of the low-coherence reflectometry in 1991 by Huang et al. [47]. In their system, they coupled a low-coherence light source to a fiber-based Michelson interferometer. The Michelson interferometer splits the beam into two different paths: a reference and a sample path. The sample path illuminates a sample and collects the backscattered light. The reference path guides the light toward a reference mirror that moves back and forth to continuously modulate the optical length of the reference path. Light that is received back from the sample and the reference mirror is combined and guided toward a photodetector. When the optical lengths of the sample and reference path are almost equal (i.e., the difference is no more than the source coherence length), the interference pattern becomes visible. The amplitude of the interference is maximum when the lengths of the two paths are exactly equal and drops as we move the reference mirror out of the coherence length. It is theoretically and experimentally proven that the axial (along the z direction) resolution of such an OCT scanner is the same as the coherence length of the source (l_c) [48]. For a Gaussian light source which produces the central wavelength of λ_0 and the bandwidth of $\Delta\lambda$, the coherence length is $l_c = \ln(2)(2/\pi)(\lambda_0^2/\Delta\lambda)$. This assembly was the first generation of an OCT imager, which is usually called a time-domain OCT (TD-OCT). The main drawback of this scanning mechanism is the need to physically move the reference mirror, which limits the speed and performance of the scanner. Development of the frequency-domain OCT technology was basically the attempt to overcome this limitation. Frequency-domain OCT devices generate images by analyzing the spectral pattern of the collected interferograms generated by combining the reflected lights from the reference and the sample arms. In frequency-domain OCT scanners, the spectrum of the interference pattern is produced by either replacing the photodetector in the conventional TD-OCT by a spectrometer in spectral-domain OCT configurations or utilizing a light source that rapidly sweeps the illumination wavelengths over a desired bandwidth in swept-source OCT (SS-OCT) devices. TD-OCT is relatively of low cost; however, the requirement for scanning path length in the reference arm severely limits the maximum speed to be typically less than a few thousand A-scans per second. Some research groups have been able to achieve scan rates as fast as 20 K A-scans per second [49]. SD-OCT can achieve imaging speed that is almost an order of magnitude faster than TD-OCT. The system is more complex and requires the design and implementation of a fast spectrometer. Therefore, the complexity and cost of a SD-OCT scanner are higher than of a TD-OCT device. SS-OCT can achieve scan rates faster than SD-OCT. By the use of signal multiplexing or full-field recording, this method can increase the imaging speed even beyond this level. A-scan rates beyond 200 K lines per second are reported for SS-OCT. While SD-OCT is typically slower than SS-OCT, it offers superior phase stability and axial resolution, which is beneficial in Doppler flowmetry or capillary imaging that require high phase stability and fine axial resolution. The current SS-OCT technology provides fast imaging capability with large imaging depth.

Here, we discuss the theory as well as design and implementation guidelines for a typical fiber-based spectral-domain OCT. Some statements are also valid for other OCT configurations.

In 1995, Fercher et al. proposed a spectral interferometry technique, based on the Fourier diffraction theorem, to measure the optical length of objects along the light beam (the z-axis) [1]. For weakly scattering objects, where the first-order Born approximation is valid, they demonstrated that, under low-coherence illumination, the spectrum of the backscattered light is proportional to the Fourier transform of the scattering potential of the sample ($F(z)$). Their work can be considered as the foundation of Fourier domain OCT systems. For this discussion, we adapt their approach to estimate the backscattered electric field from a sample when it is exposed to a monochromatic Gaussian beam of light with power spectral density of $P_S(k)$, where $k = |\mathbf{k}_i| = |\mathbf{k}_s| = 2\pi/\lambda$ is the wavenumber and $\mathbf{k}_i$ and $\mathbf{k}_s$ are the wave vectors of the illuminating and backscattering waves, respectively. When assuming that the origin of the coordinate system is placed at the surface of a specimen of thickness T, the backscattered electrical field (E^D_{Sample}) at a sufficiently large distance D along the illumination axis is estimated by [1]

$$E^D_{\text{Sample}}(\mathbf{k},t) = \alpha\sqrt{P_S(k)}e^{(ik\cdot D - i\omega t)} \times \int_0^T F(z')\,e^{-i(\mathbf{k}_s-\mathbf{k}_i)z'}dz' = \alpha\sqrt{P_S(k)}e^{(ik\cdot D-i\omega t)} \times \text{FT}\{F(z)\}, \quad (1)$$

where FT is the Fourier transform with respect to z, ω is the angular frequency of the monochromatic light, and α is a constant depending on the distances and configuration of the beam. When this beam interferes with the beam from the reference mirror which is placed at the location 0 (E^D_{Ref}), the combined electric field is

$$E^D_{\text{total}}(\mathbf{k},t) = E^D_{\text{Sample}}(\mathbf{k},t) + E^D_{\text{Ref}}(\mathbf{k},t) = \alpha\sqrt{P_S(k)}e^{(ik\cdot D-i\omega t)} \times \text{FT}\{F(z)\} + \beta\sqrt{P_S(k)}e^{(ik\cdot D-i\omega t)} \times \text{FT}\{R\delta(z)\}. \quad (2)$$

The parameter R is the reflectivity of the mirror and β is the portion of the back-reflection light that is used to produce interference. The mirror is assumed to be infinitely thin so its reflection can be modeled as a Dirac delta function at $z = 0$ ($\delta(z)$). The intensity of the interfering signal is the square of the absolute value of this electric field:

$$\begin{aligned} I(k) &= E^D_{\text{total}}(\mathbf{k},t)\,E^D_{\text{total}}(\mathbf{k},t)^* \\ &= P_S(k)\,\alpha^2\,|\text{FT}\{F(z)\}|^2 + P_S(k)\,\beta^2\,|\text{FT}\{R\delta(z)\}|^2 \\ &\quad + P_S(k)\,\alpha\beta^*\,\text{FT}\{F(z)\}\,\text{FT}\{R\delta(z)\}^* \\ &\quad + P_S(k)\,\alpha^*\beta\text{FT}\{F(z)\}^*\,\text{FT}\{R\delta(z)\}, \end{aligned} \quad (3)$$

where the superscript $*$ represents the complex conjugate. In general, if the reflection from the mirror is significantly stronger than the backscattering from the sample, then the first term ($S(k)\alpha^2|\text{FT}\{F(z)\}|^2$) has a negligible effect and can be ignored. Therefore, the intensity of interfering light at wavenumber k can be approximated by

$$\begin{aligned} I(k) &\approx P_S(k)\,\beta^2\,|\text{FT}\{R\delta(z)\}|^2 \\ &\quad + P_S(k)\,\alpha\beta^*\,\text{FT}\{F(z)\}\,\text{FT}\{R\delta(z)\}^* \\ &\quad + P_S(k)\,\alpha^*\beta\text{FT}\{F(z)\}^*\,\text{FT}\{R\delta(z)\}. \end{aligned} \quad (4)$$

By measuring the interference intensity at multiple wavenumbers and applying an inverse FT to $I(k)$, we obtain the scattering potential of the tissue along the z-axis ($F(z)$):

$$\begin{aligned} \text{IFT}\{I(k)\} &\approx \text{IFT}\left\{\beta^2 P_S(k)\right\} \star R^2\delta(z) \\ &\quad + \text{IFT}\{\alpha\beta P_S(k)\} \star F(z) \star R\delta(-z) \\ &\quad + \text{IFT}\{\alpha\beta P_S(k)\} \star F(-z)^* \star R\delta(z). \end{aligned} \quad (5)$$

The operator $\star$ is a convolution operator. Here, we assumed that α, β, and R are real numbers independent of the wavenumber. If $P_S(k)$ is a smooth function of the wavenumber, then its IFT is mostly centered at the origin and can be approximated by a Dirac delta function $\delta(z)$:

$$\text{IFT}\{I(k)\} \approx \beta^2R^2\delta(z) + \alpha\beta R \times F(z) + \alpha\beta R \times F(-z)^*. \quad (6)$$

By placing the origin of the coordinate system on the tissue surface, we know the sample is confined to positive z values. Hence, $A(z) = \text{IFT}\{I(r,k)\}$ on the positive z-axis represents the scattering potential of the object which is called the depth profile. $A(z)$ is a Hermitian function of z, and on the negative z-axis, $A(z)$ is the complex conjugate of its values on the positive axis.

In practice, the light source has a limited bandwidth which causes $\text{IFT}\{P_S(k)\}$ to have a finite width. This width is the axial resolution of the spectral-domain OCT imaging. For a given power spectral density of $P_S(k)$, it is possible to use the Fourier theorem to calculate the axial resolution of the OCT imaging. For a light source with a Gaussian spectrum of bandwidth $\Delta\lambda$ at the central wavelength of λ_0, the axial resolution is

$$\delta z = \frac{2\ln(2)}{\pi}\frac{\lambda_0^2}{\Delta\lambda} = l_c. \quad (7)$$

The theoretical resolution in spectral-domain OCT imaging is the same as the coherence length of the light source (l_c). If α and β are not uniform across the light spectrum, the axial resolution can be larger than (7). The lateral resolution in an OCT imaging system is a function of the central wavelength and the numerical aperture of the imaging system and is decoupled from its axial resolution. For more in-depth discussions and analysis of OCT image formation, we refer the readers to many articles and books including but not limited to [33, 35–43, 50–53].

In spectral-domain OCT imaging, the depth of the scatterer and the interference pattern are related via Fourier transform (see (6)). Therefore, when the depth of a scatterer (z) increases, its corresponding frequency in the interference pattern ($I(k)$) also increases. The spectrometer, however, acquires a finite number of samples from the interference pattern and, in practice, a discrete Fourier transform (DFT) is used to convert the interference pattern to a depth profile. According to the Nyquist sampling theorem, the spectral resolution of the spectrometer determines the maximum detectable frequency in the interference pattern ($I(k)$) and, hence, the imaging depth. When a scatterer falls outside this imaging range, it produces an interference pattern with a frequency larger than the Nyquist limit of the sampling rate. Due to aliasing, the depth of such a sample is misrepresented in OCT images. If the number of pixels on the camera is N and the measured spectral bandwidth (in wavenumber) is $\Delta\kappa$, the spectrometer's sampling interval is estimated by $|\delta\kappa| = |\Delta\kappa/N| \approx (2\pi/\lambda_0^2)(\Delta\lambda/N)$. Therefore, according to (6), the imaging range of the SD-OCT device is [36, 50, 52]

$$2 \times z_{\max} = \frac{1}{2}\frac{1}{(1/\lambda_0^2)(\Delta\lambda/N)} = \frac{\lambda_0^2}{2(\Delta\lambda/N)}$$

$$\text{or } z_{\max} = \frac{N\lambda_0^2}{4\Delta\lambda}, \tag{8}$$

where $z_{\max}$ is the maximum imaging depth without any aliasing effect. $z_{\max}$ is directly proportional to the number of pixels in the spectrometer. For a large imaging range, a detector with a sufficiently high number of pixels should be selected. According to (8), the imaging range can also be furthered by increasing the central wavelength or decreasing the bandwidth of the light source. However, either of these changes adversely affects the imaging resolution.

3. Materials and Methods: Design Considerations

3.1. Light Source. According to (7), a low-coherence light source is essential to obtain a fine axial resolution. Since the axial resolution of an OCT image is directly proportional to the square of the central wavelength of the light source and inversely proportional to its bandwidth, similar axial resolutions could be achieved in different regions of the spectrum. When using longer wavelengths, an increasingly wider light spectrum ($\Delta\lambda$) is required to maintain the axial resolution. The main drawback of using longer wavelengths is the effect on the lateral resolution which is discussed later in this section. On the other hand, while shorter wavelengths can provide better lateral resolution, in majority of biological samples, the scattering effect increases at those wavelengths, which limits the effective imaging depth of an OCT. Therefore, the central wavelength of the light source should be chosen based on the application and the tissue of interest. For biological samples, both visible (e.g., [21, 54–57]) and near-infrared (e.g., [58–60]) wavelengths have been used. The high absorption coefficient of water in wavelength longer than 1 μm justifies using short wavelength or visible light sources for transparent samples such as the eye. However, for typical brain imaging within the superficial 500 μm, a wavelength above 1000 nm is more popular.

The light source should be stable and should provide a low-noise output to obtain high quality images. Moreover, the emitted light should be spatially coherent. Superluminescent light emitting diodes (SLDs or SLEDs) are popular for spectral-domain OCT applications since they can provide low-noise and stable low-coherence light while they demand less maintenance in contrast to pumped lasers. With the recent advances in supercontinuum laser technology, the stability of this type of light sources has improved and their noise level has reduced. Supercontinuum laser sources provide light with an ultrawide spectral bandwidth and are used to obtain a fine axial imaging resolution. They provide more optical power in comparison with SLDs. For applications where high power and large light spectrum are needed, such light sources are a proper choice (e.g., [61, 62]). However, for applications that require high phase stability and low noise levels, such as Doppler velocimetry and OCT angiography, SLDs remain a superior choice compared to supercontinuum lasers. The spectral power distribution of the emitted light is also of great importance. A Gaussian-like power spectrum is ideal and provides an axial resolution close to the theoretical value. Deviation of the power spectrum from the Gaussian shape or sharp changes in the power spectrum can result in degradation of resolution or appearance of some vertical side lobes in OCT images. During postprocessing, some of these effects can be attenuated. Nonetheless, a Gaussian-like power spectrum is beneficial to achieve high quality OCT images.

3.2. Sample Arm. The design of a sample arm in an OCT system depends on the desired lateral resolution, maximum field of view, and the spectral bandwidth of the light source. Therefore, proper design and implementation of the sample arm are essential to achieve the desired specifications. For example, the axial resolution theoretically depends on the bandwidth of the light source; however, it still can be negatively influenced by a poor design of the sample arm if the chromatic aberrations cause significant spectral filtering in the sample arm [63]. A sample arm basically consists of four major components: a collimator, two galvanometric mirrors, intermediary optics, and an objective lens. Each component is influential in the specifications of the system.

3.2.1. Collimator. A collimator is a lens or lens mechanism that is placed in front of an optical fiber to produce a collimated beam. In combination with the rest of the arm, it forms an imaging system that images the fiber tip on the sample and, conversely, it images the focal plane of the objective on the tip of the optical fiber. In the forward path, when imaging the fiber on the sample, the collimated beam waist and the collimated beam quality have significant impact on the lateral resolution. The collimated beam radius depends on the numerical aperture of the optical fiber and also the effective focal length of the collimating lens system. When the fiber is placed at the focal plane of the collimator, the beam radius is approximated by

$$w = f \cdot \theta_{\text{fiber}} = f \cdot \frac{\lambda}{(\pi w_{\text{fiber}})}, \quad (9)$$

where f is the effective focal length of the collimator and w_{fiber} is the mode radius of the fiber at wavelength λ. The mode radius of a fiber is dependent on the wavelength and slightly changes across the spectrum of the wide-band source. In OCT applications, this dependency is neglected and the mode radius of the central wavelength is used in the calculation. By increasing the focal length of the collimator, it is possible to produce a larger collimated beam waist which leads to a smaller spot size on the sample (as discussed later in this article). When using a large focal length, the aperture of all intermediary optical components, including the galvanometric mirrors, needs to be selected accordingly to avoid beam truncation which causes signal loss or artifacts in the produced images.

In the return path, the collimator couples the photons, which are collected by the objective lens, to the single mode fiber. Optical aberrations or mismatch between the beam profile at the focal point of the collimator lens and the mode of the optical fiber causes loss in the signal and drop in image quality. The power loss is minimum when the returning beam waist matches the effective aperture of the collimator. Ideally, the collimator and the objective lens should be selected such that the loss due to mismatch between the collimator and objective lens aperture is minimum.

Optical aberrations in the collimating lens mechanism can degrade the quality of OCT images first by introducing imperfections to the collimated beam in the forward path, which can increase the sample spot size, and second by reducing the coupling efficiency to the fiber, which decreases the signal-to-noise ratio (SNR) in OCT images. Spherical and chromatic aberrations are two important imperfections that can be caused by the collimating lens system. A large spherical aberration reduces the lateral resolution of OCT images and the coupling efficiency of the backscattered photons. Longitudinal and lateral chromatic aberrations can cause the axial and lateral position of the focal spot to change as a function of wavelength. This affects the coupling efficiency across different wavelengths, which may degrade the axial resolution of the system [64] as well as its lateral resolution. Therefore, it is critical to select a proper collimator to achieve the desired imaging resolution and SNR.

A conventional biconvex or planoconvex lens may be able to produce the desired beam size; however, due to spherical and chromatic aberrations that are introduced by these types of lenses, the quality of the resulting beam is not satisfactory for high-resolution OCT applications. Typical achromatic doublets introduce slightly less chromatic and/or spherical aberrations; nonetheless, the remaining aberrations can still prevent achieving high lateral resolutions.

A better collimated beam quality can be achieved using an aspherical lens. Aspherical lenses are relatively more expensive yet usually designed to obtain short effective focal lengths. The quality of the Gaussian beam that is produced by this type of lenses is acceptable for most OCT applications. For extreme precisions, one can use a lens mechanism instead of a single lens; however, in most OCT applications, an aspherical lens can deliver the desired beam quality.

Besides the lens type, the placement of the fiber with respect to the lens can affect the quality of the beam. Small misalignments (tilt or misplacement) of the fiber and lens with respect to each other can also introduce aberrations to the beam and affect the OCT resolution. When using a lens with small aperture, such misalignments cause beam truncation, which results in loss of resolution and artifacts in the OCT images. Therefore, the fiber and collimating lens should be aligned very carefully. There are collimating packages that are offered by different vendors. These packages are designed in such a way that, for the designed wavelength and fiber radius, they can deliver a high quality beam without the need for fine alignment. Most collimating packages consist of a single aspheric lens. Some others, which are designed to produce high quality Gaussian beams (M factor close to 1), use a combination of multiple lenses to suppress aberrations. Typically, this type of collimator lenses can provide a better beam quality and higher coupling efficiency to the fiber, but they increase the cost of the system. When choosing a collimating package, it is important to select a package that is designed for angled physical contact (APC) to avoid artifacts in the OCT images due to back reflections at the fiber tip. This is discussed in more detail further in the paper.

Another important factor when choosing a collimator lens is its antireflection (AR) coating. Proper antireflection coating on the lenses in the sample arm, including the collimator, is important to acquire high quality OCT images as AR coating decreases the light reflection from the surface of the optical components and improves the image SNR. A good AR coating can reduce the reflection loss to as low as 0.2 dB, while improper coating can result in a power loss up to 3 dB or higher. More importantly, the strong light reflection from surfaces which are not covered with proper AR coating introduces image artifacts as horizontal lines across the OCT images.

3.2.2. Galvanometric Mirrors. To obtain 2- or 3-dimensional images of a specimen, generally, the OCT beam raster-scans the tissue using a laser scanning mechanism (Figure 1). Two galvanometric mirrors rapidly scan the OCT beam spot on the sample along two orthogonal directions (x and y), while the depth profile at each (x, y) location is obtained via low-coherence interferometry (see (6)). For fast and high-resolution OCT imaging, it is necessary to use a galvanometric system which can achieve high scanning speeds with accurate control over the angle of the mirrors. Mechanical vibrations or timing jitter in the mirrors can affect the accuracy of OCT images or can produce artifacts in Doppler

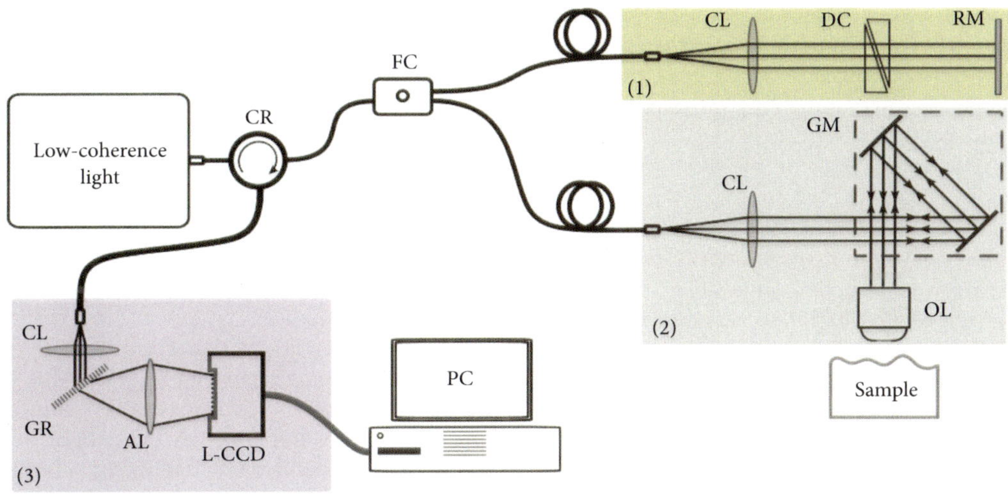

FIGURE 1: An optical fiber-based spectral-domain OCT scanner. The system uses a low-coherence light source to launch light into both sample (Path 2) and reference (Path 1) arms. In the sample arm, a set of galvanometric mirrors raster-scan the sample under test. Light that is collected back from the reference and sample arms recombines to form interference patterns which are detected by the spectrometer (Path 3).

measurements and angiography images. Therefore, a servo circuitry which is equipped with a closed-loop controller is essential to achieve high stability and accuracy in positioning and scanning speed of the mirrors. The size of the mirrors affects the maximum speed that the galvanometric scanner can achieve. Small mirrors make it possible for the scanner to operate at a higher speed; however, the mirrors must be sufficiently large to avoid beam truncation. In contrast, since the OCT resolution depends on the collimated beam waist, the OCT beam cannot be too small. When a small beam and fine OCT lateral resolution are required simultaneously, it is possible to use a small diameter collimator followed by a scan-tube lens mechanism with a proper expansion ratio after the galvanometric mirrors to expand the OCT beam prior to the objective lens. Resonant scanning mirrors can achieve very high scanning rates (~8 K B-scans per second) in comparison with the galvanometric mirrors (200 B-scans per second for small regions). However, their lack of flexibility to scan desired patterns limits their application in OCT imaging. Moreover, in most SD-OCT applications, the imaging speed is limited by the detector line rate and, in practice, the speed of SD-OCT imaging is not limited by the speed of typical galvanometric scanning mirrors.

The physical dimensions of the mirrors dictate the separation between the two mirrors necessary to avoid collision. To optimally align the galvanometer, we need to place the back-focal plane of the objective in the middle point between the mirrors. In such an arrangement, none of the mirrors satisfies the telecentric configuration [65]. Telecentricity is important in OCT imaging and is discussed in the following section. The separation between the mirrors should be limited to the minimum possible distance; otherwise, it compromises the accuracy of OCT images. If the placement of the mirrors significantly reduces the telecentricity of the optical design, the OCT images suffer from a depth dependent magnification and the magnification is different along the x and y directions. When the separation is small, the imperfection is negligible. As the distance between the mirrors increases, this deformity also becomes noticeable and can introduce errors in measurements, especially when a high magnification objective lens is used. In cases when the mirrors cannot be placed sufficiently close to each other, this problem can be avoided by the use of relay optics between the mirrors. This relay mechanism projects the image of the first mirror (x-axis) on the second mirror (y-axis) [66, 67] and the second mirror is placed at the back-focal plane of the objective. Such a relay mechanism increases the size and cost of the OCT system. Another approach is to use two-axis tilt mirrors [68, 69], which use a single mirror to scan along both x and y directions. Using current technologies, generally, galvanometric mirrors provide more linear and faster scanning speeds compared to two-axis tilt mirrors. The recent advances in microelectromechanical system (MEMS) devices have helped reduce this gap; however, currently, the speed of galvanometric mirrors outperforms that of two-axis tilt mirrors for most OCT applications.

The surface quality of galvanometric mirrors is important in low-coherence interferometry. The curvature of the mirrors' surface introduces phase-front distortion to the collimated beam which negatively impacts the image quality. The scanning mirrors should be stiff so that, during mechanical tension, at the time of large angular acceleration, the mirrors' surfaces remain flat. These considerations may affect the thickness (weight) and the material that are used in the construction process of the mirrors, which limit the speed and settling time of the galvanometric system. Therefore, when selecting the scanning mirrors, their speed as well as the mirror size and angular accuracy should match the desired parameters of the OCT system.

3.2.3. Objective Lens. Generally, in an OCT system, the lateral resolution and maximum possible field of view are limited by the objective lens. A high lateral resolution can be achieved using an objective that is designed to work at a large numerical aperture (NA). The lateral resolution of an OCT is defined as the radius of the beam spot at the focal point of the objective [33, 52]. For a Gaussian beam of radius r_0 and wavelength λ passing through a lens with the focal length of f, the spot size is inversely proportional to the NA [70]:

$$\delta x = \frac{2\lambda}{\pi}\frac{f}{r_0} = \frac{2\lambda}{\pi NA}. \tag{10}$$

Here, δx is the theoretical lateral resolution. Increasing the NA increases the divergence angle of the Gaussian beam which translates to some reduction in the depth of focus. Depth of focus (DOF) of a Gaussian beam is the range in which the beam waist radius stays smaller than $\sqrt{2}\delta x$. Generally, DOF is considered as the range where the resolution is maintained close to the desired value, and outside this range the resolution drops quickly. For a Gaussian beam, the DOF is calculated by [70]

$$b = \frac{\pi \delta x^2}{\lambda} = \frac{4\lambda}{\pi NA^2}. \tag{11}$$

Therefore, the DOF is directly proportional to the square of the spot size which poses a fundamental trade-off between the lateral resolution and the depth of focus. An optical setup with a low NA (~0.1) strikes a good balance between resolution and DOF for an OCT scanner in the infrared range of the spectrum. This is in contrast with many other microscopy technologies such as multiphoton fluorescence imaging, which demand a large NA to perform optical sectioning. For optical sectioning, a high numerical aperture is desirable since the beam outside the depth of focus diverges quickly and the effect of out-of-focus regions of a sample can be ignored. In OCT, low-coherence interferometry separates the photons that are backscattered at different depths in the tissue. Therefore, it is desirable for an OCT scanner to possess a uniform spot size across the depth of interest. This condition imposes a fundamental trade-off between the lateral resolution of an OCT scanner and its depth of focus.

Using a light source with shorter wavelengths, it is possible to strike a better balance between the lateral resolution and the depth of focus. However, a shorter wavelength experiences larger scattering in most biological tissues and has shorter penetration depth. In such samples, the scattering effect reduces the signal-to-noise ratio and ultimately the image quality. Therefore, depending on the application and the tissue properties, selecting a proper wavelength has a considerable impact on the quality and resolution of OCT images.

When a long central wavelength and a large depth of focus are necessary for an OCT application, it is recommended to use a Bessel beam. A Bessel beam with a spot size similar to a Gaussian beam offers a longer depth of focus [71–73]. The compromise here is that Bessel beams have relatively strong side lobes which can affect the accuracy of some OCT measurements. Also, implementing a system with a Bessel beam costs more due to the increase in the cost of some optical components.

Optical aberrations that are introduced by the objective lens in the sample arm can degrade the resolution of the imaging system. Low chromatic aberrations (longitudinal and lateral) are important when a light beam with a large bandwidth is used for OCT imaging. Chromatic aberrations prevent different wavelengths from coming into focus at the same focal point. Ultrawideband OCT light sources such as supercontinuum lasers are becoming more popular to achieve high axial resolution. If the objective lens is not designed for such a large bandwidth, chromatic aberrations prevent the scanner from achieving the theoretical lateral resolution. Proper coating of the objective for the desired wavelengths is also important to reduce the back reflection of the light from the objective. Since most objective lenses have a complex structure composed of several layers of different types of glass, the strength of the back reflection can be comparable to or stronger than the tissue backscattering.

One other important aberration in OCT imaging is astigmatism which may confine the area of the diffraction limited resolution to the center of the field of view. Astigmatism causes sagittal and tangential light rays to focus at different axial positions. Astigmatism reduces the lateral resolution of the OCT device when scanning the off-axis regions of a specimen. Besides the resolution, optical aberrations reduce the signal-to-noise ratio of OCT images. In a typical OCT device, the objective collects the photons that are backscattered from the tissue and sends them back toward the mirrors and the optical fiber. The fiber core acts as a pinhole and rejects the out-of-focus light. This significantly reduces the intensity of the light at the out-of-focus region significantly.

Other parameters that should be considered when selecting an OCT objective lens are magnification and telecentric property. Objective magnification determines the relationship between the change in the angle of the incident beam and the transversal displacement of the spot on the sample. A low magnification objective lens requires small mirror angles to cover a large field of view. The same scanning angles cover a smaller portion of the sample when a high magnification lens is used. For typical OCT imaging, low magnification objectives with large FOV are desirable. The design of most commercially available low magnification lenses does not allow the realization of small spot sizes, which is partially due to cost considerations. This couples the magnification of a lens to the imaging resolution which imposes a technical trade-off between the resolution and the field of view.

For laser scanning mechanisms, such as OCT, a linear relationship between the scanning angle and the transversal displacement allows the device to obtain a uniform sampling across the transverse plane by simply sweeping the angle of the galvanometric mirrors while the detector is recording at a constant rate. Such an objective is usually called an f-θ lens since the displacement of the beam is equal to the product of the effective focal length of the lens (f) and the angle of incidence of the beam (θ). In general, f-θ lenses are able to produce a flat imaging plane and uniform resolution across the imaging plane. A non-f-θ lens realizes a nonlinear relationship between the mirror angle and the beam displacement on the sample. This effect can cause a nonuniform magnification across the transverse plane and distortion in the recorded OCT images. It is important to select an objective lens that introduces a low level of distortion. For OCT applications, the uniformity of magnification at various depths is desirable. For this purpose, the lens must be used in the telecentric configuration. That condition is achieved when the geometrical center between the mirrors is placed at the back-focal plane of the objective. A telecentric f-θ lens provides a beam that is perpendicular to the flat imaging plane and its magnification is uniform within the depth of focus of the objective.

The abovementioned factors are related to the performance of an objective lens. Some other considerations are related to the mechanical properties of the objective. The distance between the back-focal plane and the surface of the lens is called back-focal length (BFL). The scanning mirrors should be placed such that the geometrical center of the mirrors collocates with the back-focal plane of the objective. If the BFL is sufficiently large, positioning the galvanometer will become more convenient. For objectives that are designed for high-resolution imaging, the BFL is short and it is almost impossible to place the mirrors in the designated space. In such conditions, a relay mechanism is necessary to image the mirrors on the desired positions [65, 67, 74]. Working distance in an objective is the distance between the focal plane and the surface of the lens. A large working distance provides more space between the sample and the lens. This is especially important for *in vivo* experiments where other devices and equipment are used simultaneously, for example, to collect other forms of data or to deliver certain chemicals to the organelle under study. In such conditions, a large working distance is very beneficial.

3.3. Beam Combination. For many applications, it is beneficial to integrate OCT imaging with other optical methodologies, such as fluorescence, white light, or laser speckle imaging. Such a combination can be achieved by placing a dichroic beam splitter in the path of the OCT beam in the sample arm. The dichroic mirror can be a long- or short-pass beam splitter; however, each additional component can introduce some imperfections or artifacts to OCT images. Most dichroic mirrors are designed to work with collimated beams. When a diverging or converging OCT beam passes through a slab of glass, different portions of the beam travel for different lengths inside the glass, which introduces optical aberrations that have a profound impact on the quality of images. Usually, this effect is more crucial when a long-pass mirror is used and the OCT beam passes through the dichroic glass.

3.4. Reference Arm. The reference arm in an OCT system replicates the light path of the sample arm. It may consist of the same components as the sample arm; however, it is possible to use dispersion-compensating (DC) blocks instead of some components, such as the objective lens, to reduce the cost. A DC component has dispersion properties similar to the original component. In the reference arm, a mirror is used in place of the sample. The path length of the reference arm should match the sample arm. In practice, it is desirable to place the reference mirror on a micromanipulator to adjust the reference arm length. In many applications, this length is adjusted such that the point of equal path length in the sample arm is placed a few hundred micrometers before the focal plane of the objective.

Since the reflection coefficient of a mirror is larger than a typical sample, the light that is collected back by the optical fiber in the reference arm is stronger than the collected light from the sample arm. A strong reference beam increases the signal-to-noise ratio (SNR) of OCT images. Also, a strong reference intensity reduces the visibility of the image artifacts caused by interference occurring between the reflected lights from different sample reflectors (see (5)). When imaging deep tissue or a weakly backscattering sample, a long exposure time is necessary to detect the weak backscattered light. In such cases, a strong reference beam saturates the detector and makes the reconstruction impossible. To adjust the intensity of the reference beam, one can place a variable attenuator or aperture in the reference arm. An alternative method of balancing the intensity of the reference beam, with respect to the sample beam, is to use a variable fiber coupler instead of the fixed ratio fiber couplers.

3.5. Spectrometer Arm. The photons that are collected in the sample and reference arms are combined by a fiber coupler in the spectrometer arm. If a 50 : 50 coupler is used, the sample and reference powers are equally divided between the output ports of the coupler and there is no preference between the ports. When a coupler with a ratio other than 50 : 50 is used to deliver more power to one of the arms, the same port that is connected to the light source should be used to detect the interference. In this condition, it is possible to use a single mode fiber optic circulator in the path of the light source to divert the path of the returned photons toward the detection (spectrometer) device.

The spectrometer arm usually consists of a single mode fiber, a collimating lens, a diffraction grating, a lens (or lens mechanism), and a line detector. The light that is emitted from the single mode fiber is collimated before illuminating the diffraction grating. The grating diffracts different wavelengths to different angles and the diffracted beam is imaged by a lens mechanism on the active area of a line CCD camera.

A line CCD camera is the detector in a SD-OCT setup. A combination of different design parameters including the source bandwidth, diffraction grating resolution, lens mechanism, and camera pixel size determines the axial imaging resolution and maximum depth that a SD-OCT can image which were partly discussed earlier in this paper. The overall bandwidth that is detected by the camera determines the axial resolution. A larger effective bandwidth results in a better axial resolution (see (7)). The maximum imaging depth depends on how finely the light spectrum is sampled by the camera. If the sampling is denser, using either a smaller detector pixel size or a lensing mechanism of longer focal length, the scanner provides a larger imaging depth; however, this mandates using a larger number of pixels to capture the same bandwidth to maintain the axial resolution. Besides the number and size of camera pixels and also the source bandwidth, the optical performance of the grating and imaging lens is important in the quality of the images. There are commercial spectrometers available which offer easy installation; however, it is not possible to adopt and tune them for every arbitrary light spectrum.

Here, we present a simple and fast guideline on how to select proper components and implement an efficient spectrometer for OCT applications.

3.5.1. Optical Grating. When selecting a grating, its designated wavelength, efficiency over the desired bandwidth, and polarization sensitivity should match the application. Transmission holographic gratings strike a good balance between efficiency, working spectrum, and insensitivity to minor misalignments, which is beneficial for OCT applications. Blazed gratings, in comparison with holographic gratings, offer a better efficiency around the working wavelength. However, their efficiency drops quickly for the wavelengths that are far from the central wavelength. This reduces the image signal-to-noise ratio (SNR) and/or degrades the OCT axial resolution by reducing the spectral bandwidth of the spectrometer.

For a light source with elliptical polarization, a grating with less polarization sensitivity is beneficial. When the efficiency is highly polarization dependent, the optical power of the diffracted light and the SNR of the generated images degrade. The grating needs to be aligned carefully so that the incident beam exposes the grating at the prescribed angle with respect to the grooving. Another advantage of transmission gratings is the lower sensitivity to the imperfections regarding the angle of incidence (AOI), compared to reflection gratings. However, to obtain the optimal performance, the alignment needs to provide the AOI that matches the grating specifications. Overall, transmission holographic gratings with small polarization preference meet the requirements of most OCT applications.

3.5.2. Lens Mechanism. The spectrometer lens mechanism focuses the light dispersed by the grating on a line detector which is placed on its focal plane. In this configuration, the spectrometer lens virtually acts as a scan lens. Different wavelengths exit the grating at different angles and are focused at different positions on a line along the direction of the light dispersion (Figure 3). Therefore, at the focal plane of the lens, different wavelengths are spatially separated and each pixel of the detector is receiving a portion of the spectrum based on its location. The lens FL should be selected such that the desired part of spectrum is projected on the extent of the sensor array. The bandwidth that is projected on the detector determines the theoretical axial resolution of the OCT system. If the FL is too large, only a portion of the bandwidth is received by the detector, which degrades the axial resolution. The SNR of OCT images reduces as well due to the power loss in the spectrometer. In this case, the spectrum is sampled more densely, which results in a larger imaging depth. In contrast, if the FL is too small, some of the camera pixels practically do not receive a sufficient number of photons and will not contribute to the reconstruction process. In such conditions, the axial resolution can still be calculated by (7) where $\Delta\lambda$ is the bandwidth of the light source. However, because of the larger sampling interval, the maximum imaging depth decreases (see (8)).

Since the pixel size is not infinitesimal, each pixel integrates over the portion of the spectrum that is projected on the sensor. Therefore, each detector's pixel acts as a low-pass filter which smoothens the interference patterns. In this process, the high frequency fringes are attenuated. Since the backscattered light from deeper objects creates higher frequency components in the interference pattern, the low-pass filtering results in a depth dependent attenuation [75]. This signal fall-off, which is sometimes called the sensitivity drop-off, causes the SNR of typical OCT images to drop quickly as depth increases.

Besides the physical dimension of the detector, the focal spot size of the lens mechanism is another factor that can cause signal fall-off. When the spectrometer spot size is smaller than the detector's pixel size, this effect can be ignored. When the spot size is comparable to or larger than the detector's pixel size, the energy of each wavelength spreads to more than one pixel which can be modeled by another low-pass filter being convolved with the interference signal, which worsens the signal fall-off. The spot size on the detector is directly proportional to the FL of the lens mechanism and inversely related to the collimated beam diameter (see (10)). If a lensing system with a large FL is required, then the size of the collimated beam that is incident on the grating should be selected accordingly to realize a small spot size on the detector. Optical aberrations such as astigmatism and Petzval field curvature result in a nonuniform spot size along the detector plane. When the aberrations are significant, the varying spot size causes an attenuation in the interference pattern, which varies as a function of the position across the detector. Therefore, these aberrations generate wavelength dependent attenuation which acts as a spectral filter that may degrade the axial resolution of the system as depth increases. Using a proper lens mechanism and collimating lens, it is possible to produce a sufficiently small spot size on the detector and reduce optical aberrations to achieve a high axial resolution and minimize the signal fall-off.

3.5.3. Detector. In recent years, several line CCD cameras were designed and manufactured particularly for SD-OCT imaging. Their spectral response, dark noise, exposure time, data transfer rate, and pixel size allow for fast, high-resolution, high-dynamic-range, and high-sensitivity OCT imaging.

The sensitivity of an OCT scanner is defined as the inverse of the sample reflectivity which produces an amplitude equal to the noise amplitude in the OCT image. This value has a great impact on the image quality and is affected by several parameters in the detector. The dynamic range is defined as the ratio between the minimum reflectivity that causes saturation and the noise level in the detector. Therefore, this value represents the span of reflectivity values in the sample arm that can be detected by the scanner. Practically, the dynamic range of a system is smaller than its sensitivity. The typical values for sensitivity and dynamic range are ~100 dB and ~60 dB, respectively [33]. The detector array records the reference and sample signals along with the interference signal. Generally, the amplitude of the reference signal is much stronger than of the interference signal; therefore, the detector should have a large photoelectron capacity to record weak interference signals that are superimposed on top of the strong reference signal. Shot noise, readout noise, thermal noise, and analog-to-digital conversion noise are the typical forms of noise that limit the sensitivity of an OCT system. It is desirable for a detector to work at the shot noise limit which is a fundamental noise in imaging techniques. A small receiver noise is essential for a SD-OCT detector to take advantage of its large photoelectron capacity and to achieve a shot noise-limited performance [37, 40, 41]. Since in SD-OCT imaging the depth profile is calculated using a discrete Fourier transform, to achieve a dynamic range of 60 dB in the image, it is not required for the detector to capture the interference signal with such a large dynamic range. A detector with a 12-bit analog-to-digital converter provides a sufficient bit depth to achieve a sufficient dynamic range and low discretization noise for SD-OCT imaging.

The scanning speed of a SD-OCT imaging system is mainly limited by the line rate of its detector. Commercially available line CCD cameras are able to provide a scan rate of >240 k lines per second at the cost of larger readout noise and a smaller dynamic range compared to lower speeds. Although such detectors are sufficiently fast for many applications, faster and more sensitive detectors are beneficial in certain *in vivo* experiments, for example, to reduce motion artifacts or perform Doppler velocimetry of fast moving particles. By employing the new state-of-the-art electron-injection based detectors [76], it is possible to improve the sensitivity and dynamic range of the OCT imaging systems at higher imaging speeds.

The spectral response of the detector acts as a filter, similar to other optical components. If the spectral response is smaller than the light source such that it deforms the spectrum of the recorded signal in comparison with the source, then it can degrade the axial resolution of the OCT imaging system. Most line detectors have fairly flat response over a large spectral range and one can ignore their effect on the axial resolution. However, to obtain an ultrahigh resolution, the spectral response of the detector should match the spectrum of the light source.

In diffraction-based spectrometers, each wavelength is diffracted to a different direction. The angle between the diffracted light beam and the optical axis of the grating is not proportional to the light frequency (κ space) [36]. Consequently, the line CCD nonuniformly samples the interference pattern in the κ space. Applying the Fourier transform (see (6)) to this sampled interference signal to obtain the depth information prior to any further processing results in low resolution images and the reconstructed image suffers from a low signal-to-noise ratio (SNR). As a result, the majority of spectral-domain OCT systems apply a resampling routine as part of the OCT signal processing procedures to convert a signal that is recorded in the λ space to a signal in the κ space. However, some hardware solution has been offered by different research groups to avoid the resampling process. For example, Hu et al. [64] and Gelikonov et al. [77] have used a custom-designed prism between the grating and the lens in their interferometer to compensate for this nonuniform frequency sampling property of the grating. However, this method involves the custom design of a specific prism. Placing the prism in the spectrometer to obtain the uniform sampling in the frequency domain is a very sensitive and time-consuming task. Moreover, there are some imperfections, like the mismatch between the dispersion in the sample and reference arms, which eventually demands a software-based resampling process and cannot be solved readily by placing a prism in the spectrometer arm. Therefore, a software-based process would be essential to obtain a uniformly sampled signal in the frequency domain [75]. The majority of spectral-domain OCT systems perform the resampling as part of the OCT signal processing procedures.

3.6. Fibers, Coupler, and Circulator. For typical fiber-based OCT applications, single mode optical fibers and single mode fiber-based components are required. The core diameter of such fibers is in the range of a few microns to 10 μm depending on the central wavelength. The designed spectral bandwidth of the fibers should match the light source bandwidth. Otherwise, the coupling efficiency might be low or the fiber may act as a multimode fiber. Although there are some studies that demonstrate the benefit of using multimode fibers in certain applications [78–81], in a typical OCT design, using a multimode fiber can cause numerous artifacts in the image.

The fibers should also provide angled physical contact (APC) with the rest of the system to avoid image artifacts. Noticeable back reflection of light from any object can contaminate OCT images. The back reflection from the surface of the fiber tip is sufficiently strong to cause image artifacts. For this reason, the fibers need to be polished to have an angled surface which reduces the power of the back-reflected light that is coupled back to the fiber. When using single mode fibers with angled polish, other fiber-based components (collimators, fiber coupler, light source, and optical fiber circulator) need to be selected accordingly. The components that are compatible with angled polish fibers are declared by "APC" in their naming typically.

Light polarization may change when it passes through a typical single mode fiber. If the sample polarization and reference polarization are different, this potentially reduces the interference amplitude between them and reduces the SNR of OCT images. To achieve high SNR imaging, it is recommended to use polarization controllers in sample, reference, and spectrometer arms. This allows the user to adjust the polarization of the sample and reference arms to achieve a high imaging SNR. However, for most biological samples, this step is not necessary to obtain OCT images. Polarization

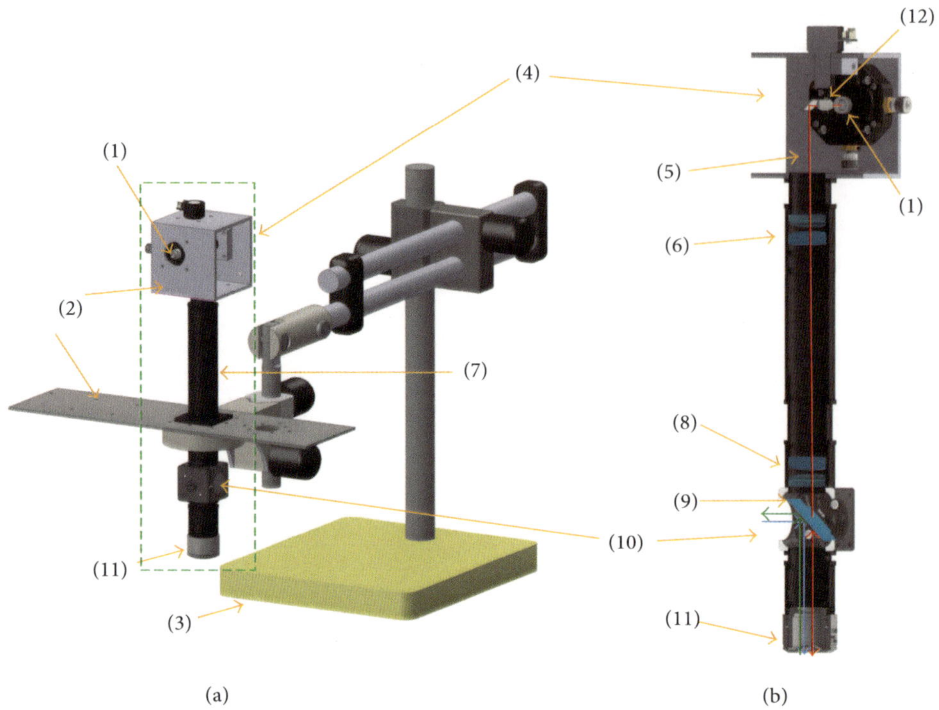

FIGURE 2: Sample arm. This image demonstrates the design of the sample arm that is mounted on a typical microscope stand. This system uses two identical semi-Plössl lenses as scan and tube lenses. The cross section shows the optics including galvanometric mirrors, scan and tube lens mechanisms, and the infrared dichroic mirror. (1) OCT collimator, (2) custom mounts, (3) commercial microscope stand, (4) scan head, (5) OCT beam, (6) scan lens, (7) lens tube, (8) tube lens, (9) dichroic mirror, (10) visible port, (11) objective lens, and (12) galvanometric mirrors. The SOLIDWORKS designs of lens tubes and cage cube-mounted beam splitters are adapted from Thorlabs (Thorlabs, NJ, USA). The microscope stand model has been modified from [45].

maintaining fibers should be used for polarization-sensitive OCT (PS-OCT) imaging. These fibers are designed to not alter the light polarization and are suitable for PS-OCT applications.

4. Materials and Methods: OCT Scanner Development

In this section, we discuss details of a SD-OCT system that we implemented for brain imaging applications. With slight modifications, the same design can be used for other applications that may require a different resolution, imaging depth, or light spectrum.

4.1. Light Source. To image up to several hundred micrometers deep inside a scattering medium, such as the brain tissue, with axial and lateral resolutions of ~5 μm or better, a light source with a long central wavelength is preferred. According to (7), the axial resolution of a SD-OCT system is linearly proportional to the square of central wavelength of the light source and inversely proportional to its spectral bandwidth. To achieve a balance between the penetration depth and resolution, a light source with a central wavelength of 1300 nm and FWHM bandwidth of >170 nm (LS2000B, Thorlabs, NJ, USA) was used. With this light source, the theoretical axial resolution is about 4.4 μm in air, which meets our requirement.

Since the infrared light is not visible to the human eye, for alignment purposes, an aiming red laser was utilized in this design. The beam of this laser is coupled to the infrared fiber path via a fiber coupler. The same fiber coupler splits the OCT light beam between the reference and the sample arms.

4.2. Sample Arm. Our goal was to obtain a lateral resolution of 5 μm, so we accordingly selected an OCT objective lens (LSM02, Thorlabs, NJ, USA) with an effective focal length of 18 mm. In combination with an incident beam diameter of 3.4 mm, this lens provides a theoretical FWHM spot size of ~5.2 μm. The collimated beam was produced by connecting a single mode optical fiber to a collimating package with the effective focal length of 18.86 mm (F280APC-C, Thorlabs, NJ, USA).

Our sample arm uses a set of galvanometric mirrors (GMs) to raster-scan the tissue. The GMs are designed for beam diameters smaller than 5 mm (GVSM002, Thorlabs, NJ, USA). The geometrical middle point between the mirrors should be positioned at the back-focal plane of the objective lens (OL). In this OL, as in most objective lens compounds, the back-focal plane is very close to the objective, which makes it challenging to place the mirrors at the designated positions. We installed the mirrors on a custom-made holder (Figure 2), with separation of ~10 mm between the mirrors to avoid accidental collision. A scan-tube

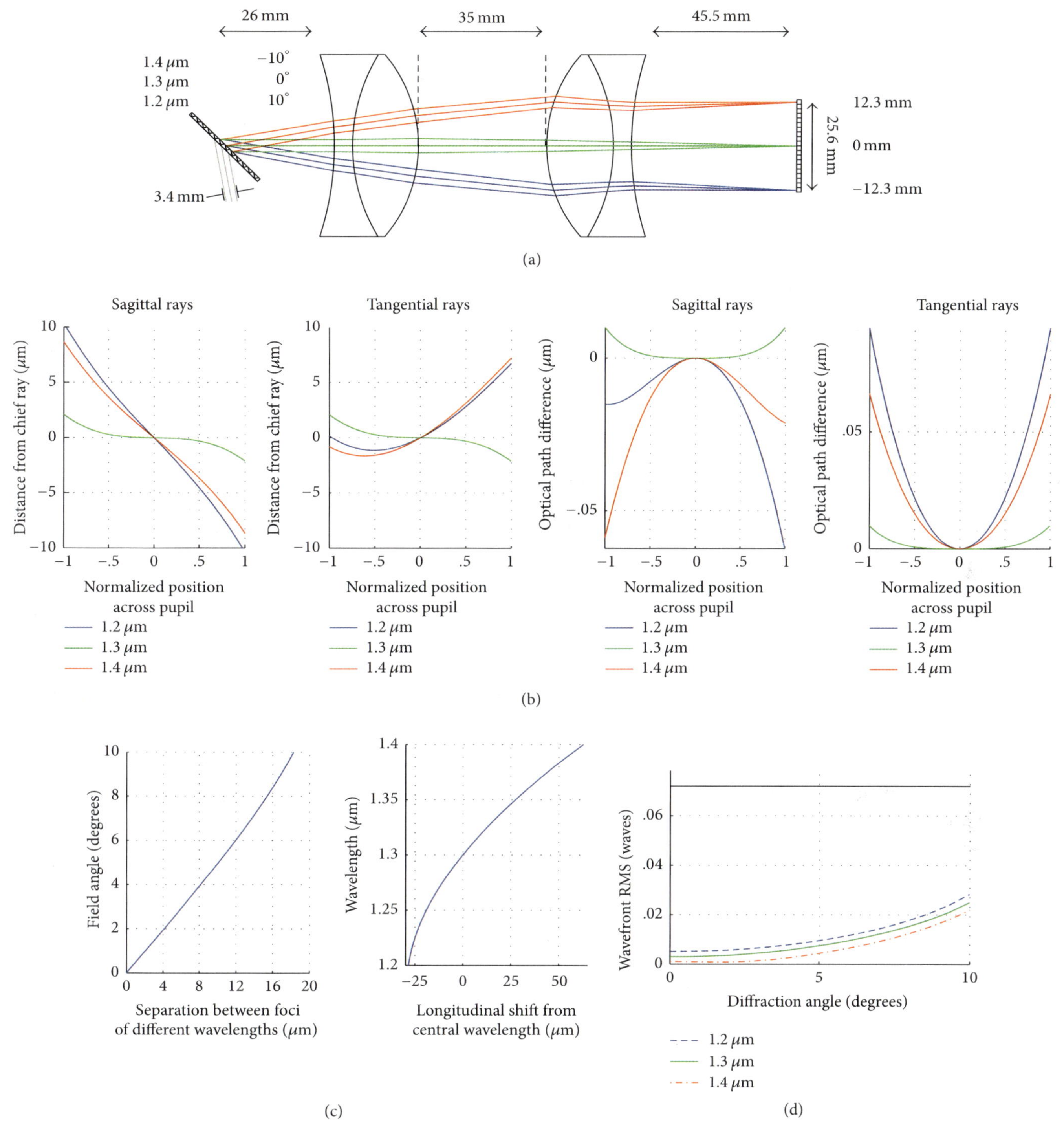

FIGURE 3: A schematic of the spectrometer. The grating diffracts the broadband collimated beam. The lens system acts as a scan lens and focuses the diffracted light beams on the focal plane array of a line scan camera. The focal length of the lens mechanism is designed such that the desired spectrum (1200 nm to 1400 nm) is covering the extent of the detector array. Optical aberrations in this scan lens system are minimized to reduce the sensitivity roll-off of the OCT images.

lens mechanism images the middle point between the mirrors to the back-focal plane of the objective [74]. With the use of the scan-tube lens, there is space to place a dichroic beam splitter in the sample arm. This dichroic beam splitter can separate OCT and visible light paths (visible port in Figure 2) to allow simultaneous OCT and bright-field or fluorescence imaging. The current design of the system is for a long-pass dichroic beam splitter; however, a custom mount can be slightly modified to accommodate a short-pass dichroic mirror instead. When using a long-pass dichroic, the OCT

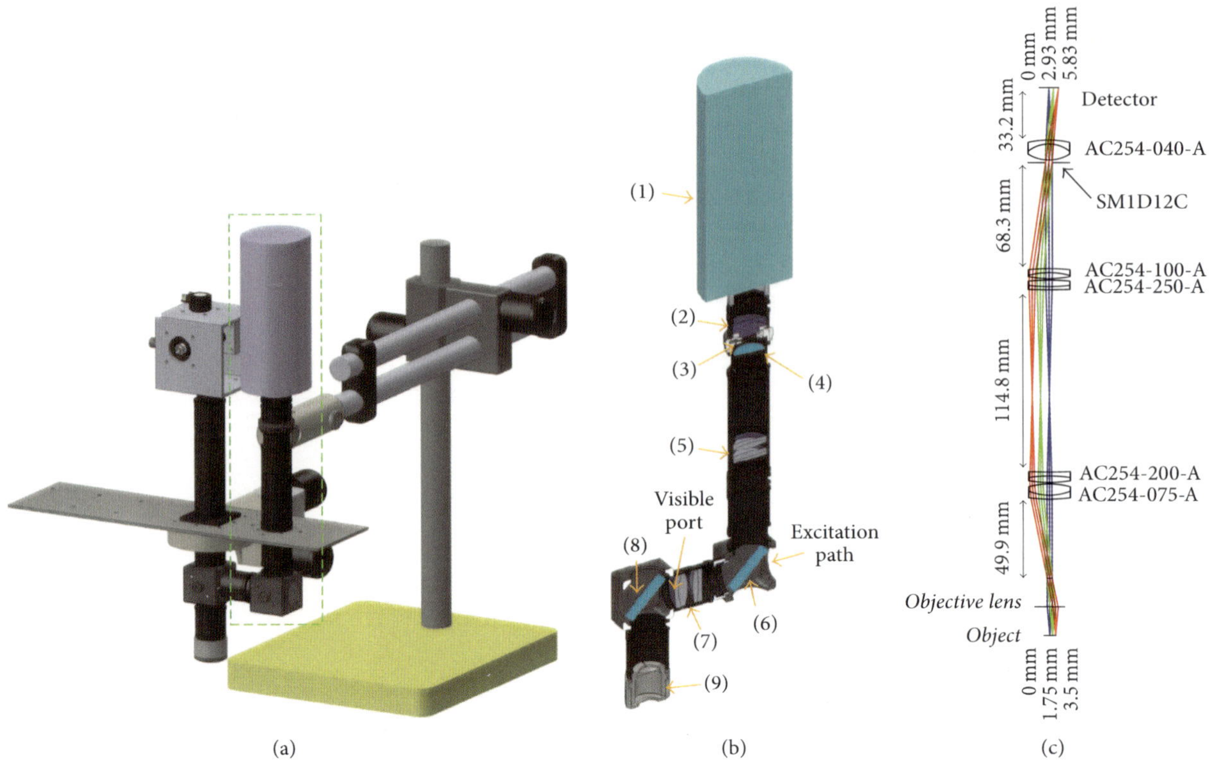

FIGURE 4: An example of fluorescence imaging integrated with the OCT sample arm. The system uses the same objectives for OCT and fluorescence imaging for simultaneous OCT and fluorescence recordings. (a) The placement of fluorescence imaging path in the OCT sample arm. (b) The SOLIDWORKS design of the fluorescence imaging path. (c) The schematic of the optical design of the path. The fluorescence imaging path uses two semi-Plössl lenses to image the back-focal plane of the objective to the entrance pupil of an imaging lens. There is an adjustable iris in this path to adjust the aperture diameter of the imaging system. This is helpful to reduce the aberration in the optical path and improve the imaging resolution. (1) Fluorescence camera (EXi Aqua, QImaging, BC, Canada), (2) adjustable iris (SM1D12CZ, Thorlabs, NJ, USA), (3) imaging lens (AC254-040-A, Thorlabs, NJ, USA), (4) optical filter, (5) semi-Plössl lens (AC254-100-A + AC254-250-A), (6) visible light dichroic mirror (FF440-520-Di01, Semrock Inc., NY, USA), (7) semi-Plössl lens (AC254-075-A + AC254-200-A), (8) long-pass infrared dichroic mirror (FF875-Di01, Semrock Inc., NY, USA), and (9) objective lens.

beam passes through the glass slab. The length that light travels via the glass varies as a function of each scanning angle. This causes a spurious change in the depth of reconstructed OCT images across the transversal plane as the beam raster-scans the tissue. This spurious change in the depth may appear as a bias in OCT Doppler velocity measurements. A short-pass dichroic beam splitter does not introduce such artifacts; however, multiple reflections from different layers of reflectors in a short-pass filter can cause strong artifacts in the OCT image that appear as replications of the original image. A short-pass dichroic mirror should be selected carefully to avoid such artifacts. We used a long-pass beam splitter (FF875-Di01, Semrock Inc., NY, USA). This dichroic mirror reflects wavelengths that are shorter than 900 nm. A simple design of a fluorescence/bright-field imaging path is shown in Figure 4. To avoid bias in Doppler measurements, when using a long-pass dichroic mirror, we suggest using a bidirectional scanning mechanism. In the bidirectional scanning mechanism, the same cross section is scanned from two opposite directions, and after the Doppler calculations, the measurements are averaged. The Doppler artifacts that are introduced by structural imperfections, like the dichroic glass, are additively negative in the two recordings and are canceled out.

The sample arm (including the fluorescence imaging) is mounted on a custom-made adapter and a typical microscope double-arm boom stand. This allows for adjustment of the OCT head angle to approach the specimen from the desired angle.

4.3. Visible Path. The design of the system allows for installation of a visible path on the imaging device. The visible path can be used for delivery of optogenetic stimulation light pulses or incorporating a secondary imaging mechanism such as optical intrinsic signal imaging (OICI) or fluorescence imaging. In this part, we offer a sample fluorescence imaging design that has been added and tested on this device. Figure 4 represents the optical design and hardware realization of the imaging path.

4.4. Reference Arm. Ideally, in an OCT system, the reference arm should replicate the length that light travels and the dispersion that it experiences in the sample arm. If the light dispersion in the sample arm is not properly replicated in

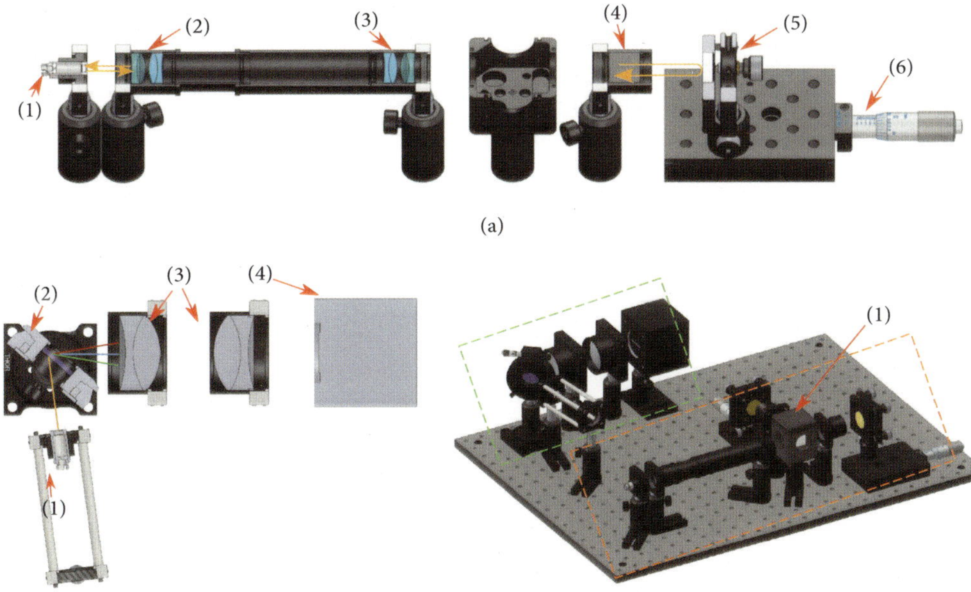

Figure 5: SOLIDWORKS design of the reference and spectrometer arms. (a) The CAD design of the reference arm. (1) Fiber connector, (2) scan lens, (3) tube lens, (4) dispersion compensator for the objective, (5) a gold mirror installed on a kinematic mount with adjustable angles, and (6) translational stage to precisely control the position of the reference mirror. (b) The CAD design of the spectrometer arm. The grating has an optimal incident angle of 48.6° and diffraction angle of 48.6°. It is mounted on rotatory mounts to adjust the angles of the grating with respect to the OCT beam and the lens mechanism. (1) Fiber connection, (2) grating installed on rotatory mounts, (3) Plössl lens, and (4) line CCD camera. (c) The reference and spectrometer arms are installed on an optical breadboard to facilitate transportation. There is a movable mirror mount, component (1), in the reference arm that can change the beam path to match different objective lenses in the sample arm. The reference and spectrometer arms are encircled with orange and green rectangles, respectively.

the reference arm, it may greatly affect the axial resolution and quality of the OCT images. To minimize the dispersion mismatch between sample and reference arms, we assembled the reference arm with identical components which were used in our sample arm, except the scan head and the objective lens. In the reference arm, we used a single glass dispersion compensator (DC) (LSM02DC, Thorlabs, NJ, USA) to replicate the optical path of the beam inside the objective lens (Figure 5). This design reduces the cost. At the end of our reference arm, we use a protected gold mirror (PF10-03-M01, Thorlabs, NJ, USA) to obtain a flat reflection spectrum within the wavelength range of our light source. The light power that is back-reflected from a mirror is typically much stronger than the backscattered light from tissue in the sample arm. We installed an adjustable iris in the reference arm to tune the reference power. A removable right-angle mirror is used in the reference arm to incorporate multiple optical paths in the reference arm. This is useful when more than one objective lens mechanism is used during an experiment.

4.5. Spectrometer. To record interference patterns, we used a custom-made spectrometer which consists of a transmission grating, a Plössl compound lens, and a line charge-coupled device (L-CCD). We chose a volume phase holographic transmission grating (Wasatch Photonics, UT, USA) with 1145 grooves per mm to obtain a large bandwidth and a balanced performance for different light polarization. The optimal incident and diffraction angles at $\lambda = 1.31\,\mu$m for this grating are both ~48.6°. These requirements were accommodated by installing the grating on two rotation mounts (PRM1 and RP01, Thorlabs, NJ, USA) for the fine adjustment of yaw and roll angles (Figure 5). The beam was delivered to the spectrometer via an SMF-28*e*+ single mode fiber (CIR-1310-50-APC, Thorlabs, NJ, USA) and a collimating package (F280APC-C, Thorlabs, NJ, USA).

The diffracted beam is imaged on the L-CCD by two achromatic doublet lenses (AC508-150-C, Thorlabs, NJ, USA) which form a Plössl compound (Figure 3) with an effective focal length of ~72 mm. The L-CCD detector in this system (SU1024-LDH2, Sensors Unlimited, NJ, USA) has 1024 pixels with a 25 μm pixel pitch and aperture height of 500 μm. The length of the focal plane array in this camera is 25.6 mm which is sufficient to receive at least 200 nm of the spectrum (1200 nm to 1400 nm) in combination with our lensing mechanism. For an incident beam diameter of 3.4 mm, this lens compound forms a theoretical FWHM spot size of ~20 μm which is smaller than the camera pixel pitch. Optical aberrations that are introduced by this lens system are sufficiently small (Figures 3(b) and 3(c)) and the root-mean-square wavefront error stays below the diffraction limit at different diffraction angles (Figure 3(d)). The small spot size reduces the sensitivity roll-off by reducing the spread of each wavelength over multiple pixels.

4.6. Fiber-Based Components. The light source was directly connected to a fiber optic circulator (CIR-1310-50-APC, Thorlabs, NJ, USA). The circulator illuminates a 50/50 fiber coupler (FC1310-70-50-APC, Thorlabs, NJ, USA) which splits the OCT beam between the sample and reference arms via two single mode fibers with APC cuts (P3-980A-FC-2, Thorlabs, NJ, USA). The backscattered or back-reflected light for the sample and reference arms is recombined by the fiber coupler and guided toward the spectrometer arm via the fiber circulator. The fiber coupler possesses two input ports where one is used for OCT beam and the other is connected to a red laser diode. The red light is used for alignment purposes and to help the operator guide the beam to cover the region of interest during OCT imaging. When no visible light for alignment is used and a fiber coupler with 50/50 split ratio is being used, no fiber circulator is needed as the free coupler port can be connected to the spectrometer directly.

4.7. Animal Preparation. We used OCT to visualize the vascular network in the cerebral cortex of Sprague-Dawley rats. All animal procedures were approved by the Institutional Animal Care and Use Committee (IACUC) at the University of Wisconsin and Medical College of Wisconsin. During all surgical and imaging procedures, the animals were anesthetized using isoflurane and body temperature was regulated using a heated water pad. For intravital imaging of the cerebrovasculature, a closed cranial window of ~5 mm diameter was prepared. Briefly, a craniotomy was made and the dura mater covering the brain within the window was removed to enhance resolution and depth of imaging. Prior to imaging, the craniotomy was sealed with a glass coverslip.

5. Results

By placing a gold mirror (PF10-03-M01, Thorlabs, NJ, USA) in the sample arm, we measured the axial resolution of this device. This was done by imaging the mirror and then measuring the FWHM of the recorded A-scans along the axial direction. The mirror surface acts as a Dirac delta input to the imaging system along the depth direction; however, due to the finite axial resolution of the system, we observed that the A-scan broadens along the depth (Figure 6). The practical axial resolution of the system is ~5.6 μm, which is slightly larger than the theoretical resolution. The imperfections in the components and optical aberrations in lensing mechanisms can contribute to the degradation of axial resolution. The non-Gaussian distribution of the source power is another important factor that negatively affects the resolution.

Using a 3D phantom with embedded subresolution gold nanoshells [74], we measured the point spread function (PSF) of this OCT scanner at its focal plane. The results verify a lateral full width at half maximum (FWHM) spot size of ~5.6 μm and ~6.4 μm (or equally $1/e^2$ spot diameters of 9.5 μm and 10.9 μm) along the fast and slow galvanometric axes, respectively. The lateral resolution was consistent across the field of view of 4 mm in diameter (Figure 7). These OCT resolution measurements are in close agreement with the empirical spot size measurements provided by the objective

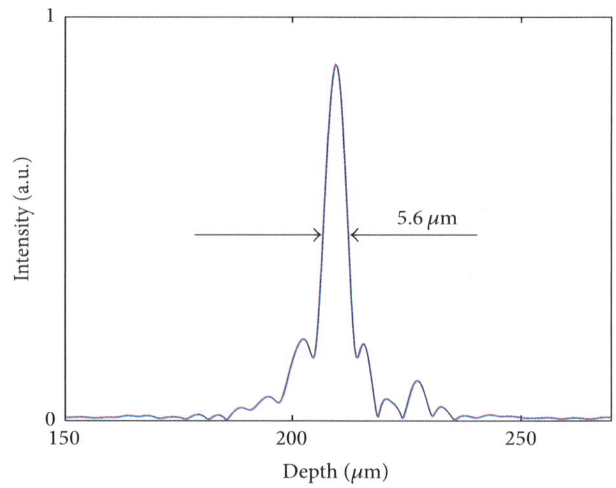

Figure 6: The measured axial resolution of the implemented SD-OCT system.

vendor. When used with an incident beam diameter of 4 mm, they measured a $1/e^2$ spot size of ~11 μm at the focal point of the objective.

During our resolution measurements, the focal plane of the image was set at ~390 μm below the surface of the phantom to allow for recording the PSF of the OCT scanner at different distances from the focal plane. At the focal plane, we observed a quasi-Gaussian and localized distribution of the intensity (Figure 8). At deeper regions, the spread of the Gaussian increases and its intensity drops. At regions above the focal plane, we see that the spot size shrinks while a ring starts to form around it. This ring is visible for particles that are located ~100 μm above the focal plane. At shallower regions (e.g., 120 μm above the focal plane), the ring becomes almost as intense as the central spot. We anticipate that the combination of spherical aberrations in the objective and the beam cut at the collimating packages are contributing to the formation of the ring which appears in our PSF measurements. To investigate the effect of tube/scan lens mechanism on the formation of the ring, we conducted similar measurements by modifying the design of our sample arm to place the objective lens directly after the scanning mirrors. When the scan/tube lens mechanism was removed from the OCT path, our measurements (not included in this manuscript) revealed a similar ring pattern above the focal plane of the OCT images. This observation confirmed that the intermediary lens mechanism does not significantly contribute to the formation of the surrounding rings in the OCT PSF.

The intensity change in the OCT images across the field of view was measured using the 3D phantom. At each desired location in the field of view, a 400 μm × 400 μm block of the phantom was imaged and the average intensity of the detected nanoshells at the focal plane of the system was calculated. This calculation was repeated at different locations across the field of view and results are presented in Figure 9. The system is able to provide almost a flat intensity (less than 2 dB drop in the intensity) for a field of view as wide as 4 mm in diameter. However, when approaching the boundary of the diffraction limited field of view (~4.7 mm), the intensity

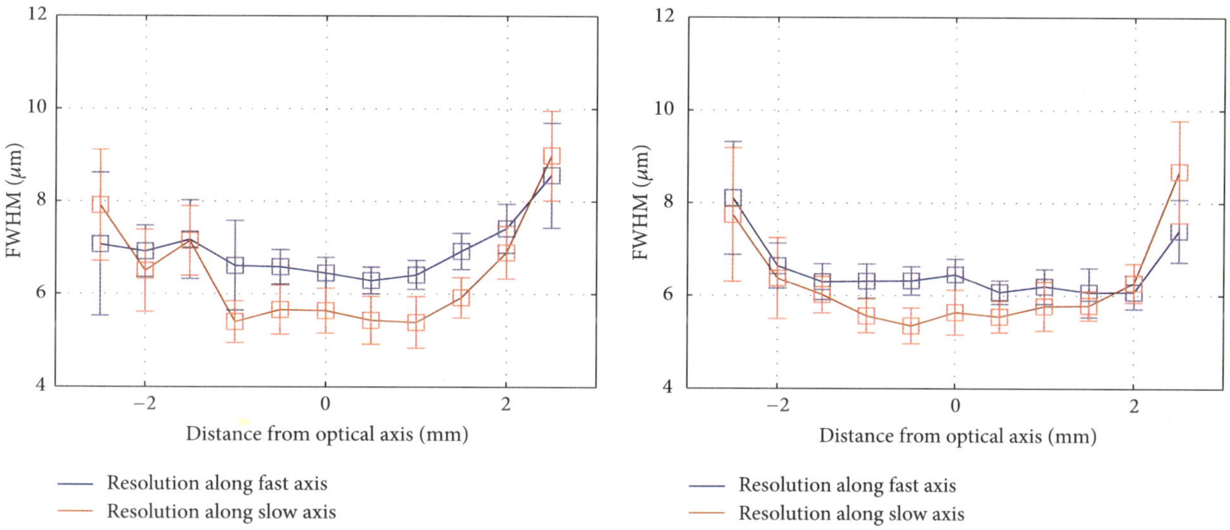

FIGURE 7: Variation in the lateral resolution of the implemented SD-OCT. Within the radius of 2 mm, the FWHM resolution stays below 7.1 μm. The maximum diffraction limited field of view for this objective is ~4.7 mm in diameter. When approaching this limit, the FWHM increases to ~8 μm. The error bars mark the standard deviations. There are at least 70 measurements at each location.

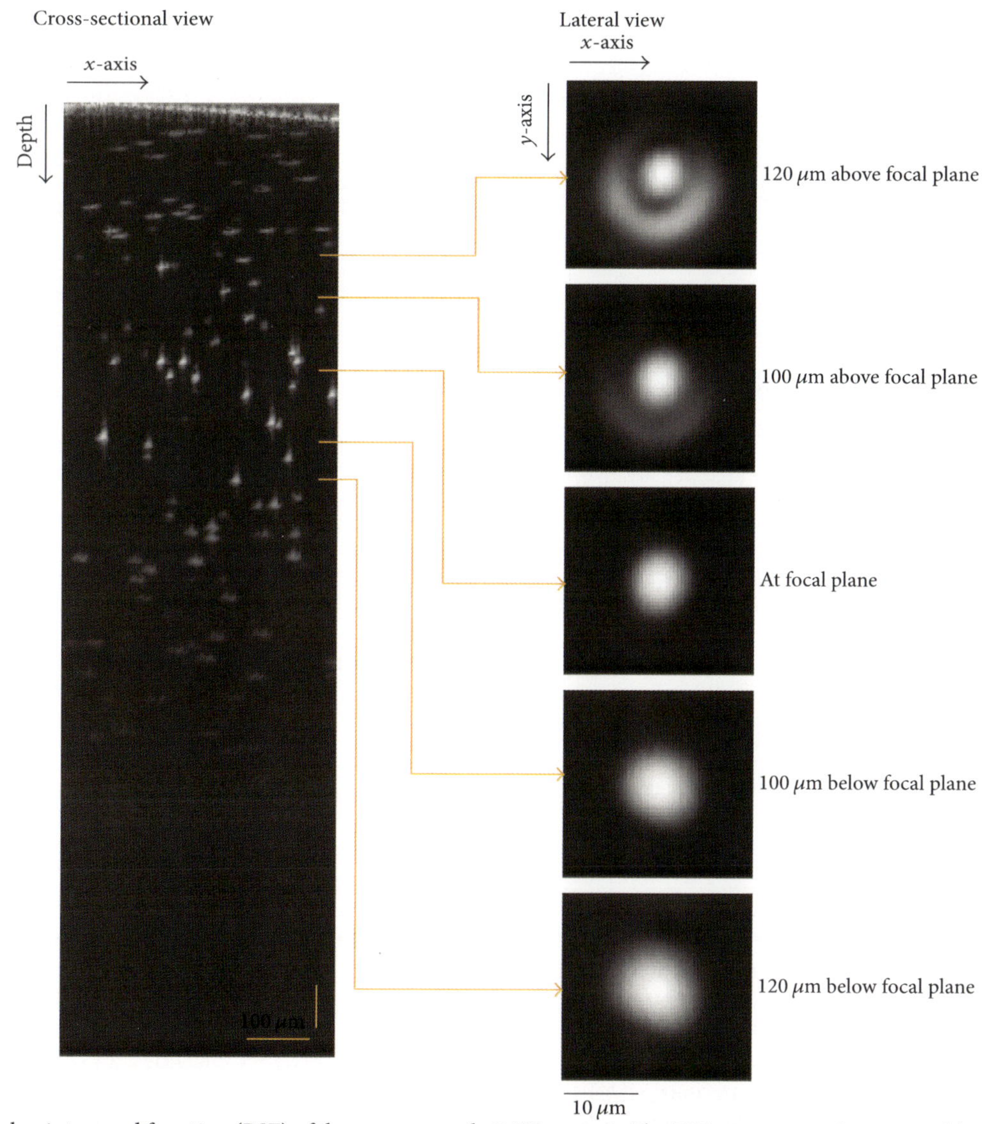

FIGURE 8: The lateral point spread function (PSF) of the custom-made OCT scanner. The PSF was measured using a gold nanoshell embedded phantom when the focal plane was placed ~390 μm below the surface of the phantom. The lateral PSF includes some rings at regions above the focal plane.

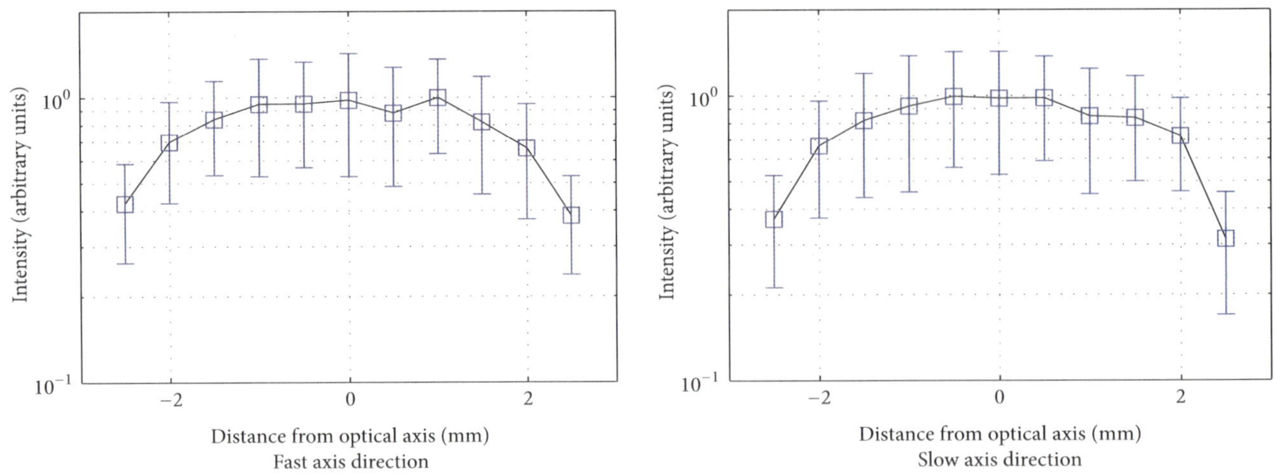

FIGURE 9: Variation in the intensity of OCT images across the field of view. At the borders of our field of view (4.7 mm in diameter), the intensity drops rapidly.

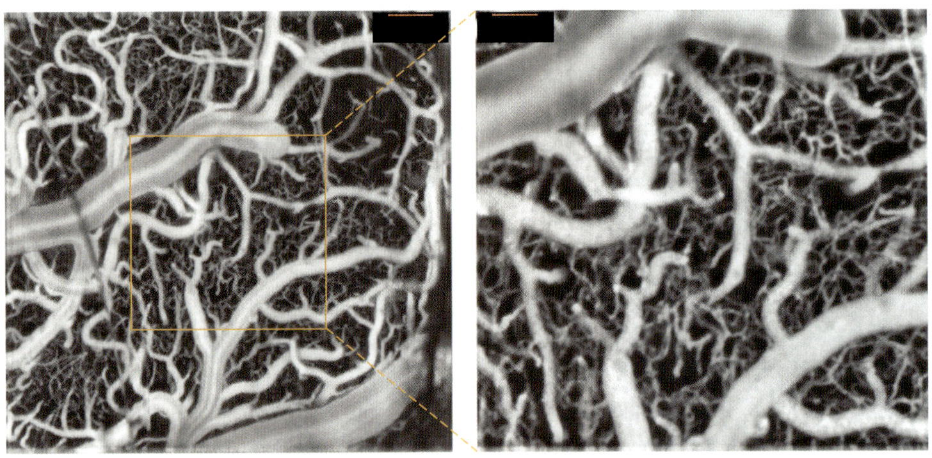

FIGURE 10: OCT angiography of capillaries in a rat brain. The imaging resolution system allows for imaging vasculature and small capillaries in the cortex of a rat. The image is the maximum intensity projection of a thickness of ~220 μm after averaging 5 angiography recordings [46].

declines noticeably. A combination of beam truncation and different optical aberrations can result in such an intensity drop in the imaging system.

We have successfully used this setup for *in vivo* brain imaging in adult rodents. The system was able to image cerebral arteries, veins, and capillaries. Figure 10 demonstrates the vasculature morphology that was captured by the device using OCT angiography [13, 14, 46]. This image shows the maximum intensity projection (MIP) obtained from a 3D OCT angiography that was captured from a rat. The depth that was used for MIP was from 60 μm to 280 μm below the brain surface. In this recording, the focal plane of the OCT scanner was positioned ~100 μm below the surface.

By adjusting the focal plane of the OCT, it is possible to record high-resolution OCT images at deeper regions in the brain. An example of vascular network which was captured at three different depths is demonstrated in Figure 11. In this figure, some penetrating arterioles are marked with red arrows. The circle on the image marks an arteriole that penetrates up to 500 μm without branching.

Typical imaging depth of a SD-OCT system in the brain is limited to less than 1 mm. When the camera exposure time was set to ~25 μs, the proposed system was able to image vessels at a depth of up to 900 μm (Figure 12). To the best of our knowledge, this is the deepest reported SD-OCT vascular imaging at a central wavelength of 1300 nm.

6. Conclusion

This paper is aimed at assisting researchers in optimizing the optical design of spectral-domain OCT systems according to their specific needs and research goals. In this paper, a brief overview of the fundamental theory of image formation in SD-OCT scanners was presented. Then, a broad category of theoretical and practical considerations for the selection and design of optical and optomechanical components were discussed. The paper provides an analysis on each of the critical components and explains how different parameters of each piece can affect the ultimate performance of a SD-OCT scanner. We also provided some details to design

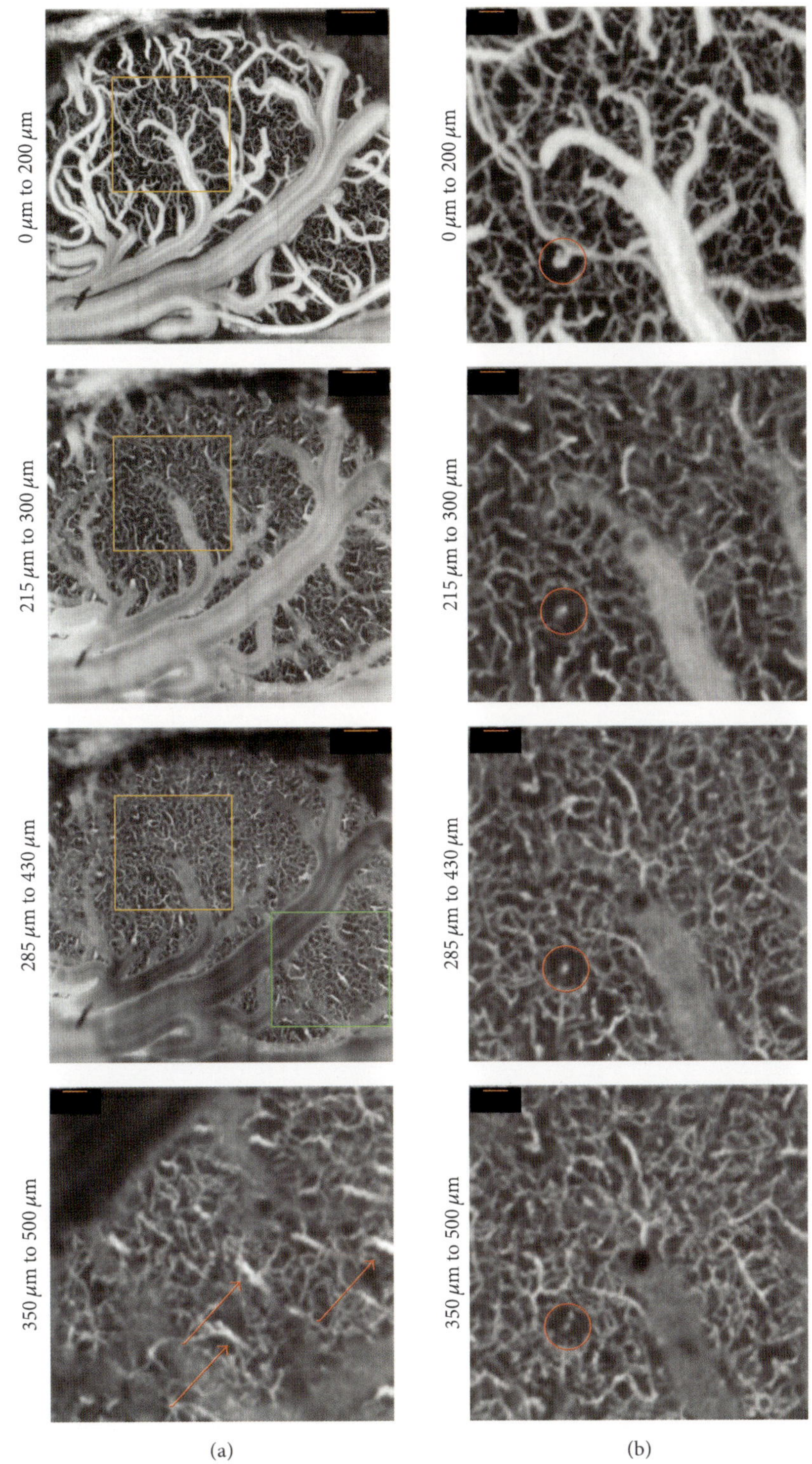

FIGURE 11: Imaging at various depths. By adjusting the position of the focal plane inside the brain, it is possible to analyze the topology and distribution of vascular networks at different depths in the brain. The red arrows mark some penetrating arterioles. The red circle on (b) depicts an arteriole that penetrates up to 500 μm deep in the brain before branching.

a mechanism to combine the OCT with other imaging modalities such as fluorescence imaging or optogenetic neurostimulation technology.

The components and optical design of a custom-made SD-OCT system that we used in our research for our *in vivo* brain imaging applications were discussed in further detail in this paper. This device is a relatively low cost imaging system (~$34 K) which was constructed mostly using off-the-shelf components. Making the scanner head and a few adapters requires some level of machining; however, the

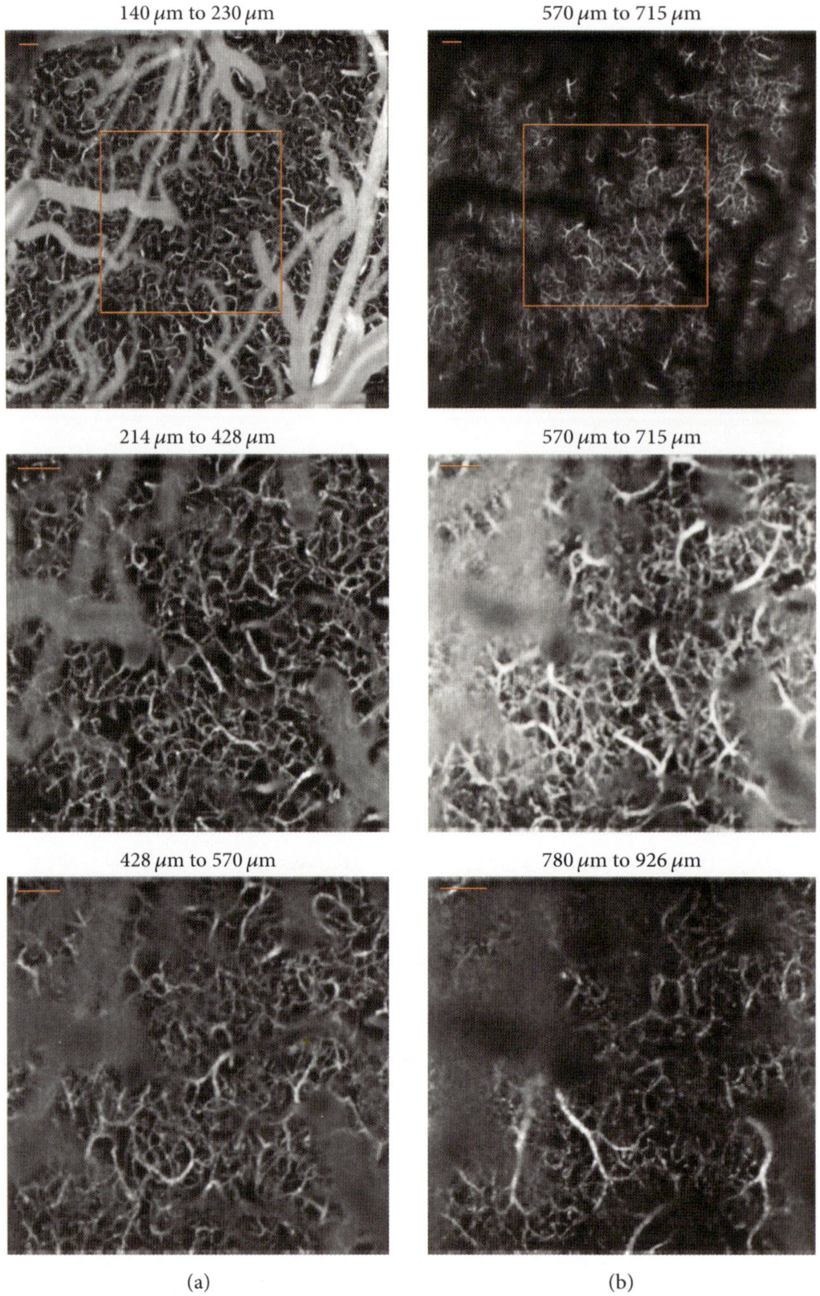

FIGURE 12: Deep brain imaging. The system is able to image vasculature up to 900 μm deep in the brain of a rat. When imaging deep regions of a brain, the large pial vessels on the surface block the OCT beam and create dark and noisy areas in the OCT angiography images. (a) and (b) are obtained from the same region but at different depths. The shadow of the superficial vessels on (a) prevents imaging the capillaries at deeper regions that are shown on (b). The scale bars are 200 μm.

designs are fairly simple and easy to implement. Adaptation of the minimum number of custom-made parts reduced the time and cost of our system development.

Our design included the fabrication of a scan and tube lens system to incorporate a dichroic beam splitter in the path of the OCT beam and addition of a second imaging modality through the same objective lens. A simple design of a fluorescence imaging path is discussed in Figure 4.

The designed SD-OCT imaging system possesses a close-to-diffraction-limit lateral resolution and its axial resolution is comparable to the theoretical value. The system provides the diffraction limited resolution across a large FOV (4 mm in diameter) without major intensity drop across this area. In the design, we considered the possibility of two different optical paths in the reference arm. Therefore, it is possible to change the objective lens during the experiment. Alternating between two objectives may require some adjustments in the objective collar.

Although the presented design and discussions facilitate the process of building new custom-made SD-OCT scanners, the authors would like to emphasize that such system development projects are delicate time-consuming efforts. Proper alignment of the components is crucial to achieve the diffraction limited performance to obtain theoretical specifications. While for an experienced person the process can take a few days, for a nonexpert designer with no or minimum background in optical design, the process can be tedious and much more time-consuming to assemble and align the same platform. The Supplementary Materials summarize some steps to help nonexpert users achieve proper optical alignment.

Obviously, such an imaging system requires a proper data acquisition system and analysis software to become a fully functional setup. While for two-photon microscopy there are already open-source software packages to operate a newly developed system, similar software packages are not available for custom-made OCT scanners. Development of publicly available software packages that are modular and adjustable to general needs in the field of OCT imaging seems necessary.

Conflicts of Interest

The authors declare that there are no conflicts of interest regarding the publication of this paper.

Acknowledgments

This research was supported by the Army Research Office (ARO) Grant no. 68750-LS-II, the National Science Foundation (NSF) Career Award Grant no. 1454300, the Brain and Behavior Research Foundation (NARSAD) Young Investigator Award Grants nos. 20610 and 23620, and the University of Wisconsin Research Growth Initiative (RGI) Grants nos. 101X213 and 101X254.

References

[1] A. F. Fercher, C. K. Hitzenberger, G. Kamp, and S. Y. El-Zaiat, "Measurement of intraocular distances by backscattering spectral interferometry," *Optics Communications*, vol. 117, no. 1-2, pp. 43–48, 1995.

[2] J. Men, Y. Huang, J. Solanki et al., "Optical Coherence Tomography for Brain Imaging and Developmental Biology," *IEEE Journal of Selected Topics in Quantum Electronics*, vol. 22, no. 4, 2016.

[3] C. Magnain, J. C. Augustinack, M. Reuter et al., "Blockface histology with optical coherence tomography: a comparison with Nissl staining," *NeuroImage*, vol. 84, pp. 524–533, 2014.

[4] V. J. Srinivasan, H. Radhakrishnan, J. Y. Jiang, S. Barry, and A. E. Cable, "Optical Coherence Microscopy for deep tissue imaging of the cerebral cortex with intrinsic contrast," *Optics Express*, vol. 20, no. 3, pp. 2220–2239, 2012.

[5] C. Magnain, J. C. Augustinack, E. Konukoglu et al., "Optical coherence tomography visualizes neurons in human entorhinal cortex," *Neurophotonics*, vol. 2, no. 1, Article ID 14063RR, 2015.

[6] N. A. Nassif, B. Cense, B. H. Park et al., "In vivo high-resolution video-rate spectral-domain optical coherence tomography of the human retina and optic nerve," *Optics Express*, vol. 12, no. 3, pp. 367–376, 2004.

[7] M. Wojtkowski, V. Srinivasan, J. G. Fujimoto et al., "Three-dimensional retinal imaging with high-speed ultrahigh-resolution optical coherence tomography," *Ophthalmology*, vol. 112, no. 10, pp. 1734–1746, 2005.

[8] Y. Ikuno and Y. Tano, "Retinal and choroidal biometry in highly myopic eyes with spectral-domain optical coherence tomography," *Investigative Ophthalmology & Visual Science*, vol. 50, no. 8, pp. 3876–3880, 2009.

[9] V. J. Srinivasan, Y. Chen, J. S. Duker, and J. G. Fujimoto, "In vivo functional imaging of intrinsic scattering changes in the human retina with high-speed ultrahigh resolution OCT," *Optics Express*, vol. 17, no. 5, pp. 3861–3877, 2009.

[10] Q. Zhang, R. Lu, B. Wang, J. D. Messinger, C. A. Curcio, and X. Yao, "Functional optical coherence tomography enables in vivo physiological assessment of retinal rod and cone photoreceptors," *Scientific Reports*, vol. 5, article no. 9595, 2015.

[11] S. Makita, Y. Hong, M. Yamanari, T. Yatagai, and Y. Yasuno, "Optical coherence angiography," *Optics Express*, vol. 14, no. 17, pp. 7821–7840, 2006.

[12] R. K. Wang, S. L. Jacques, Z. Ma, S. Hurst, S. R. Hanson, and A. Gruber, "Three dimensional optical angiography," *Optics Express*, vol. 15, no. 7, pp. 4083–4097, 2007.

[13] B. J. Vakoc, R. M. Lanning, J. A. Tyrrell et al., "Three-dimensional microscopy of the tumor microenvironment in vivo using optical frequency domain imaging," *Nature Medicine*, vol. 15, no. 10, pp. 1219–1223, 2009.

[14] V. J. Srinivasan, J. Y. Jiang, M. A. Yaseen et al., "Rapid volumetric angiography of cortical microvasculature with optical coherence tomography," *Optics Expresss*, vol. 35, no. 1, pp. 43–45, 2010.

[15] Y. Zhao, Z. Chen, C. Saxer, S. Xiang, J. F. De Boer, and J. S. Nelson, "Phase-resolved optical coherence tomography and optical Doppler tomography for imaging blood flow in human skin with fast scanning speed and high velocity sensitivity," *Optics Expresss*, vol. 25, no. 2, pp. 114–116, 2000.

[16] R. Leitgeb, L. Schmetterer, M. Wojtkowski, C. K. Hitzenberger, M. Sticker, and A. F. Fercher, "Flow velocity measurements by frequency domain short coherence interferometry," in *Proceedings of the Coherence Domain Optical Methods in Biomedical Science and Clinical Applications VI*, pp. 16–21, USA, January 2002.

[17] B. R. White, M. C. Pierce, N. Nassif et al., "In vivo dynamic human retinal blood flow imaging using ultra-high-speed spectral domain optical doppler tomography," *Optics Express*, vol. 11, no. 25, pp. 3490–3497, 2003.

[18] V. X. D. Yang, M. L. Gordon, B. Qi et al., "High speed, wide velocity dynamic range Doppler optical coherence tomography (Part I): System design, signal processing, and performance," *Optics Express*, vol. 11, no. 7, pp. 794–809, 2003.

[19] U. Morgner, W. Drexler, F. X. Kärtner et al., "Spectroscopic optical coherence tomography," *Optics Expresss*, vol. 25, no. 2, pp. 111–113, 2000.

[20] D. J. Faber, E. G. Mik, M. C. G. Aalders, and T. G. Van Leeuwen, "Toward assessment of blood oxygen saturation by spectroscopic optical coherence tomography," *Optics Expresss*, vol. 30, no. 9, pp. 1015–1017, 2005.

[21] F. E. Robles, C. Wilson, G. Grant, and A. Wax, "Molecular imaging true-colour spectroscopic optical coherence tomography," *Nature Photonics*, vol. 5, no. 12, pp. 744–747, 2011.

[22] M. R. Hee, D. Huang, E. A. Swanson, and J. G. Fujimoto, "Polarization-sensitive low-coherence reflectometer for birefringence characterization and ranging," *Journal of the Optical Society of America B: Optical Physics*, vol. 9, no. 6, pp. 903–908, 1992.

[23] J. F. De Boer, T. E. Milner, M. J. C. Van Gemert, and J. S. Nelson, "Two-dimensional birefringence imaging in biological tissue by polarization-sensitive optical coherence tomography," *Optics Expresss*, vol. 22, no. 12, pp. 934–936, 1997.

[24] J. F. De Boer, T. E. Milner, and J. Stuart Nelson, "Determination of the depth-resolved Stokes parameters of light backscattered from turbid media by use of polarization-sensitive optical coherence tomography," *Optics Expresss*, vol. 24, no. 5, pp. 300–302, 1999.

[25] E. Götzinger, M. Pircher, and C. K. Hitzenberger, "High speed spectral domain polarization sensitive optical coherence tomography of the human retina," *Optics Express*, vol. 13, no. 25, pp. 10217–10229, 2005.

[26] B. H. Park, M. C. Pierce, B. Cense et al., "Real-time fiber-based multi-functional spectral-domain optical coherence tomography at 1.3 μm," *Optics Express*, vol. 13, no. 11, pp. 3931–3944, 2005.

[27] M. Yamanari, S. Makita, V. D. Madjarova, T. Yatagai, and Y. Yasuno, "Fiber-based polarization-sensitive Fourier domain optical coherence tomography using B-scan-oriented polarization modulation method," *Optics Express*, vol. 14, no. 14, pp. 6502–6515, 2006.

[28] B. Baumann, E. Götzinger, M. Pircher et al., "Segmentation and quantification of retinal lesions in age-related macular degeneration using polarization-sensitive optical coherence tomography," *Journal of Biomedical Optics*, vol. 15, no. 6, Article ID 061704, 2010.

[29] J. M. Schmitt, "OCT elastography: Imaging microscopic deformation and strain of tissue," *Optics Express*, vol. 3, no. 6, pp. 199–211, 1998.

[30] J. Rogowska, N. A. Patel, J. G. Fujimoto, and M. E. Brezinski, "Optical coherence tomographic elastography technique for measuring deformation and strain of atherosclerotic tissues," *Heart*, vol. 90, no. 5, pp. 556–562, 2004.

[31] S. G. Adie, X. Liang, B. F. Kennedy, R. John, D. D. Sampson, and S. A. Boppart, "Spectroscopic optical coherence elastography," *Optics Express*, vol. 18, no. 25, pp. 25519–25534, 2010.

[32] B. F. Kennedy, X. Liang, S. G. Adie et al., "In vivo three-dimensional optical coherence elastography," *Optics Express*, vol. 19, no. 7, pp. 6623–6634, 2011.

[33] W. Drexler and J. G. Fujimoto, *Optical Coherence Tomography*, Springer International Publishing, Cham, 1st edition, 2008.

[34] J. Kim, W. Brown, J. R. Maher, H. Levinson, and A. Wax, "Functional optical coherence tomography: Principles and progress," *Physics in Medicine and Biology*, vol. 60, no. 10, pp. R211–R237, 2015.

[35] C. Dorrer, N. Belabas, J.-P. Likforman, and M. Joffre, "Spectral resolution and sampling issues in fourier-transform spectral interferometry," *Journal of the Optical Society of America B: Optical Physics*, vol. 17, no. 10, pp. 1795–1802, 2000.

[36] M. Wojtkowski, R. Leitgeb, A. Kowalczyk, T. Bajraszewski, and A. F. Fercher, "In vivo human retinal imaging by Fourier domain optical coherence tomography," *Journal of Biomedical Optics*, vol. 7, no. 3, pp. 457–463, 2002.

[37] M. A. Choma, M. V. Sarunic, C. Yang, and J. A. Izatt, "Sensitivity advantage of swept source and Fourier domain optical coherence tomography," *Optics Express*, vol. 11, no. 18, pp. 2183–2189, 2003.

[38] S. H. Yun, G. J. Tearney, J. F. De Boer, N. Iftimia, and B. E. Bouma, "High-speed optical frequency-domain imaging," *Optics Express*, vol. 11, no. 22, pp. 2953–2963, 2003.

[39] A. F. Fercher, W. Drexler, C. K. Hitzenberger, and T. Lasser, "Optical coherence tomography—principles and applications," *Reports on Progress in Physics*, vol. 66, no. 2, pp. 239–303, 2003.

[40] R. Leitgeb, C. K. Hitzenberger, and A. F. Fercher, "Performance of fourier domain vs. Time domain optical coherence tomography," *Optics Express*, vol. 11, no. 8, pp. 889–894, 2003.

[41] J. F. de Boer, B. Cense, B. H. Park, M. C. Pierce, G. J. Tearney, and B. E. Bouma, "Improved signal-to-noise ratio in spectral-domain compared with time-domain optical coherence tomography," *Optics Expresss*, vol. 28, no. 21, pp. 2067–2069, 2003.

[42] B. Cense, N. A. Nassif, T. C. Chen et al., "Ultrahigh-resolution high-speed retinal imaging using spectral-domain optical coherence tomography," *Optics Express*, vol. 12, no. 11, pp. 2435–2447, 2004.

[43] M. Wojtkowski, V. J. Srinivasan, T. H. Ko, J. G. Fujimoto, A. Kowalczyk, and J. S. Duker, "Ultrahigh-resolution, high-speed, Fourier domain optical coherence tomography and methods for dispersion compensation," *Optics Express*, vol. 12, no. 11, pp. 2404–2422, 2004.

[44] B. F. Kennedy, K. M. Kennedy, and D. D. Sampson, "A review of optical coherence elastography: Fundamentals, techniques and prospects," *IEEE Journal of Selected Topics in Quantum Electronics*, vol. 20, no. 2, 2014.

[45] "A quickly made model of a microscope stand," https://grabcad.com/library/microscope-stand-1.

[46] F. Atry, S. Frye, T. J. Richner et al., "Monitoring cerebral hemodynamics following optogenetic stimulation via optical coherence tomography," *IEEE Transactions on Biomedical Engineering*, vol. 62, no. 2, pp. 766–773, 2015.

[47] D. Huang, E. A. Swanson, C. P. Lin et al., "Optical coherence tomography," *Science*, vol. 254, no. 5035, pp. 1178–1181, 1991.

[48] J. A. Izatt, M. D. Kulkarni, H.-W. Wang, K. Kobayashi, and M. V. Sivak Jr., "Optical coherence tomography and microscopy in gastrointestinal tissues," *IEEE Journal of Selected Topics in Quantum Electronics*, vol. 2, no. 4, pp. 1017–1028, 1996.

[49] Y.-K. Kim and Y. P. Kim, "High-speed time-domain optical coherence tomography with an imaging speed of ten frames per second with 2000 A-scan," *Optical Engineering*, vol. 49, no. 5, Article ID 055601, 2010.

[50] G. Häusler and M. W. Lindner, "'Coherence radar' and 'spectral radar'—new tools for dermatological diagnosis," *Journal of Biomedical Optics*, vol. 3, no. 1, pp. 21–31, 1998.

[51] J. M. Schmitt, "Optical Coherence Tomography (OCT): a review," *IEEE Journal of Selected Topics in Quantum Electronics*, vol. 5, no. 4, pp. 1205–1215, 1999.

[52] P. H. Tomlins and R. K. Wang, "Theory, developments and applications of optical coherence tomography," *Journal of Physics D: Applied Physics*, vol. 38, no. 15, pp. 2519–2535, 2005.

[53] J. Kalkman, "Fourier-Domain Optical Coherence Tomography Signal Analysis and Numerical Modeling," *International Journal of Optics*, vol. 2017, Article ID 9586067, 2017.

[54] F. E. Robles, S. Chowdhury, and A. Wax, "Assessing hemoglobin concentration using spectroscopic optical coherence tomography for feasibility of tissue diagnosticcs," *Biomedical Optics Express*, vol. 1, no. 1, pp. 310–317, 2010.

[55] J. Yi and X. Li, "Estimation of oxygen saturation from erythrocytes by high-resolution spectroscopic optical coherence tomography," *Optics Expresss*, vol. 35, no. 12, pp. 2094–2096, 2010.

[56] J. Yi, Q. Wei, W. Liu, V. Backman, and H. F. Zhang, "Visible-light optical coherence tomography for retinal oximetry," *Optics Expresss*, vol. 38, no. 11, pp. 1796–1798, 2013.

[57] S. P. Chong, C. W. Merkle, D. F. Cooke et al., "Noninvasive, in vivo imaging of subcortical mouse brain regions with 1.7 μm optical coherence tomography," *Optics Express*, vol. 40, no. 21, pp. 4911–4914, 2015.

[58] S. Ishida, N. Nishizawa, T. Ohta, and K. Itoh, "Ultrahigh-resolution optical coherence tomography in 1.7 μm region with fiber laser supercontinuum in low-water-absorption samples," *Applied Physics Express*, vol. 4, no. 5, 2011.

[59] S. Ishida and N. Nishizawa, "Quantitative comparison of contrast and imaging depth of ultrahigh-resolution optical coherence tomography images in 800-1700 nm wavelength region," *Biomedical Optics Express*, vol. 3, no. 2, pp. 282–294, 2012.

[60] Y. Hattori, H. Kawagoe, Y. Ando, M. Yamanaka, and N. Nishizawa, "High-speed ultrahigh-resolution spectral domain optical coherence tomography using high-power supercontinuum at 0.8 μm wavelength," *Applied Physics Express*, vol. 8, no. 8, Article ID 082501, 2015.

[61] M. Szkulmowski, M. Wojtkowski, T. Bajraszewski et al., "Quality improvement for high resolution in vivo images by spectral domain optical coherence tomography with supercontinuum source," *Optics Communications*, vol. 246, no. 4-6, pp. 569–578, 2005.

[62] J. Barrick, A. Doblas, M. R. Gardner, P. R. Sears, L. E. Ostrowski, and A. L. Oldenburg, "High-speed and high-sensitivity parallel spectraldomain optical coherence tomography using a supercontinuum light source," *Optics Expresss*, vol. 41, no. 24, pp. 5620–5623, 2016.

[63] Z. Hu and A. M. Rollins, "Theory of two beam interference with arbitrary spectra," *Optics Express*, vol. 14, no. 26, pp. 12751–12759, 2006.

[64] Z. Hu, Y. Pan, and A. M. Rollins, "Analytical model of spectrometer-based two-beam spectral interferometry," *Applied Optics*, vol. 46, no. 35, pp. 8499–8505, 2007.

[65] Z. Hu and A. M. Rolling, "Quasi-telecentric optical design of a microscope-compatible OCT scanner," *Optics Express*, vol. 13, no. 17, pp. 6407–6415, 2005.

[66] E. H. K. Stelzer, *The Intermediate Optical System of Laser-Scanning Confocal Microscopes*, Springer US, Boston, MA, 2006.

[67] A. Negrean and H. D. Mansvelder, "Optimal lens design and use in laser-scanning microscopy," *Biomedical Optics Express*, vol. 5, no. 5, pp. 1588–1609, 2014.

[68] M. Hafez, T. Sidler, and R.-P. Salathé, "Study of the beam path distortion profiles generated by a two-axis tilt single-mirror laser scanner," *Optical Engineering*, vol. 42, no. 4, pp. 1048–1057, 2003.

[69] J. Y. Kim, C. Lee, K. Park, G. Lim, and C. Kim, "Fast optical-resolution photoacoustic microscopy using a 2-axis water-proofing MEMS scanner," *Scientific Reports*, vol. 5, article no. 7932, 2015.

[70] B. E. A. Saleh and M. C. Teich, *Fundamentals of Photonics*, Wiley-Interscience, 1991.

[71] R. A. Leitgeb, M. Villiger, A. H. Bachmann, and L. Steinmann, "Extended focus depth for Fourier domain optical coherence microscopy," *Optics Expresss*, vol. 31, no. 16, pp. 2450–2452, 2006.

[72] K.-S. Lee and J. P. Rolland, "Bessel beam spectral-domain high-resolution optical coherence tomography with micro-optic axicon providing extended focusing range," *Optics Expresss*, vol. 33, no. 15, pp. 1696–1698, 2008.

[73] S. Tamborski, H. C. Lyu, H. Dolezyczek et al., "Extended-focus optical coherence microscopy for high-resolution imaging of the murine brain," *Biomedical Optics Express*, vol. 7, no. 11, Article ID #269710, pp. 4400–4414, 2016.

[74] F. Atry and R. Pashaie, "Analysis of intermediary scan-lens and tube-lens mechanisms for optical coherence tomography," *Applied Optics*, vol. 55, no. 4, pp. 646–653, 2016.

[75] T. Bajraszewski, M. Wojtkowski, M. Szkulmowski, A. Szkulmowska, R. Huber, and A. Kowalczyk, "Improved spectral optical coherence tomography using optical frequency comb," *Optics Express*, vol. 16, no. 6, pp. 4163–4176, 2008.

[76] V. Fathipour, A. Bonakdar, and H. Mohseni, "Advances on Sensitive Electron-Injection Based Cameras for Low-Flux, Short-Wave Infrared Applications," *Frontiers in Materials*, vol. 3, no. 33, 2016.

[77] V. M. Gelikonov, G. V. Gelikonov, and P. A. Shilyagin, "Linear-wavenumber spectrometer for high-speed spectral-domain optical coherence tomography," *Optics and Spectroscopy (English translation of Optika i Spektroskopiya)*, vol. 106, no. 3, pp. 459–465, 2009.

[78] J. Kim, D. T. Miller, E. Kim, S. Oh, J. Oh, and T. E. Milner, "Optical coherence tomography speckle reduction by a partially spatially coherent source," *Journal of Biomedical Optics*, vol. 10, no. 6, Article ID 064034, 2005.

[79] A.-H. Dhalla, J. V. Migacz, and J. A. Izatt, "Crosstalk rejection in parallel optical coherence tomography using spatially incoherent illumination with partially coherent sources," *Optics Expresss*, vol. 35, no. 13, pp. 2305–2307, 2010.

[80] M. Sato, D. Saito, K. Shouji, R. Kurotani, H. Abe, and I. Nishidate, "Ultrathin forward-imaging short multimode fiber probe for full-field optical coherence microscopy," *Optics Communications*, vol. 381, pp. 296–308, 2016.

[81] P. Eugui, D. J. Harper, A. Lichtenegger et al., "Multimode fiber for angle-resolved optical coherence tomography," in *Proceedings of the Bio-Optics: Design and Application*, p. BoM4A.3, San Diego, California, USA, 2017.

14

Study on Quantitative Phase Imaging by Dual-Wavelength Digital Holography Microscopy

Lin Hu, Yunxiu Shui, Yaohui Dai, Haiyu Wu, Gang Zhu, and Yan Yang

College of Mechanical Engineering, Chongqing University of Technology, Chongqing 400054, China

Correspondence should be addressed to Yan Yang; yangyan@cqut.edu.cn

Guest Editor: Xiaowei Li

A dual-wavelength digital holographic microscopy with premagnification is proposed to obtain the object surface measurements over the large gradient. The quantitative phase images of specimens are captured in high precision by the processing of filtering and phase compensation. The phase images are acquired without phase unwrapping, which is necessary in traditional digital holographic microscopy; thereby the proposed system can greatly increase the speed of reconstruction. The results of numerical simulation and optical experiments demonstrated that the reconstructed speed increased by 37.9 times, and the relative error of measurement is 4% compared with the traditional holographic microscopy system. It means that the proposed system can directly acquire the higher quality quantitative phase distribution for specimens.

1. Introduction

Digital holographic microscopy (DHM) is a powerful technology for the measurement of microscopic samples by recording and reconstructing the amplitude and phase of the wave. The dual-wavelength digital holographic microscopy uses two different wavelength lasers to simultaneously record the hologram and numerically reconstruct the phase information according to phase distribution under two wavelengths. As early as 1998, a new method for the extraction of quantitative phase imaging by using partially coherent illumination and an ordinary transmission microscope was proposed by A. Barty et al., which can recover a phase even in the presence of amplitude modulation [1]. A solution for absolute phase measurements was presented by Cuche Etienne et al. [2]. They introduced digital reference wave and phase mask, which were applied in phase-contrast imaging and optical meteorology. This property of holograms offers phase-contrast techniques, which can be used in quantitative 3D imaging. The compensation of the inherent wave front curvature by subtracting the reference hologram in DHM for quantitative phase-contrast imaging is introduced by Ferraro, Pietro, et al. [3]. This simple method can be implemented efficiently under the ideal experimental conditions. The dual-wavelength phase-shifting digital holography that selectively extracts wavelength information from five wavelength-multiplexed holograms is presented by Tahara, Tatsuki, et al. [4]. The color hyperchaotic image encryption method by the DHM and CA encoding algorithm is proposed by Sichuan University, which has great security and robustness [5, 6]. It is an important application about digital holography. The enhanced quantitative three-dimensional measurement system is introduced by JaeYong Lee. It can simplify the configuration by a dual-peak quantum dot wavelength converter and a blue LED [7]. There are great researches for reducing speckle noise and increasing the speed of reconstruction [8–10]. We have successfully achieved the near real-time three-dimensional surface measurement by digital holography [11, 12].

If the optical path difference is less than the equivalent wavelength, the real phase distribution of object can be directly obtained without phase unwrapping by the dual-wavelength digital holographic interferometry. If the optical path difference is greater than the equivalent wavelength, the phase unwrapping process will be simplified by dual-wavelength digital holographic interferometry. The noise immunity and the scope of phase unwrapping algorithms will be improved and expanded, respectively [13–15]. C. J. Mann obtained the measurement results of the surface height over several microns of range by three-wavelength digital holography [16]. The optical path length can be converted to physical thickness, and the sample height information will be provided. The dual-wavelength DHM imaging experiment of a stepped phase plate and the smoke particles were conducted by P. Song et al. [17]. M.K. Kim et al. successfully achieved three-dimensional imaging of resolution plates and cancer cells using dual-wavelength DHM [18]. A phase-imaging technique to quantitatively study the three-dimensional structure of cells by simultaneous dual-wavelength reflection digital holography was presented by A. Khmaladze et al. [19]. It proves that the dual-wavelength DHM allows a faster imaging, which does not rely on the surrounding pixels to correct the phase discontinuities, but simply compares two single-wavelength phases. Dual-wavelength DHM has been widely applied in the fields of surface topography measurement [20–22], cell imaging [23, 24], 3D particle imaging [25, 26], transparent medium physical quantity measurement [27–29], and so on. Due to the limitation of the dual-wavelength DHM technique, more noise will be introduced with the increase in synthesis wavelength and the expanding of the sample range. The accuracy of measurement decreased as the noise increased. The setup of dual-wavelength DHM consists of two laser machines, which makes it difficult to ensure the concentricity of two laser beams in an optical assembly. All the above have the adverse effects on obtaining high precision phase information.

A system of dual-wavelength DHM with premagnification, which can directly and accurately obtain quantitative phase images, is presented. The principle of surface topography measurement and phase unwrapping method based on dual wavelength digital holography is introduced. The effectiveness of the system is verified by computer simulations and optical experiments using the 1951 USAF target and the standard groove object. Compared with the traditional single wavelength DHM, the system can not only obtain the phase information without phase unwrapping, but also get the low noise and high precision quantitative phase images.

2. Experimental Principle

2.1. Principle. In dual-wavelength digital holographic microscopy, two laser beams in different wavelengths from separated laser sources are coupled into one beam, and the optical assemblies are shown in Figure 1. The two lasers were used as coherent light sources. Both beams are collimated by spatial filter. Beam splitters (BS1) divide the beams into the reference and the object arms. The object beam passes through the sample and microscope objective. There are different tilts in orthogonal directions for reference waves of two lasers, which allows us to capture both wavelengths simultaneously. Then the interference pattern between the reference waves and object wave is recorded by a CCD camera.

The dual-wavelength composite digital hologram will be acquired on the CCD, and the interference pattern can be expressed as

$$I_H(x,y) = |O_1|^2 + |R_1|^2 + |O_2|^2 + |R_2|^2 + R_1^*O_1 + R_1O_1^* + R_2^*O_2 + R_2O_2^* \quad (1)$$

where I_H is the intensity of the composite digital hologram, x, y are the coordinates of the holographic plane, $*$ is the complex conjugate, O_1, O_2 are object beams, and R_1, R_2 are reference beams.

The interferometric phase can be extracted by the spatial filtering method and shifted to the center position to perform the Fourier transform. The complex amplitude distribution of the reproducing light field can be obtained as follows:

$$O_i(x,y) = R_i(x,y)\left[R_i^*(x,y)O_i(x,y)\right] \quad i = 1,2 \quad (2)$$

The intensity and the wavefront phase distribution of optical field can be calculated according to the following expressions:

$$I_i(x,y) = O_i(x',y')R_i^*(x',y') \quad (3)$$

$$\Phi_i(x,y) = \arctan\left[\frac{\operatorname{Im} O_i(x',y')}{\operatorname{Re} O_i(x',y')}\right] \quad (4)$$

In order to overcome the issue of phase ambiguity produced by single wavelength approach, a synthetic beat-wavelength is used and expressed as follows:

$$\Phi = \Phi_1(x,y) - \Phi_2(x,y) = 2\pi\frac{nh}{\lambda_1} - 2\pi\frac{nh}{\lambda_2} = 2\pi\frac{nh}{\Lambda} \quad (5)$$

where h is the height distribution of the specimens, which means the optical path difference of the twice of the topography for reflection scheme. nh is the optical path difference of the homogeneous sample in the air. Λ is the equivalent wavelength defined as follows:

$$\Lambda = \frac{\lambda_1\lambda_2}{|\lambda_1 - \lambda_2|} \quad (6)$$

If the optical path difference is less than the equivalent wavelength, the real phase of specimens can be directly obtained without unwrapping. Otherwise, the phase distribution is wrapped between $-\pi$ and $+\pi$. In this case, the package phase difference is compensated to obtain its equivalent phase.

$$\Phi \begin{cases} \Phi_1(x,y) - \Phi_2(x,y) & \Phi > 0 \\ \Phi_1(x,y) - \Phi_2(x,y) + 2\pi & \Phi < 0 \end{cases} \quad (7)$$

In summary, the dual-wavelength digital holographic microscopy to measure the large gradient of specimens can solve the problem of phase unwrapping with single-wavelength digital holographic microscopy. As the range of measurement increases, the noise of the phase distribution in the single-wavelength digital holographic microscope also increases. Therefore, the reasonable choice of wavelength and the method of noise reduction are key points in the dual-wavelength holographic microscopy.

3. Experimental Results

3.1. Experimental Setup. The experiments were conducted in a DHM developed by our team. The schematic diagram and

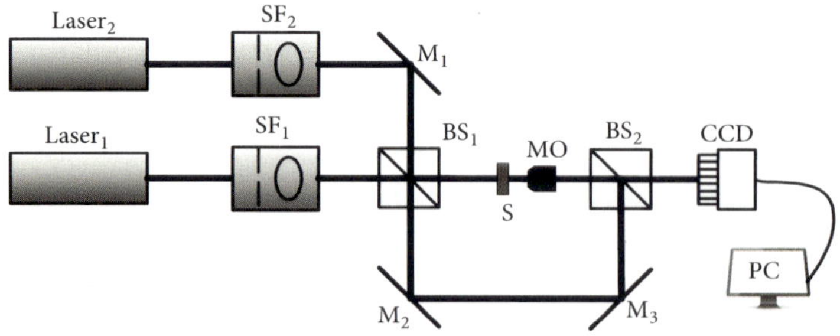

FIGURE 1: The schematic diagram of dual-wavelength digital holographic microscopy. SF, spatial filter; M, mirrors; BS, beam splitters; S, sample; MO, microscope objective; CCD, charge-coupled devices; PC, personal computer.

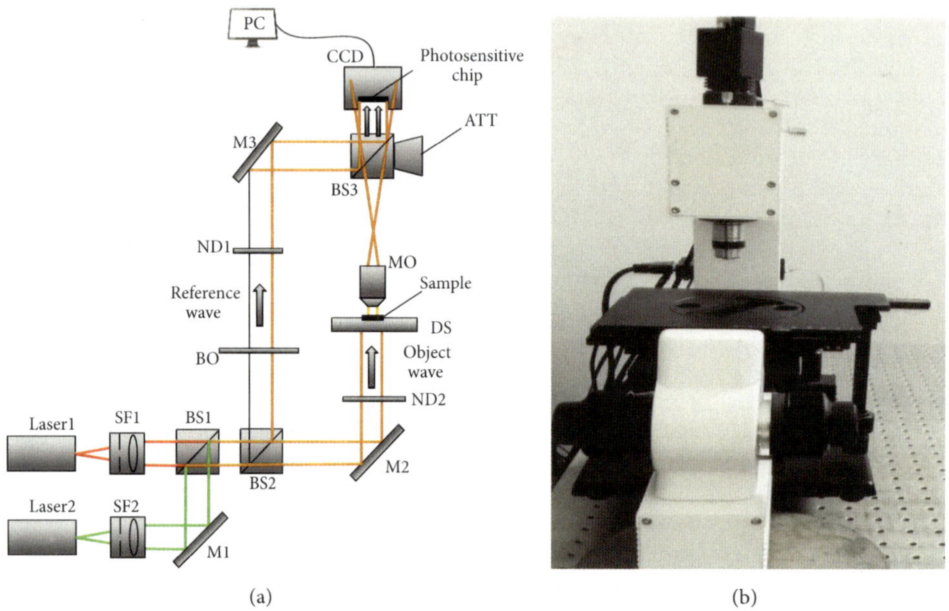

FIGURE 2: (a) The schematic diagram of the dual-wavelength DHM system. SF, spatial filter; BS, beam splitters; M, mirrors; ND, neutral density; DS, displacement stage; S, sample; MO, microscope objective; ATT, adjustable tilting stage; BO, beam obstacle. (b) Real picture of DHM.

the setup are shown in Figure 2, where $\lambda_1 = 632.8\ nm$ and $\lambda_2 = 532\ nm$, the equivalent wavelength, $\Lambda = 3.34\ \mu m$. The beams emitted by the two lasers pass through the beam expanders and collimation systems (SF), the mirror and the cube beam splitting prism (BS). Then the beam is divided into two beams by beam splitter; one is object wave, and the other is the reference wave. In the premagnification optical path, the object wave, which is reflected by mirror, passes through the microscope objective (MO, 10×, $NA = 0.25$) and the sample. On the other hand, the reference wave passes through beam obstacle and neutral density. It should be noted that beam obstacle is turned on so that the optical path is the one of off-axis digital holography (in this experiment, the beam obstacle is turned on); otherwise, it will become in-line digital holography. The two beams are combined by the cube splitting prism. Finally, the interference hologram is recorded by CCD (SENTECH STC-SBS241POE; the pixel area is $1296 pixel \times 966 pixel$). It is worth noting that the dual-wavelength preamplification digital holographic microscopy systems with large numerical aperture microscope objectives can acquire more sample information, which can be compressed to amplify the CCD acceptable sampling frequency [30].

3.2. Numerical Simulation. In order to illustrate that the phase image of sample whose optical path difference is less than the equivalent wavelength can be obtained directly and quickly by dual-wavelength DHM; a numerical simulation experiment is carried out. The computer-generated cone is a phase object with a maximum phase height of $2.5\mu m$. The simulation parameters of cone are as follows: $\lambda_1 = 632.8\ nm$, $\lambda_2 = 532 nm$, in which each single wavelength is less than the optical path difference. According to (6), the equivalent wavelength is $3.34\mu m$, which means that the dual-wavelength phase unwrapping will obtain a continuous phase distribution. Based on the designed height, the wrapping phase distribution of single-wavelength and the phase distribution of equivalent wavelength are shown in Figure 3, respectively.

To demonstrate the advantages and compare the unwrapping speed of the dual-wavelength DHM and traditional

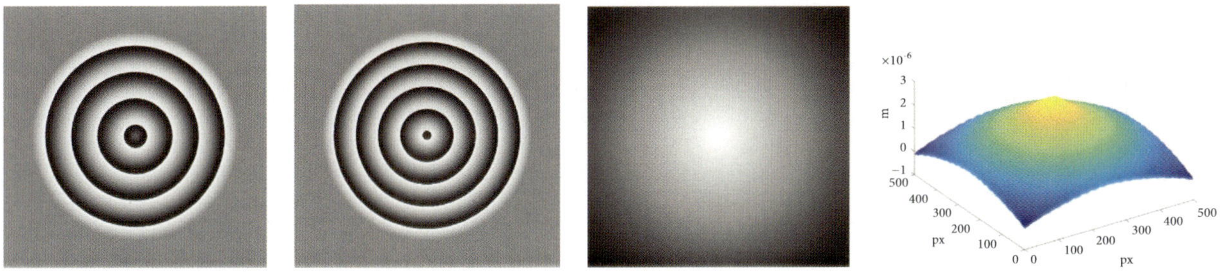

(a) (b) (c) (d)

FIGURE 3: Simulation results for slope: (a) phase distribution for 632.8 *nm*, (b) phase distribution for 532 *nm*, (c) phase distribution for equivalent wavelength, and (d) 3D view of dual-wavelength phase distribution.

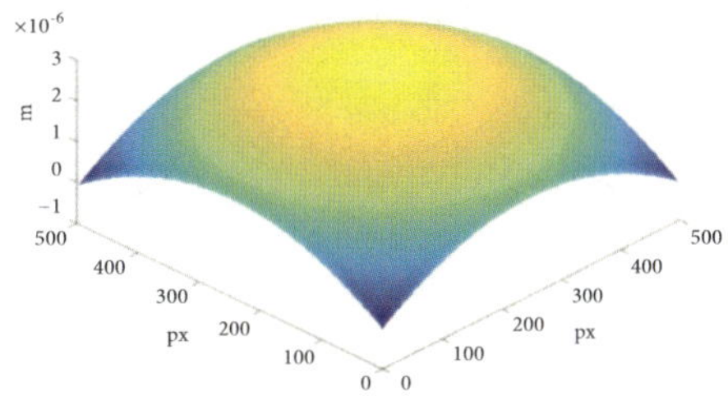

FIGURE 4: 3D view of dual-wavelength phase distribution.

single-wavelength DHM, a 632.8 *nm* wrapped phase image is chosen to obtain the final height distribution by unwrapping procedure, which is named the quality-guided phase unwrapping algorithm. And the results for 632.8 *nm* through the quality-guided phase unwrapping algorithm [31, 32] are shown in Figure 4.

The height distributions of cone by dual-wavelength DHM and quality-guided phase unwrapping method are compared. And the results are shown in Table 1. The results show that both of them can accurately obtain the height distribution of the cone, but in terms of time, the dual-wavelength holographic microscopy in the system is faster. The speed of reconstruction by the proposed system is 37.9 times more than the value of the quality-guided phase unwrapping method.

3.3. Optical Experiment

3.3.1. Quantitative Phase Imaging of USAF 1951 Target. Based on the dual-wavelength phase imaging system, the surface of an USAF 1951 target was measured. At the same time, the target was also measured by the scanning three-dimensional profiler (NanoMap 500LS, AEP Technology Inc., USA) and the results showed that the height was about $0 \sim 100nm$. Since the sample height is much smaller than the single-wavelength, the phase distribution of the sample can be obtained directly. Figures 5 and 6 present the measurement results using traditional single-wavelength and dual-wavelength, respectively.

In the reconstruction process, filtering and secondary phase distortion compensation processing are performed [33–35]. The hologram is spectrally separated by Fourier transform. The microscope objective lens will introduce a secondary phase distortion factor in the system. In order to obtain the real phase, two holograms are recorded, which are the reference holograms for the background of the hologram and the hologram of measured sample. The real phase can be obtained by subtracting the two reconstruction phases of hologram.

The comparison of the height distribution curve along the middle symmetrical line of phase with different wavelength (in Figures 5(a), 5(b), and 6(a)) is shown in Figure 7. The height distribution along the middle symmetrical line of sample was scanned by NanoMap 500LS and its absolute height was measured to be 54 *nm*. The average absolute height and the relative error between the measurement result and scanning result using three measurement methods are shown in Table 2.

Table 2 shows that the result of the equivalent wavelength is much closer to scanning value. From the experimental results, it can be seen that both the traditional DHM and the dual-wavelength DHM can obtain the phase distribution of the object surface. It is not hard to see that the system of dual-wavelength DHM can accurately obtain the three-dimensional morphology distribution of the object surface. The results prove that the measurement accuracy and stability of the proposed system are better than the traditional single wavelength.

3.3.2. Quantitative Phase Imaging of Groove. The transparent groove standard plate, which is artificially designed, was used as an experimental sample to measure its three-dimensional

Table 1: Comparison of two methods.

Method	Maximum phase (μm)	Time (s)
Quality-guided phase unwrapping	2.5749	716.61
Dual-wavelength DHM (in the system)	2.5971	18.93

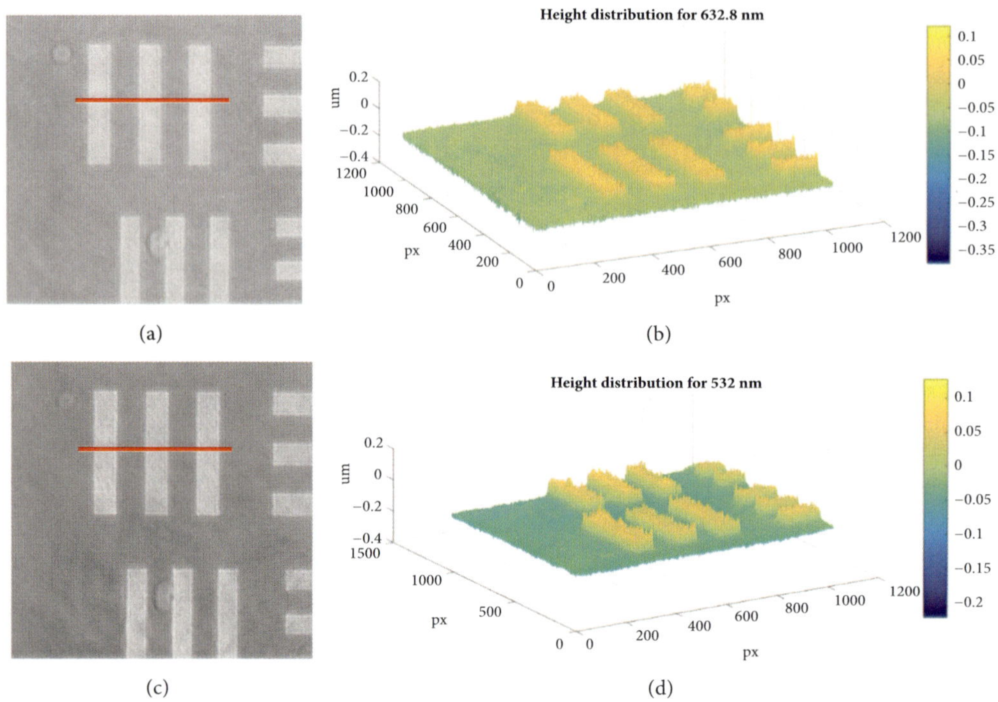

Figure 5: Experimental results with single wavelength measurement. (a) Phase image for 632.8 nm, (b) height distribution for 632.8 nm, (c) phase image for 532 nm, and (d) height distribution for 532 nm.

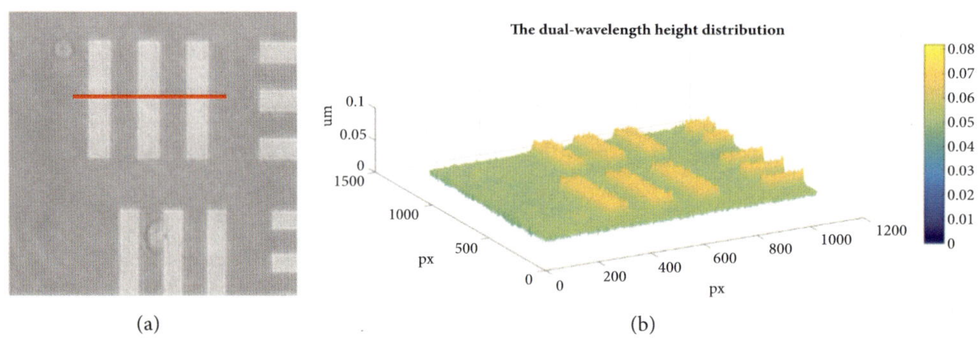

Figure 6: Experimental results for equivalent wavelength. (a) Phase image for equivalent wavelength, (b) 3D view of dual-wavelength height distribution.

Table 2: The comparison of the height distributions.

Wavelength(nm)	Absolute height (nm)	Relative error (compared with the scanned value)	Variance
632.8	61.8	14.4%	1.925×10^{-4}
532	99.3	83.8%	3.188×10^{-4}
equivalent wavelength	47.5	13.8%	1.504×10^{-5}

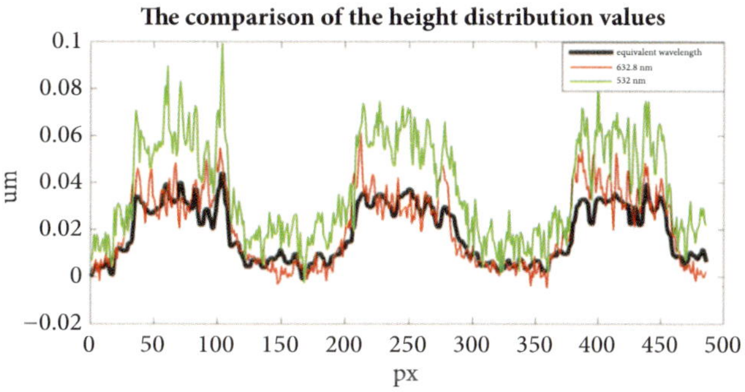

FIGURE 7: The comparison of the height distribution values along the middle symmetrical line of sample in pane Figures 5(a), 5(b), and 6(a).

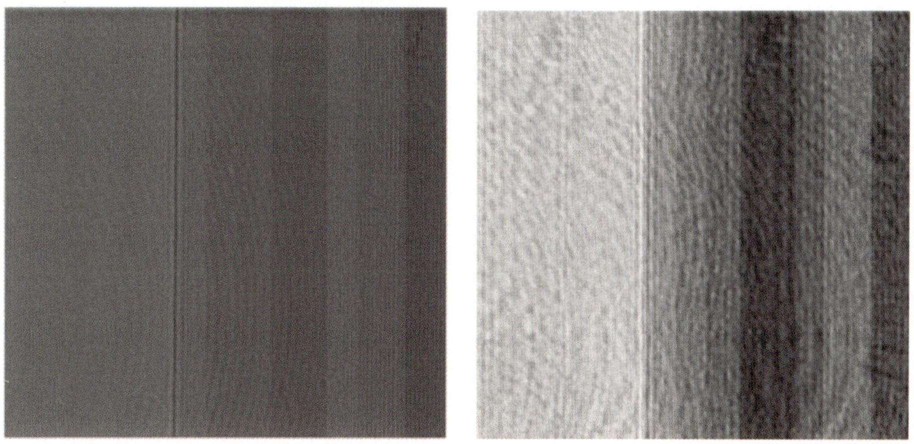

(a) (b)

FIGURE 8: Experimental results with single wavelength measurement. (a) Phase image for 632.8 nm, (b) phase image for 532 nm.

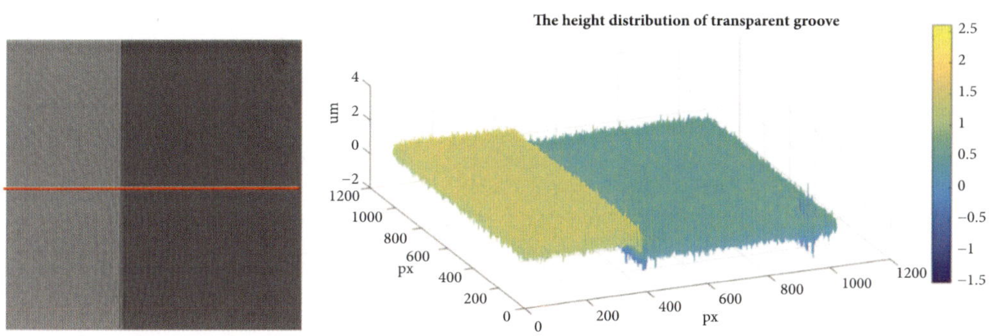

(a) (b)

FIGURE 9: Results for equivalent wavelength. (a) Phase distribution, (b) three-dimensional height distribution.

appearance. The width of groove is 1 mm and the depth is 9 μm. Figure 8(a) shows the phase distribution measured by wavelength, 632.8 nm. Figure 8(b) shows the phase distribution measured by wavelength, 532 nm. It is clear to see that there is phase folding in the phase distribution at each single wavelength. The phase distribution measured by the dual wavelength interference is shown in Figure 9.

The height distribution of the middle symmetrical line of the groove (Figure 9(a)) is obtained by the median filtering, which is shown in Figure 10. According to the formula of the surface profile distribution and phase distribution (in (5)), the height difference between the base and the bottom of the groove $b = 0.936$ μm. The relative error between the actual groove height and experimental result is 4%, which is in good agreement with the theoretical values.

4. Conclusions

In order to solve the quantitative phase imaging problem of traditional digital holographic microscopy system for the large gradient of object surface, a dual-wavelength preamplification digital holographic microscopy optical system is

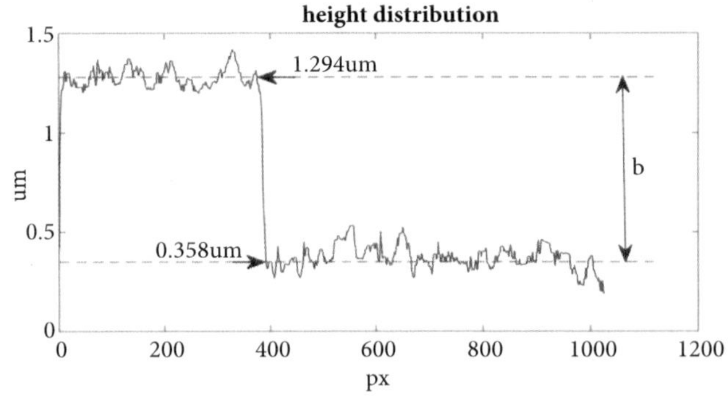

FIGURE 10: Height distribution values along the middle symmetrical line of groove in Figure 8(a).

proposed. The numerical simulation and experiments are carried out and the effectiveness of the system is verified. The experimental results of the sample phase imaging of the traditional DHM and dual-wavelength preamplification DHM system are compared. The results show that the experimental system can effectively overcome the limitations of the single-wavelength method in the imaging of complex surface objects, increase the observation speed, and simplify the reconstruction process, which further validates the effectiveness of the experimental system for quantitative phase imaging.

Conflicts of Interest

The authors declare that they have no conflicts of interest.

Acknowledgments

This work has been supported by the National Natural Science Foundation of China (Nos. 11272368 and 51875068), the International Cooperation Special Project in Science and Technology of China (No. 2015DFR70480), the Chongqing Municipal Education Commission's science and technology research project (KJ1600929), and the Graduate Innovation Foundation of Chongqing University of Technology (No. ycx2018216).

References

[1] A. Barty, K. A. Nugent, D. Paganin, and A. Roberts, "Quantitative optical phase microscopy," *Optics Expresss*, vol. 23, no. 11, pp. 817–819, 1998.

[2] E. Cuche, F. Bevilacqua, and C. Depeursinge, "Digital holography for quantitative phase-contrast imaging," *Optics Expresss*, vol. 24, no. 5, pp. 291–293, 1999.

[3] A. Finizio, C. Magro, G. Pierattini et al., "Compensation of the inherent wave front curvature in digital holographic coherent microscopy for quantitative phase-contrast imaging," *Applied Optics*, vol. 42, no. 11, p. 1938, 2003.

[4] T. Tahara, R. Mori, S. Kikunaga, Y. Arai, and Y. Takaki, "Dual-wavelength phase-shifting digital holography selectively extracting wavelength information from wavelength-multiplexed holograms," *Optics Express*, vol. 40, no. 12, pp. 2810–2813, 2015.

[5] X. Li, Y. Wang, and Q. H. Wang, "Modified integral imaging reconstruction and encryption using an improved SR reconstruction algorithm," *Optics & Lasers in Engineering*, pp. 112–162, 2019.

[6] X. Li, D. Xiao, and Q.-H. Wang, "Error-free holographic frames encryption with CA pixel-permutation encoding algorithm," *Optics and Lasers in Engineering*, vol. 100, pp. 200–207, 2018.

[7] S. Jeon, J. Lee, J. Cho, S. Jang, Y. Kim, and N. Park, "Wavelength-multiplexed digital holography for quantitative phase measurement using quantum dot film," *Optics Express*, vol. 26, no. 21, p. 27305, 2018.

[8] H. Cho, D. Kim, Y. Yu, W. Jung, and S. Shin, "Dual-wavelength Digital Holography Microscope for BGA Measurement Using Partial Coherence Sources," *Journal of the Optical Society of Korea*, vol. 15, no. 4, pp. 352–356, 2011.

[9] J. Cho, J. Lim, S. Jeon et al., "Dual-wavelength off-axis digital holography using a single light-emitting diode," *Optics Express*, vol. 26, no. 2, pp. 2123–2131, 2018.

[10] V. Bianco, P. Memmolo, and M. Leo, "Strategies for reducing speckle noise in digital holography," *Light Science & Applications*, p. 7, 2018.

[11] Z. Gang, Z. Zhixiong, W. Huarui, and Y. Yan, "Near real-time digital holographic microscope based on GPU parallel computing," in *Proceedings of the SPIE- 2017 International Conference on Optical Instruments and Technology: Optical Systems and Modern Optoelectronic Instruments*, p. 10616, 2018.

[12] Z. Zhao, "A holographic display method combing amplitude of real picture and phase of hologram," *Pro. SPIE*, vol. 10255, Article ID 1025503, 2017.

[13] J. Kühn, J. Conchello, C. J. Cogswell et al., "Real-time dual-wavelength digital holographic microscopy for extended measurement range with enhanced axial resolution," *Biomedical Optics International Society for Optics and Photonics*, 2008.

[14] Z. Gao, Z. Jiang, Y. Wang et al., "Phase imaging for photorefractive holographic gratings with dual-wavelength digital holography," in *Proceedings of the Photonics Asia*, p. 85561G, Beijing, China.

[15] Y. Zeng, F. Wang, H. Lei, X. Hu, and X. Hu, "Surface profile measurement of microstructures based on dual-wavelength digital microscopic image-plane holography," *Guangxue Xuebao/Acta Optica Sinica*, vol. 33, no. 10, 2013.

[16] C. J. Mann, P. R. Bingham, V. C. Paquit, and K. W. Tobin, "Quantitative phase imaging by three-wavelength digital holography," *Optics Express*, vol. 16, no. 13, pp. 9753–9764, 2008.

[17] S. Peiyong, *3D Topography Recovery of Objects Based on Digital Holographic Technology*, Nanjing Normal University, 2011.

[18] D. Parshall and M. K. Kim, "Digital holographic microscopy with dual-wavelength phase unwrapping," *Applied Optics*, vol. 45, no. 3, pp. 451–459, 2006.

[19] A. Khmaladze, M. Kim, and C.-M. Lo, "Phase imaging of cells by simultaneous dual-wavelength reflection digital holography," *Optics Express*, vol. 16, no. 15, pp. 10900–10911, 2008.

[20] Y. Kou, E. Li, J. Di, Y. Zhang, M. Li, and J. Zhao, "Surface morphology measurement of tiny object based on dual-wavelength digital holography," *Zhongguo Jiguang*, vol. 41, no. 2, 2014.

[21] D. G. Abdelsalam and D. Kim, "Two-wavelength in-line phase-shifting interferometry based on polarizing separation for accurate surface profiling," *Applied Optics*, vol. 50, no. 33, pp. 6153–6161, 2011.

[22] W. Yunxin, W. Dayong, Y. Yishu et al., "Application and Analysis in the Biomedicine Field Using Digital Holographic Technology," *Chinese Journal of Lasers*, vol. 41, no. 2, Article ID 002, 2014.

[23] A. Khmaladze, E. Seeley, J. Jasensky et al., "Dual wavelength digital holographic imaging of cells with phase background subtraction," 2012.

[24] D. G. Abdelsalam, R. Magnusson, and D. Kim, "Single-shot, dual-wavelength digital holography based on polarizing separation," *Applied Optics*, vol. 50, no. 19, pp. 3360–3368, 2011.

[25] S. Grare, S. Coetmellec, D. Allano, G. Grehan, M. Brunel, and D. Lebrun, "Dual wavelength digital holography for 3D particle image velocimetry," *Journal of the European Optical Society: Rapid Publications*, vol. 10, no. 3, 2015.

[26] Z. Yanan, L. Junshen, L. Yuan et al., "Three-Dimensional Displacement Tracking Technique of Particle Based on Digital Holographic Microscopy," *Chinese Journal of Lasers*, pp. 44-12, 2017.

[27] B. Rappaz, P. Marquet, E. Cuche, Y. Emery, C. Depeursinge, and P. J. Magistretti, "Measurement of the integral refractive index and dynamic cell morphometry of living cells with digital holographic microscopy," *Optics Express*, vol. 13, no. 23, pp. 9361–9373, 2005.

[28] D. Boss, P. T. So, E. Beaurepaire et al., "Dual-wavelength Digital Holography for quantification of cell volume and integral refractive index (RI)," in *Proceedings of the European Conferences on Biomedical Optics*, p. 808608, Munich, Germany.

[29] M. Frómeta, G. Moreno, J. Ricardo et al., "Optimized setup for integral refractive index direct determination applying digital holographic microscopy by reflection and transmission," *Optics Communications*, vol. 387, pp. 252–256, 2017.

[30] Z. Yanan, L. Hai, L. Yuan et al., "Compensation of the Phase Aberrationsin Digital Holographic Microscopy Based on Reference Lens Method," *Acta Photonica Sinica*, vol. 47, no. 1, pp. 216–222, 2018.

[31] Y. Lu, X. Wang, and X. Zhang, "Weighted least-squares phase unwrapping algorithm based on derivative variance correlation map," *Optik - International Journal for Light and Electron Optics*, vol. 118, no. 2, pp. 62–66, 2007.

[32] S. Li, W. Chen, and X. Su, "Reliability-guided phase unwrapping in wavelet-transform profilometry," *Applied Optics*, vol. 47, no. 18, pp. 3369–3377, 2008.

[33] Z. Wenjing, X. Qiangsheng, and Y. Yingjie, "Compensation for the Quadratic Phase Aberration in Off2axis Digital Micro2holography," *Acta Photonica Sinica*, vol. 38, no. 8, pp. 1972–1976, 2009.

[34] Z. Wang, Y. Chen, and Z. Jiang, "Dual-wavelength digital holographic phase reconstruction based on a polarizationmultiplexing configuration," *Chinese Optics Letters*, vol. 14, no. 1, pp. 31–35, 2016.

[35] S. Liu, "Dynamic phase-contrast microscopy of living cells by digital holography," *Journal of Beijing University of Aeronautics & Astronautics*, vol. 38, no. 9, pp. 1186–1188, 2012.

15

Targeted Photodynamic Therapy: A Novel Approach to Abolition of Human Cancer Stem Cells

Anine Crous, Elvin Chizenga, Natasha Hodgkinson, and Heidi Abrahamse

Laser Research Centre, Faculty of Health Sciences, University of Johannesburg, P.O. Box 17011, Doornfontein, Johannesburg 2028, Room 5308, John Orr Building, South Africa

Correspondence should be addressed to Heidi Abrahamse; habrahamse@uj.ac.za

Academic Editor: Giancarlo C. Righini

Cancer is a global burden that has prompted extensive research into prevention and treatment, over many decades. Scientific studies have shown that subset of cells within a tumour, known as cancer stem cells (CSCs), are responsible for tumourigenesis, metastasis, drug resistance, and recurrences. CSCs have characteristic features of enhanced self-renewal, proliferation, and limited but multidirectional differentiation capacity. The discovery of CSCs has initiated extensive research into novel cancer treatment regimes. Evidence indicates that CSCs are resistant to conventional chemo- and radiation therapy leading to treatment failures, cancer metastasis, secondary cancer formation, and relapse. Because of the observed phenomena in the course of cancer prognosis, a need for treatment modalities targeting CSCs is important. Photodynamic therapy (PDT) is a clinically approved, minimally invasive, therapeutic procedure that can exert a selective cytotoxic activity toward cancerous cells while reducing toxicity to normal cells. It uses a photosensitizer (PS) that becomes excited when subjected to light at a specific wavelength, and the PS forms reactive oxygen species (ROS) killing malignant cells. Currently, PDT is being investigated as a target specific treatment for CSCs by the addition of carrier molecules and antibody conjugates bound to the PS. Targeted PDT (TPDT) may be able to not only eradicate the tumour mass but kill CSCs as well.

1. Introduction

Cancer is a global burden, affecting people from all socioeconomic backgrounds, in all geographic regions of the world. This disease is responsible for 1 in 7 deaths worldwide and is predicted to affect 21.6 million people and 13 million deaths by the year 2030 [1]. To effectively combat cancer in modern medicine, much attention is being directed toward cancer prevention and treatment, while concurrently investigating causes of cancer and cancer dynamics [2]. Cancer is caused by a genetic mutation, on a cellular level, where errors in DNA instructions lead to cells evading normal cell cycle functions specifically cell cycle arrest [3]. Along with cancer cells evading cell cycle arrest cancer can have the ability to metastasize and recur even after conventional therapy. Recent evidence has shown that a subset of cells called cancer stem cells (CSCs) can initiate tumour formation [4]. This small subpopulation of cells resides within the tissue mass and has characteristics that include self-renewal, differentiation, and tumourigenicity when transplanted into an animal host. A number of cell surface markers such as CD44, CD24, and CD133 are often used to identify and enrich CSCs. The clinical relevance of CSCs has been strengthened by emerging evidence demonstrating that CSCs are resistant to conventional chemotherapy and radiation treatment and that CSCs are very likely to be the origin of cancer metastasis. CSCs are believed to be an important target for novel anticancer drug discovery [5].

Photodynamic therapy (PDT) is a well-documented therapy that has emerged as an effective treatment modality of cancer [6]. It involves the use of a nontoxic dye, or photosensitizer (PS), which is effectively activated by light at a specific wavelength [7]. This then generates singlet oxygen, or other reactive oxygen species (ROS), causing degradation of cellular components by damaging different biomolecules including proteins, DNA, and lipids leading to cancer cell destruction and death [8, 9]. According to Abrahamse et al. (2016) optimum standards of a good PS include being a single pure compound, having a strong absorption peak in the red to near infrared spectral region (between 650 and 800 nm), possessing a substantial triplet quantum yield, and having no dark toxicity and a relatively rapid clearance rate from normal tissues [6]. Along with a PS having optimum standards, what distinguishes PDT from conventional therapies such as chemotherapy is its affinity for malignant cells that results in selective cytotoxic effects, killing cancer cells with minimal effects on normal cells, unlike chemotherapy and genotoxic drugs that destroy the cancer cells and its surrounding normal tissue [10].

PDT can further be enhanced by making it target definitive. This can be done with the addition of nanoparticles and monoclonal antibodies to enhance tumour selectivity, PS absorption, and cell specificity. Nanoparticles (NPs) are currently being used for tumour imaging *in vivo*, profiling of cancer biomarkers, and targeted drug delivery [11]. They have also been used to increase solubility and improve the transcytosis of PSs through membranes and endothelial barriers. The combined use of NPs and monoclonal antibodies concurrently has potential for a very high therapeutic efficacy. This review will focus on the aspect of the stem cell theory, CSCs, and the potential role of targeted PDT using nanoparticles and monoclonal antibodies.

2. Cancer

The National Cancer Institute defines cancer as a collection of diseases in which cells of the body begin to grow uncontrollably and spread to other tissues. Although there are different theories of how cancer develops and many derangements involved, they are all attributed to DNA damage that alters the normal cellular growth pathways. Mutations in genes that control normal cellular functions result in a dysregulated growth of cells, followed by uncontrolled replication of cells that finally outgrow normal cells to form a tumour mass [12]. Biologically, cancer is a genetic disease but not, with a few exceptions, inheritable. Only a few cases where mutations are inherited as a single gene disorder can cause cancer.

2.1. Molecular Biology of Cancer. Since the beginning of the era of oncology, a multitude of research has been done to understand the molecular mechanisms of cancer. Hippocrates was among the first scientists to describe the different forms of tumours and called them "cancer" that described the crablike shape of the tumour invading normal tissue. Research is still ongoing in hopes to elucidate on cancer biology at the molecular level and new discoveries are documented on regularly [13].

A deep understanding of the molecular mechanisms of cancer is, however, essential for the development of novel therapeutic modalities. As described previously, cancer is a result of dysregulated cell growth. In a normal cell, all cellular processes from cell birth, growth, proliferation, division, and death respond to stimulatory or inhibitory signals transmitted from regulatory genes and signalling molecules, both intracellularly and extracellularly [13, 14]. Thus, the cell cycle is an extensively regulated process. In addition to the key proteins in cell cycle regulation, Cyclins, and Cyclin Dependent Kinases, there are genes that normally stimulate cell growth called proto-oncogenes which when mutated become dominant-acting stimulatory genes that cause cancer. These cancer causing genes are, therefore, termed oncogenes. A cell also has inhibitory genes that suppress cell growth, called tumour-suppressor genes, which code for regulatory proteins, the most common of which are p53 and the Retinoblastoma [14]. When a cell suffers a dysfunction of these proteins, unregulated growth issues occur. Mutations in regulatory genes may result in the activation of oncogenes and downregulation of genes that suppress tumour formation; this alongside other mutations inactivating genes responsible for maintaining genomic integrity initiates cancer [15].

Table 1: Characteristics of cancer cells.

Impairment of cell cycle control
Replicative senescence
Accumulated mutations
Dedifferentiation
Lack of Contact inhibition
Formation of blood vessel network
Invasiveness
Metastasis

2.2. Characteristics of Cancer Cells. Biologically, cells in the same tumour differ in virtually all phenotypic features. Cancer cells, at different levels of development arising from the same clone of cells, have variations in morphology, gene expression, metabolism, motility, proliferative, immunogenic, and metastatic potential. Cancer begins at a cellular level in a multistep process involving the activation of cells that produce cancer cells commonly referred to as Cancer Stem Cells (CSCs) [16, 17]. These cells lose their cellular features of terminal differentiation (dedifferentiation) and increase in fraction of tissue outgrowing normal cells alongside faulty tissue repair. Finally, for a cancer cell to proliferate it overrides the process of replicative senescence to become immortal and obtains a supply of nutrients and oxygen that maintains the high rate of proliferation. This phenomenon is achieved by the cells ability to secrete a protein, called vascular endothelial growth factor in response to hypoxia. A cancer cell also secretes its own hormones and signal transduction molecules [14]. The molecules that are not expressed in normal cells are a characteristic feature that makes it possible to target cancer cells for drug delivery. These and other characteristics of cancer cells are summarised in Table 1.

3. Cancer Stem Cells

3.1. The CSC Hypothesis. The cancer stem cell hypothesis is grounded by the following conditions: a tumour is defined, on a cellular level, by uncontrolled proliferation caused by the addition of random mutations in critical genes that control cell growth. Although gene mutations cause cancer, it has been reported that malignancy can be caused by a

Table 2: Specific CSC markers and function [adapted from [25]].

CSC	Marker	Function
Glioma	CD 49f	Cell adhesion
Ovary	CD 117	SC-factor receptor
Liver	CD 13	Kidney disease marker
Lung	CD 56	Cell adhesion
Renal	CD105	Co-receptor TGF-β
Neuroblastoma	CD 114	Colony stimulating factor 3 receptor
Gastric	CD 54	Cell adhesion
Leukaemia	TIM-3	Immune checkpoint receptor
	CD 96	T-cell specific receptor
	CD 9	Cell adhesion
Breast	CD 55	Complement inhibitor
Melanoma	CD 20	B-cell lineage
	ABCB 5	ABC transporter

subpopulation (SP) of cells found within the tumour, called CSCs, thought to be responsible for tumorigenesis, tumour maintenance, tumour spread, and tumour relapse [18]. Adult stem cells (ASCs) are responsible for tissue renewal and repair. Stem cell characteristics include their ability for self-renewal without loss of proliferation capacity; they are perdurable and drug resistant and capable of differentiating into any type of adult cell in the body due to the influence of micro environmental and some other factors. Asymmetric division allows for stem cell (SC) survival giving one daughter cell the ability for self-renewal and the other differentiation and proliferation capacity. Cancer stem cells share characteristics similar to normal SCs and express SC representing markers [19]. These CSCs make up as few as 1% of the cells in a tumour, making them difficult to detect and study. Like normal SCs, CSCs have a number of properties permitting them to survive traditional cancer chemotherapy and radiation therapy. These cells express high levels of ATP-binding cassette drug transporters, providing for a level of resistance, are relatively quiescent, and have higher levels of DNA repair and a lowered ability to enter apoptosis [20]. The mechanism by which CSCs originate is (either) by normal SCs that underwent genetic and epigenetic changes and (or) by dedifferentiation from somatic tumour cells [19].

3.2. CSC Identification and Enrichment. The first evidence of CSCs was seen in a study conducted on acute myeloid leukaemia in 1994. This SP of cells was transplanted into immunocompromised mice. The cells were enriched using surface markers (CD34+/CD38−) and were identified as tumour initiating [21]. Cancer stem cells have been identified in solid tumours, including breast, brain, colon, pancreas, lung, prostate, melanoma, and glioblastoma, all capable of inducing malignancy in nonobese diabetic/severe combined immune deficient (NOD/SCID) mice [22, 23]. A variety of cell surface markers have been used to isolate and enrich CSCs. The expressions of CSC surface markers are tissue type and tumour type-specific. Enhanced identification and targeting of CSCs involve the use of multiple and a combination of surface markers [24]. Kim et al. (2017) conducted a study where 40 CSC surface markers were classified into 3 different categories, depending on their expression on human embryonic SCs, ASCs, and normal tissue cells. More than 80% of the markers were shown not to be expressed on normal tissue cells, making the CSC surface markers useful therapeutic targets against CSCs due to their low cross-reactivity to normal tissue cells [25]. The similarity between normal SC and CSC surface markers suggests that CSCs predominantly originate from normal SCs via the accumulation of epigenetic and genetic alterations [26]. Table 2 highlights some of the CSC markers, and the markers are specific to the type of CSC, as well as their function.

3.3. CSCs and Conventional Cancer Therapy. The role CSCs play in cancer relapse and metastasis is seen in their ability to be chemo- and/or radiation-therapy resistant (Figure 1). This has been proven by different researchers, whereby the proportion of cell SPs was examined before and after chemotherapy. The SP consisting of CSCs had a significant increase in their mean value after treatment [5]. However, controversial results have also been reported, where two patients being treated for CSC therapy resistance were assessed. The radiation resistance of breast CSCs was tested using early passage, patient-derived xenografts from two separate patients. The CD44+CD24−/low Lineage− and ALDH+ breast CSCs from one patient showed depletion of the SP of cells after radiation therapy as well as decreased tumour sphere frequency and tumorigenic capacity. In contrast, CSCs from the other patient displayed enrichment after irradiation and resistance to therapy [27].

4. Photodynamic Therapy

Photodynamic therapy is a well-studied therapy for cancer. It uses PSs (nontoxic dyes) that are activated by absorption of visible light to initially form the excited singlet state, followed by transition to the long-lived excited triplet state. This triplet state can undergo photochemical reactions, in the presence of oxygen, to form ROS (including singlet oxygen) that can destroy cancer cells [6]. The therapeutic effect of PSs lies in

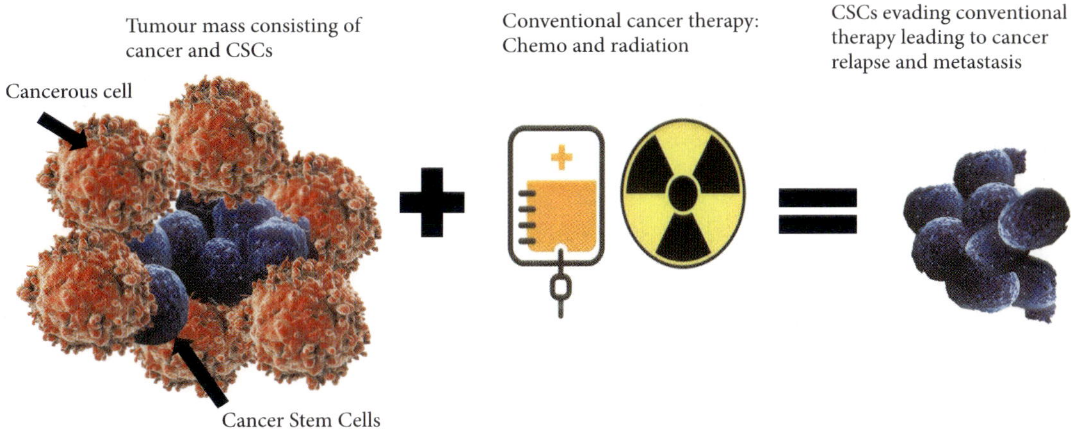

FIGURE 1: CSCs and conventional cancer therapy. Chemo- and radiation therapy fail to eradicate CSCs, which can lead to cancer relapse and metastasis.

their binding capacity, as they preferentially bind to intracellular organelles causing photo-oxidative damage to proteins and lipids that reside within a few nanometres of the PS binding sites [28]. Selective cell destruction of PDT is achieved through its enhanced permeability retention (EPR), seen in malignant cells, minimizing destruction of noncancerous tissue [29]. Other theories document localization to increased expression of certain receptors on tumour cells, decreased intratumoural pH, or tumour-associated macrophages that phagocytise PS molecules [30]. Depending on the structural characteristics of the PS it will be localized in different organelles. It has been shown that overall charge, charge distribution, lipophilicity, and overall structure predominantly determine cellular uptake and subcellular localization of a PS and ultimately determine its therapeutic effect [31].

4.1. PDT and Cancer. Photodynamic therapy proves to be operative as an anticancer treatment as well as improving patient survival. Efficacy is seen when PDT is part of a multimodal approach and is used as the first-line treatment for premalignant or early disease and as stand-alone palliative treatment. Even though PDT shows great potential, there are still some limitations that prevent a firm position for PDT in the standard care regimen of cancer [21]. A drawback commonly seen is decreased efficacy of PDT for larger lesions, due to inadequate tissue penetration of light or PS [32]. Besides larger lesions, PDT is also not indicated for metastasizing tumours. Metastasis remains one of the largest challenges in cancer therapy and PDT is no exception. Tumour recurrence is often reported where inadequate tumour eradication is noted. This is said to be due to insufficient penetration, but also the presence of PDT resistant tumour tissues [33]. The limitations of PDT mentioned above can be ascribed to CSCs having characteristics significant to PDT evasion, such as metastasis and drug evasion through their multidrug efflux pumps. This has been proven in a study conducted by Morgan et al. (2010), where a SP of cells containing the ATP-dependent transporter ABCG2 gene, a multidrug resistant pump, was responsible for initiating tumour regrowth after PDT treatment. The ABCG2 may lower intracellular levels of substrate PS below the threshold for cell death in tumours treated by PDT, leaving resistant cells to repopulate the tumour [34]. Hence the need to revolutionise PDT by making it more target specific.

4.2. Targeted Photodynamic Therapy Using Immunoconjugates and Carrier Molecules. In cell biology, the nucleus is shown to be the most hypersensitive intracellular organelle and hence therapies targeting the nucleus tend to be more effective than those that are limited to the cytoplasm [35]. To achieve a targeted drug delivery system, suitably engineered biomolecules have been the field of focus for oncology researchers. A cell targeting biomolecule can be synthesized by combining two or more particles each having specific functions to target biological components of living systems, including antigens and molecules inside, or on the surface of cells.

Studies have been focused on developing antibody-directed phototherapy (ADP), where antibody conjugation is used to deliver PSs to the tumour via tumour-associated surface markers. Targeting improves both specificity and efficiency and overcomes some of the current limitations of nontargeted PDT. The PS, when passing into the tumour microenvironment, can diffuse into cells or reside in the extracellular/cell surface environment depending on their physical properties. This in turn may also depend on the antibody. Photosensitizers that are hydrophobic in nature may be able to cross cellular membranes directly along their concentration gradient, and PS-drugs that have associated with lipid binding proteins such as low-density lipoprotein or human serum albumin can get taken up, specifically via cell surface receptors. Antibody-directed phototherapy refines this receptor delivery by using antibodies (Abs) to deliver PS-drugs to cell surface markers overexpressed on tumour cells; these can then specifically induce ADP-drug internalisation [36]. This theory was proven by Vrouenraets et al., where the PS aluminium (III) phthalocyanine tetrasulfonate [AlPc(SO3H)4] a hydrophilic compound was used. This PS in its free form does not reach the critical intracellular target, making it ineffective. A monoclonal antibody was conjugated to the PS and had shown preservation of integrity and immunoreactivity and full stability [37]. Internalisation of this ADP-drug showed increased cytotoxicity and selective tumour targeting making it an ideal approach to enhance PDT. Another form of ADP is Near Infrared-Photoimmunotherapy (NIR-PIT), which is cancer therapy utilizing an antibody-photoabsorber conjugate (APC) and specifically near infrared light (NIR-light) to activate the PS. The significance of using a PS that is activated by NIR is the properties of the light being harmless to surrounding normal tissue and its penetration depth of 1-2 cm, making it specifically useful on surface tumours [38]. This treatment also exploits the overexpression of specific antigens on a variety of cancer cells, such as mesothelin. The Ab directed against mesothelin (hYP218) shows a high binding affinity between the Ab and antigen pair, supporting highly selective cell killing when applied as an ADP. A study that demonstrated the use of NIR-PIT was conducted by Nagaya et al. (2016), where they used a PS (IR700) conjugated to hYP218 on an epidermoid carcinoma cell line. The results showed high tumour accumulation and a high tumour-background ratio as well as a significant decrease in cancer cell survival when compared to their controls [39]. Another study that evaluated the use of NIR-PIT *in vivo* was conducted by Mitsunaga et al. (2011), where they conjugated IR700 to epidermal growth factor (EGF) receptors. This form of molecular-targeted cancer therapy showed tumour shrinkage, where cells expressing the EGF showed membrane damage leading to tumour cell death [40]. The results obtained from the studies indicate that ADP whether it be internal or external targeting using an Ab directed at specific antigens expressed by cancer cells can successfully be implemented as a method for advancing PDT.

Along with ADP, the anticancer response of a PS can further be enhanced using a carrier molecule. Carrier molecules in the form of nanoparticles (NPs) can accumulate in a tumour because of the well-known enhanced permeability retention (EPR) effect. These molecules can be adapted specifically to their application by changing its composition that will enable receptor targeting through overexpression of surface markers on the malignant cells, and they can enhance photon absorption and improve singlet oxygen by the surface plasmon resonance effect specifically gold nanoparticles [41], which can lead to increased ROS and cancer cell destruction. Biodegradable and nonbiodegradable NPs can be used in this application. Biodegradable NPs degrade in their biological environment through a hydrolytic process, releasing the PS.

Using a NP that is internalised by the cell has an advantage over external NPs loaded with PSs. It can be said that the photoactivity of a PS loaded NP depends on its photochemical and cell penetrating properties; furthermore the NP activity *in vivo* is governed by pharmacokinetics and tissue distribution. Extravasation of the NP is dependent on its size, also affecting photodynamic activity. A smaller NP has greater photodynamic effect and clears out faster from the bloodstream. Nonbiodegradable NPs have the advantage of actively delivering singlet oxygen. This can be achieved externally through membrane contact, as well as internalisation through its cell penetrating properties giving photosensitivity to the targeted tumour tissue. They have enhanced particle residence time allowing for the photo action to be monitored over time. Nonbiodegradable NPs also have the following advantages over biodegradable NPs: they are stable in fluctuating pH and temperature and they are easily manipulated to specificities during development, evading bacterial contamination and they allow for PS retention and singlet oxygen diffusion due to its pore size [41].

Figure 2 graphically represents the advantages of using multicomponent photosensitizing molecules and their different roles in TPDT.

In a study by Stuchinskaya et al., a 4-component photosensitizing compound composed of phthalocyanine PS, an antibody (anti-HER2 monoclonal Ab), polyethylene glycol and an Au-NP, was used as a potential drug for targeted photodynamic cancer therapy. This multicomponent drug proved to be more stable toward aggregation and efficiently produced cytotoxic singlet oxygen under irradiation with visible red light. This study showed that the coupling of PSs, NPs, and antibodies in one biomolecule selectively targeted breast cancer cells that showed overexpression of HER2 EGF cell surface receptor [42].

A study conducted on mice for central- and peripheral-type early-stage lung cancer using tissue factor to improve the selectivity and effectiveness of nontargeted PDT (ntPDT) showed that factor VII-targeted PDT (fVII-TPDT) using fVII-Sn(IV) chlorin e6 Dihydroxide Trisodium Salt significantly enhanced (up to 25-fold) the *in vitro* effect, destroying A549 and H460 lung cancer cells via the rapid induction of apoptosis and necrosis. *In vivo* administration of fVII-TPDT significantly inhibited or eliminated subcutaneous A549 and H460 tumour xenografts in an athymic nude mouse model without any obvious side effects [43].

Work conducted by Yu et al. [44] using an effective PS, methylene blue (MB), combined with Au-NPs, was prepared using an intermolecular interaction between a polystyrene-alt-maleic acid layer on the Au-NPs and MB. The Au@polymer/MB NPs produced a high quantum yield of singlet oxygen molecules, 50% more that that of free MB, and they were then excited by laser with a wavelengths of 660 nm, having little to no dark toxicity. Furthermore, transferrin was conjugated on the Au@polymer/MB NPs via an EDC/NHS reaction to enhance the selectivity to HeLa cells compared to 3T3 fibroblasts. With a hand-held single laser treatment (32 mW/cm) for 4 min, the new Au@polymer/MB-Tf NPs showed a 2-fold enhancement of PDT efficiency toward HeLa cells over the use of free MB at 4 times dosage. Biological experiments showed that the HeLa cells reacted well with Au@polymer/MB-Tf NPs and PDT was achieved leading to apoptosis.

One of the first nanoplatforms to be applied in drug delivery systems are liposomes. Their ability lies within their containment of hydrophilic drugs in their aqueous core and hydrophobic agents within their lipid bilayers, making them efficient delivery vehicles. 5-Aminolevulinic acid (ALA) prodrugs for PDT were encapsulated in dipalmitoylphosphatidyl choline–based liposomes. *In vitro* experiments demonstrated an increased uptake of the conjugated molecule into human cholangiocarcinoma HuCC-T1 cells

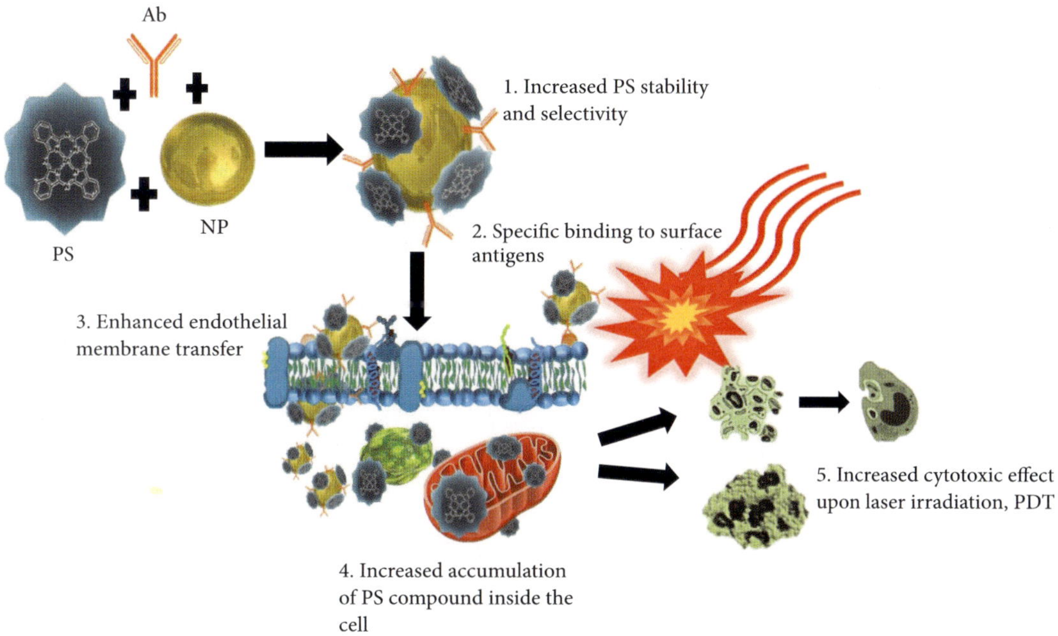

Figure 2: Advantages of using multicomponent photosensitizing molecules and their different roles in TPDT: (1) Conjugating a PS to NPs and antibodies increases its stability in the circulation and tumour selectivity. (2) Antibodies recognise specific antigens on the surface of cells enabling specific binding. (3) Nanoparticles enhance endothelial membrane by increasing solubility. (4) Because of increased solubility, there is maximum absorption resulting in efficient accumulation of photosensitizer inside the cancer cell. (5) Ultimately, there is increased cytotoxic effect in PDT.

compared to ALA alone having an increased photocytotoxic effect [45]. Magnetoliposomes (MLs) loaded with zinc phthalocyanine (ZnPc)/cucurbituril (CB) complexes (CB:ZnPc-MLs) were synthesized for TPDT and magnetohyperthermia in malignant melanoma cells. Significant reduction in cell viability was observed with melanoma cells treated with CB: ZnPc-MLs after application of both 670 nm light and AC magnetic field. The combined PDT and magnetohyperthermia by CB:ZnPcMLs were much more effective than each therapy alone [46].

In a study conducted by Governatore et al., they explored the use of ADP constructed from mAbs targeted against colorectal cancer antigens, to increase the selectivity of the PS for tumour over normal tissue. The PS chlorine6 (ce6) was conjugated to anticolon cancer monoclonal antibody 17.1A. Polylysine linkers bearing several ce6 molecules were covalently attached in a site-specific manner to partially reduced IgG molecules, which allowed photoimmunoconjugates to bear either cationic or anionic charges. The conjugates retained immunoreactivity. The cationic photoimmunoconjugate delivered 4 times more ce6 to the cells than the anionic photoimmunoconjugate, and both 17.1A conjugates showed, in comparison to nonspecific rabbit IgG conjugates, selectivity for antigen positive target cells. Illumination with only 3 J cm^2 of 666 nm light reduced the cell viability over 90% for the cationic 17.1A conjugate and by 73% for the anionic 17.1A conjugate after incubation with 1 μM ce6 equivalent of the respective conjugates. By contrast, 1 μM free ce6 gave only a 35% reduction in viability [47].

The results of this study strongly suggest the significance of using NPs and antibodies in TPDT.

Table 3 is a representation of where targeted therapy through conjugation of NPs or antibodies has advanced conventional cancer therapies.

This technology can be applied to CSCs as well, allowing for not only the cytotoxicity of the tumour but the SP of cells (CSCs) residing within it, sparing side effects in surrounding normal tissues. A representation of the above mentioned therapy is illustrated in Figure 3.

5. Conclusion

Conclusive evidence indicates that CSCs play the major roles in therapeutic resistance due to their significant characteristics of self-renewal, drug resistance, differentiation, and metastasis. Other studies also suggest that the proportion of CSCs within a tumour may correlate with the severity of the cancer. Because CSCs are able to evade the effects of conventional therapies there is, therefore, a need for targeted therapies with the purpose of enhancing patient survival by minimizing cancer recurrence, side effects, and metastasis. Scientific evidence suggests that improved cancer treatment and targeted therapies are expected to yield the highest cure rates. However, current methods are limited by their low sensitivity to early disease and a lack of specificity for targeted cell killing, as well as therapy resistance due to CSCs. PDT using targeted drug delivery systems comprised of PSs coupled to nanoparticles and monoclonal antibodies has potential to successfully treat cancer. Antibodies can assist in the accurate detection of tumour cells and provide increased PS accumulation in CSCs, along with nanoparticles that can increase EPR and cellular uptake, making TPDT

Table 3: Cancer and TPDT.

Cancer	TPDT	Reference
Breast Cancer	Phthalocyanine conjugated to anti-HER2 monoclonal Ab, polyethylene glycol and Au-NP	[42]
Lung Cancer	Sn(IV) Chlorin e6 Dihydroxide Trisodium Salt conjugated to factor VII	[43]
Cervical Cancer	Methylene Blue conjugated to Au-NPs and Transferrin	[44]
Cholangiocarcinoma (Bile Duct Cancer)	5-Aminolevulinic acid conjugated to dipalmitoyl-phosphatidyl choline–based liposomes	[45]
Melanoma	zinc phthalocyanine (ZnPc)/cucurbituril (CB) conjugated to Magnetoliposomes	[46]
Colon Cancer	chlorine6 (ce6) conjugated to anti-colon cancer monoclonal antibody 17.1A	[47]

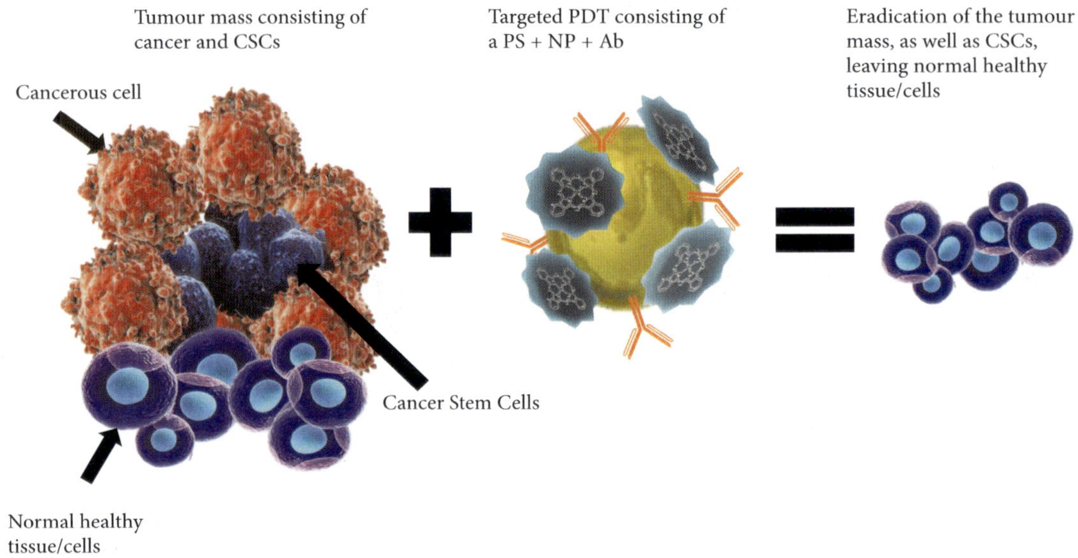

Figure 3: CSCs and targeted PDT using antibodies and nanoparticles. TPDT targeting cancer as well as CSCs, eradicating the entire tumour that advocates for normal healthy tissue growth.

a study of interest due to its therapeutic enhancements including efficacy, specificity, marginal toxicity to normal cells, biocompatibility, and minimal side effects.

Conflicts of Interest

The authors indicate no potential conflicts of interest.

Acknowledgments

This work is based on the research supported by the South African Research Chairs Initiative of the Department of Science and Technology and National Research Foundation of South Africa (Grant no. 98337). The authors sincerely thank the University of Johannesburg, the National Laser Centre, and the National Research Foundation of South Africa (CSIR-DST) for their financial grant support.

References

[1] American Cancer Society, "Cancer Facts & Figures 2017," https://www.cancer.org/research/cancer-facts-statistics/all-cancer-facts-figures/cancer-facts-figures-2017.html.

[2] B. S. Sylla and C. P. Wild, "A million africans a year dying from cancer by 2030: What can cancer research and control offer to the continent?" *International Journal of Cancer*, vol. 130, no. 2, pp. 245–250, 2012.

[3] Mayo Clinic, *Diseases and Conditions Cancer*, 2015, http://www.mayoclinic.org/diseases-conditions/cancer/basics/causes/con-20032378.

[4] E. Garcion, P. Naveilhan, F. Berger, and D. Wion, "Cancer stem cells: Beyond Koch's postulates," *Cancer Letters*, vol. 278, no. 1, pp. 3–8, 2009.

[5] Z. Yu, T. G. Pestell, M. P. Lisanti, and R. G. Pestell, "Cancer stem cells," *The International Journal of Biochemistry & Cell Biology*, vol. 44, no. 12, pp. 2144–2151, 2012.

[6] H. Abrahamse and M. R. Hamblin, "New photosensitizers for photodynamic therapy," *Biochemical Journal*, vol. 473, no. 4, pp. 347–364, 2016.

[7] M. T. Wan and J. Y. Lin, "Current evidence and applications of photodynamic therapy in dermatology," *Clinical, Cosmetic and Investigational Dermatology*, vol. 7, pp. 145–163, 2014.

[8] L. B. De Paula, F. L. Primo, M. R. Pinto, P. C. Morais, and A. C. Tedesco, "Combination of hyperthermia and photodynamic therapy on mesenchymal stem cell line treated with chloroaluminum phthalocyanine magnetic-nanoemulsion," *Journal of*

Magnetism and Magnetic Materials, vol. 380, pp. 372–376, 2015.

[9] G. M. F. Calixto, J. Bernegossi, L. M. de Freitas, C. R. Fontana, M. Chorilli, and A. M. Grumezescu, "Nanotechnology-based drug delivery systems for photodynamic therapy of cancer: A review," *Molecules*, vol. 21, p. 342, 2016.

[10] P. Agostinis, K. Berg, K. A. Cengel et al., "Photodynamic therapy of cancer: an update," *CA: A Cancer Journal for Clinicians*, vol. 61, no. 4, pp. 250–281, 2011.

[11] M. V. Yezhelyev, X. Gao, Y. Xing, A. Al-Hajj, S. Nie, and R. M. O'Regan, "Emerging use of nanoparticles in diagnosis and treatment of breast cancer," *The Lancet Oncology*, vol. 7, no. 8, pp. 657–667, 2006.

[12] R. Lewis, *Human Genetics; Concepts and Application*, McGraw–Hill Primis, 9th edition, 2009.

[13] J. S. Bertram, "The molecular biology of cancer," *Molecular Aspects of Medicine*, vol. 21, no. 6, pp. 167–223, 2001.

[14] B. A. Pierce, *Genetics Essentials; Concepts and Connections*, W. H. Freeman and Company, New York, NY, USA, 2010.

[15] D. Hanahan and R. A. Weinberg, "The hallmarks of cancer," *Cell*, vol. 100, no. 1, pp. 57–70, 2000.

[16] V. Plaks, N. Kong, and Z. Werb, "The cancer stem cell niche: how essential is the niche in regulating stemness of tumor cells?" *Cell Stem Cell*, vol. 16, no. 3, pp. 225–238, 2015.

[17] S. Sell, "On the stem cell origin of cancer," *The American Journal of Pathology*, vol. 176, no. 6, pp. 2584–2594, 2010.

[18] A. Allegra, A. Alonci, G. Penna et al., "The cancer stem cell hypothesis: A guide to potential molecular targets," *Cancer Investigation*, vol. 32, no. 9, pp. 470–495, 2014.

[19] A. Soltysova, V. Altanerova, and C. Altaner, "Cancer stem cells," *Neoplasma*, vol. 52, no. 6, pp. 435–440, 2005.

[20] H. Lou and M. Dean, "Targeted therapy for cancer stem cells: the patched pathway and ABC transporters," *Oncogene*, vol. 26, no. 9, pp. 1357–1360, 2007.

[21] T. Lapidot, C. Sirard, J. Vormoor et al., "A cell initiating human acute myeloid leukaemia after transplantation into SCID mice," *Nature*, vol. 367, no. 6464, pp. 645–648, 1994.

[22] M. Al-Hajj, M. S. Wicha, A. Benito-Hernandez, S. J. Morrison, and M. F. Clarke, "Prospective identification of tumorigenic breast cancer cells," *Proceedings of the National Acadamy of Sciences of the United States of America*, vol. 100, no. 7, pp. 3983–3988, 2003.

[23] S. K. Singh, I. D. Clarke, M. Terasaki et al., "Identification of a cancer stem cell in human brain tumors," *Cancer Research*, vol. 63, no. 18, pp. 5821–5828, 2003.

[24] V. Tirino, V. Desiderio, F. Paino et al., "Cancer stem cells in solid tumors: an overview and new approaches for their isolation and characterization," *The FASEB Journal*, vol. 27, no. 1, pp. 13–24, 2013.

[25] W.-T. Kim and C. J. Ryu, "Cancer stem cell surface markers on normal stem cells," *BMB Reports*, vol. 50, no. 6, pp. 285–298, 2017.

[26] F. Islam, V. Gopalan, R. A. Smith, and A. K.-Y. Lam, "Translational potential of cancer stem cells: a review of the detection of cancer stem cells and their roles in cancer recurrence and cancer treatment," *Experimental Cell Research*, vol. 335, no. 1, pp. 135–147, 2015.

[27] S. P. Zielske, A. C. Spalding, M. S. Wicha, and T. S. Lawrence, "Ablation of breast cancer stem cells with radiation," *Translational Oncology*, vol. 4, no. 4, pp. 217–223, 2011.

[28] D. van Straten, V. Mashayekhi, H. S. de Bruijn, S. Oliveira, and D. J. Robinson, "Oncologic photodynamic therapy: Basic principles, current clinical status and future directions," *Cancers*, vol. 9, no. 2, p. 19, 2017.

[29] J.-O. Yoo and K.-S. Ha, "New insights into the mechanisms for photodynamic therapy-induced cancer cell death," *International Review of Cell and Molecular Biology*, vol. 295, pp. 139–174, 2012.

[30] R. W. Boyle and D. Dolphin, "Structure and biodistribution relationships of photodynamic sensitizers," *Photochemistry and Photobiology*, vol. 64, no. 3, pp. 469–485, 1996.

[31] L. Benov, "Photodynamic therapy: current status and future directions," *Medical Principles and Practice*, vol. 24, pp. 14–28, 2015.

[32] T. Yano, M. Muto, K. Minashi et al., "Photodynamic therapy as salvage treatment for local failure after chemoradiotherapy in patients with esophageal squamous cell carcinoma: A phase II study," *International Journal of Cancer*, vol. 131, no. 5, pp. 1228–1234, 2012.

[33] A. Casas, G. Di Venosa, T. Hasan, and A. Batlle, "Mechanisms of resistance to photodynamic therapy," *Current Medicinal Chemistry*, vol. 18, no. 16, pp. 2486–2515, 2011.

[34] J. Morgan, J. D. Jackson, X. Zheng, S. K. Pandey, and R. K. Pandey, "Substrate affinity of photosensitizers derived from chlorophyll-a: The ABCG2 transporter affects the phototoxic response of SP stem cell-like cancer cells to photodynamic therapy," *Molecular Pharmaceutics*, vol. 7, no. 5, pp. 1789–1804, 2010.

[35] R. Vankayala, C.-L. Kuo, K. Nuthalapati, C.-S. Chiang, and K. C. Hwang, "Nucleus-targeting gold nanoclusters for simultaneous in vivo fluorescence imaging, gene delivery, and NIR-light activated photodynamic therapy," *Advanced Functional Materials*, vol. 25, no. 37, pp. 5934–5945, 2015.

[36] H. Pye, I. Stamati, G. Yahioglu, M. Butt, and M. Deonarain, "Antibody-Directed Phototherapy (ADP)," *Antibodies*, vol. 2, pp. 270–305, 2013.

[37] M. B. Vrouenraets, G. W. M. Visser, M. Stigter, H. Oppelaar, G. B. Snow, and G. A. M. S. Van Dongen, "Targeting of aluminum (III) phthalocyanine tetrasulfonate by use of internalizing monoclonal antibodies: Improved efficacy in photodynamic therapy," *Cancer Research*, vol. 61, no. 5, pp. 1970–1975, 2001.

[38] S. Okuyama, T. Nagaya, K. Sato et al., "Interstitial near-infrared photoimmunotherapy: Effective treatment areas and light doses needed for use with fiber optic diffusers," *Oncotarget* , vol. 9, no. 13, pp. 11159–11169, 2018.

[39] T. Nagaya, Y. Nakamura, K. Sato et al., "Near infrared photoimmunotherapy with an anti-mesothelin antibody," *Oncotarget* , vol. 7, no. 17, pp. 23361–23369, 2016.

[40] M. Mitsunaga, M. Ogawa, N. Kosaka, L. T. Rosenblum, P. L. Choyke, and H. Kobayashi, "Cancer cell-selective in vivo near infrared photoimmunotherapy targeting specific membrane molecules," *Nature Medicine*, vol. 17, no. 12, pp. 1685–1691, 2011.

[41] L. Colombeau, S. Acherar, F. Baros et al., "Inorganic nanoparticles for photodynamic therapy," *Topics in Current Chemistry*, vol. 370, pp. 113–134, 2016.

[42] T. Stuchinskaya, M. Moreno, M. J. Cook, D. R. Edwards, and D. A. Russell, "Targeted photodynamic therapy of breast cancer cells using antibody–phthalocyanine–gold nanoparticle conjugates," *Photochemical & Photobiological Sciences*, vol. 10, no. 5, pp. 822–831, 2011.

[43] J. Cheng, J. Xu, J. Duanmu, H. Zhou, C. J. Booth, and Z. Hu, "Effective treatment of human lung cancer by targeting tissue

factor with a factor vii-targeted photodynamic therapy," *Current Cancer Drug Targets*, vol. 11, no. 9, pp. 1069–1081, 2011.

[44] J. Yu, C.-H. Hsu, C.-C. Huang, and P.-Y. Chang, "Development of therapeutic Au-methylene blue nanoparticles for targeted photodynamic therapy of cervical cancer cells," *ACS Applied Materials & Interfaces*, vol. 7, no. 1, pp. 432–441, 2015.

[45] K. H. Choi, C.-W. Chung, C. H. Kim, D. H. Kim, Y.-I. Jeong, and D. H. Kang, "Effect of 5-aminolevulinic acid-encapsulate liposomes on photodynamic therapy in human cholangiocarcinoma cells," *Journal of Nanoscience and Nanotechnology*, vol. 14, no. 8, pp. 5628–5632, 2014.

[46] G. C. Bolfarini, M. P. Siqueira-Moura, G. J. F. Demets, P. C. Morais, and A. C. Tedesco, "In vitro evaluation of combined hyperthermia and photodynamic effects using magnetoliposomes loaded with cucurbit," *Journal of Photochemistry and Photobiology B: Biology*, vol. 115, pp. 1–4, 2012.

[47] M. Del Governatore, M. R. Hamblin, E. E. Piccinini, G. Ugolini, and T. Hasan, "Targeted photodestruction of human colon cancer cells using charged 17.1A chlorin(e6) immunoconjugates," *British Journal of Cancer*, vol. 82, no. 1, pp. 56–64, 2000.

16

Extended Reach 10 Gb/s Transmission with an Optical I/Q Modulator using VCSELs over OFDM-Based Multimode Fiber Link

Deepak Kedia and Monika Nehra

Department of Electronics and Communication Engineering, Guru Jambheshwar University of Science & Technology, Hisar, Haryana 125001, India

Correspondence should be addressed to Monika Nehra; ssmonikanehra@gmail.com

Academic Editor: Chenggen Quan

This paper investigates the possibilities of extending the reach of a 10 Gb/s orthogonal frequency division multiplexing- (OFDM-) based multimode fiber (MMF) link with an optical I/Q modulation technique through vertical cavity surface emitting lasers (VCSELs). The proposed I/Q modulation technique comprises two VCSELs in a 90° hybrid combination to produce I/Q modulated OFDM signal. The results show the excellent performance (in terms of BER and Q factor) of the direct-detection optical- (DDO-) OFDM system with the proposed I/Q modulation in comparison to the direct modulation case. The proposed system is able to achieve worst-case BER of about 1.8616×10^{-3} and Q factor of about 10.9949 dB over a 5 km MMF link. The I/Q modulation technique in the DDO-OFDM system has further been investigated for extending the transmission reach of the MMF link using multispan configuration.

1. Introduction

To date, the implementation of MMF links has gained significant attention for high-speed transmission over local area networks (LANs). The main reasons that led to the widespread use of MMF in short-reach applications rely on the availability of necessary bandwidth at low expenses and simplification of the component-to-fiber alignment. In MMF, the signal propagates on different fiber modes with different group velocities that result in differential modal delay (DMD). Over a short distance, the DMD gives rise to intersymbol interference (ISI) with conventional modulation formats. The ISI imposes a limitation on the maximum data speed for a specific length of fiber and causes dispersion limited performance of MMF links rather than noise limited one [1]. Different dispersion compensation techniques have been proposed in the literature like (i) Decision-Feedback Equalizers (DFE) and (ii) Feed-Forward Equalizers (FFE) [2]. No doubt, the DFE/FFE schemes are performing well in single mode fiber (SMF) system [3], but these have limitations to compensate for different characteristics (e.g., chromatic dispersion and modal dispersion) of the MMF system.

However, using OFDM with high-order modulation formats in existing MMF links, the effect of ISI can be minimized due to frequency selective attenuation of the orthogonal channels [4]. Secondly, the partially overlapped spectra of OFDM subcarriers are helpful in achieving high optical spectral efficiency. Thirdly, the RF-to-optical up-down conversion helps in reducing the electrical bandwidth requirement for direct-detection system which is an extremely attractive feature for any high-speed transmission system using OFDM.

With the advancements in laser source technology, the directly modulated VCSELs have become the most preferable laser source for short-range applications due to their attractive attributes, for example, low manufacturing cost, low drive voltage, high modulation bandwidth, low power consumption, and capability of array integration [5]. The short wavelength VCSELs operating at 850 nm wavelength have been proposed for MMF links also that specify high link bandwidth and low ISI penalty [6]. The large active

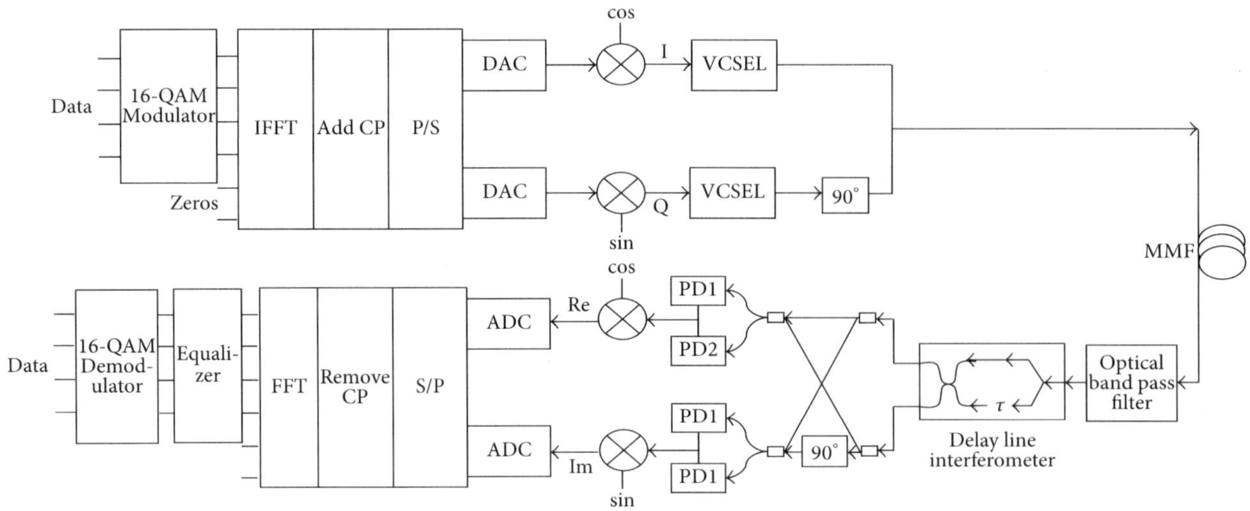

Figure 1: Schematic diagram of the DDO-OFDM system with an optical I/Q modulator using VCSELs for MMF links.

area (~15 μm) and low-source power of VCSELs offered high coupling efficiency for MMF links [7].

Recent extensive studies on MMF links demonstrated enhanced data rates but have limitations in terms of transmission reach (~1000 m). Lowery and Armstrong [8] demonstrated a 10 Gb/s data rate over a 300 m MMF link with the transmission of OFDM signals using zero bias. Further, Deng et al. [9] investigated experimentally 16.375 Gb/s dual-band optical OFDM (O-OFDM) transmission over 200 m MMF using directly modulated VCSELs. The dual-band O-OFDM offers large passband frequency tunability and excellent performance robustness that helps achieve the desired transmission performance with VCSEL modulation bandwidth for cost sensitive applications. Caspar et al. [10] demonstrated 25 Gb/s transmission over 1000 m MMF based on an optimized VCSEL with a single primary mode.

Thus, recent works on MMF links (≥10 Gb/s) using optical OFDM have been able to support transmission up to 1000 m only. Further, the VCSELs have been directly modulated in these researches. In this paper, an optical I/Q modulator using VCSELs is being proposed for 10 Gb/s data rate MMF link with a transmission reach of 5 km. To the best of the authors' knowledge, this paper is a maiden attempt to use an optical I/Q modulation technique for MMF links based on direct detection for extending the transmission reach. As compared to the direct modulation technique using single VCSEL, a considerably improved transmission performance is achieved through the proposed optical I/Q modulation.

2. DDO-OFDM System with I/Q Modulation Using VCSELs

Figure 1 shows a direct-detection system for MMF links which uses an optical I/Q modulator using VCSELs. The OFDM transmitter performs various functions as (i) mapping of data bits into OFDM symbols through inverse fast Fourier transform (IFFT), (ii) inserting cyclic prefix (CP) to combat ISI, and (iii) generating real-time signals through digital-to-analog conversion (DAC).

Thus, data is transmitted in parallel over several subcarriers resulting in a longer symbol period as compared to a serial data transmission for the same total bit rate. Here, equalization is simplified due to the reduced impact of ISI on a longer symbol period. The baseband OFDM signal is defined as [11]

$$S_B(t) = \sum_{k=-(N_{sc}/2)+1}^{k=N_{sc}/2} C_k \exp\left(j2\pi f_k t\right), \tag{1}$$

$$f_k = (k-1)\, t_s, \tag{2}$$

where $S_B(t)$ denotes the baseband OFDM signal, N_{sc} denotes the number of subcarriers, C_k denotes the information symbol or pilot symbol of the k_{th} subcarrier, and f_k is the frequency of the subcarrier. Equation (1) shows one OFDM symbol for simplicity. The resulting baseband OFDM signal is optically modulated with I/Q modulator which comprises two VCSELs to separately modulate the real (I) and imaginary (Q) parts of $S_B(t)$. The use of direct modulation laser is beneficial to improve the capabilities of the optical link in terms of distance-capacity product [12]. Both VCSEL sources are operated at the same wavelength (1550 nm) and the outputs of both VCSELs are combined in a 90° hybrid to produce the I/Q modulated optical OFDM signal.

The VCSELs are the surface emitting semiconducting lasers provided with a laser cavity formed by several epitaxial layers. The output power (P_{out}) of the VCSEL can be calculated as [13]

$$P_{out} = n_i n_o \left(\frac{h\nu}{q}\right)\left(I - I_{th}\right), \tag{3}$$

where n_i represents the injection efficiency, n_o represents the optical efficiency, h represents Planck's constant, ν represents the frequency, q represents the elementary charge, I represents the injected current, and I_{th} represents the threshold

TABLE 1: VCSEL laser emitter parameters.

Parameter	Value
Test wavelength	850 nm
Operating wavelength	1550 nm
Operating temperature	25°C
Threshold current	1.49538 mA
FWHM linewidth	10 MHz
Linear output power	0.35 mW
Turn on delay	1.73287 ns
P-I slope	0.23405 mW/mA
Relative intensity noise	−150 dB/Hz

TABLE 2: Different parameters for MMF.

Parameter	Value
Fiber length	5 km
Attenuation	3.0 dB/km
Dispersion	17 ps/nm/km
Gamma cutback factor	0.75
Intermodal bandwidth	−150 dB/Hz

current. The electric field spectrum of the transmitted optical signal is a replica of the baseband RF OFDM signal, with no need for any optical carrier component to be transmitted.

At the receiver, the I/Q modulated optical OFDM signal is directly detected by delay line interferometer with balanced detection. Here, the received signal is first passed through a delay line interferometer before balanced detection of both in-phase and quadrature tributaries. The outputs of the interferometer are further processed with the help of adequate phase shifts. The interferometer helps in converting the phase difference information of two consecutive symbols into intensity information [14, 15]. The delay of interferometer is adjusted to one-symbol duration. This intensity information can be easily detected by the photodiodes, where the resulting photocurrent shows the nonlinear mapping of baseband OFDM signal. The OFDM receiver performs the reverse operation to that of OFDM transmitter as (i) analog-to-digital conversion (ADC), (ii) removal of CP, and (iii) fast Fourier transform (FFT) to convert the OFDM symbols back into data bits [16].

3. Simulation Results

The system shown in Figure 1 has been simulated and 16-QAM modulation technique is being used to map the data bits onto the information symbols. The sampling frequency is set to 10 GHz. The number of OFDM subcarriers is set to 64, while IFFT covers 256 samples and zero padding is set in the middle of the OFDM symbol. The CP length (equal to 12.5%) is also inserted in front of the OFDM symbol to ensure the guard interval. To ensure efficient chromatic dispersion compensation, it is necessary to keep the length of the cyclically extended guard interval longer than the delay spread caused by chromatic dispersion. The baseband OFDM signal is directly modulated with an optical I/Q modulator using VCSELs. The I/Q modulated optical OFDM signal is then launched into the MMF. The different parameters of VCSEL laser emitters and MMF are listed in Tables 1 and 2, respectively.

The simulation results thus obtained are presented in Figures 2–5, respectively. Figure 2 shows the launch and received spectra of 10 Gb/s optical OFDM signal over 5 km MMF link using an optical I/Q modulator and a direct one. The optical spectra significantly show the impact of fiber dispersion on the baseband signal recovery. Figures 2(a) and 2(c), respectively, show the launch spectra and received spectra in an MMF link using optical I/Q modulation. On the other hand, Figures 2(b) and 2(d) demonstrate the respective launch and received spectra in the case of direct modulation using single VCSEL. It is clear that the use of the proposed optical I/Q modulation provides an improved performance with respect to modal dispersion as compared to direct modulation.

In the simulation setup, a single PIN photodiode is employed at the receiver to detect the optical OFDM signal. The proposed 10 Gb/s DDO-OFDM system has been implemented with transmission reach of MMF link varying from 0.1 km to 5 km. The resulting eye and frequency spectra of the system (5 km MMF link with −20 dBm launch power) are shown in Figure 3. Figures 3(a) and 3(b) demonstrate the eye spectra in case of an optical I/Q modulator and direct modulator, respectively. The eye opening is 4.4373×10^{-2} (a.u.) in case of I/Q modulated signal and jitter is 5.5792×10^{-4} ns, whereas the eye opening is 4.2742×10^{-3} (a.u.) in case of direct modulated signal and jitter is 4.9525×10^{-3} ns. Figures 3(c) and 3(d), respectively, demonstrate the frequency spectra in case of an optical I/Q modulator and direct modulator. The frequency spectra illustrate maximum allowable frequency along with the corresponding power. The maximum power in case of I/Q modulated signal is −17.801876 dB, whereas it is −29.079815 dB in case of direct modulated signal.

The performance of the proposed DDO-OFDM system has further been analyzed in terms of bit error rate (BER) and Q factor. These parameters have been assessed with respect to the length of the MMF link. A comparative analysis has been presented for the two cases, that is, using an optical I/Q modulator and using a direct modulator. The corresponding results are shown in Figures 4(a) and 4(b), respectively. It is clear from the results that the DDO-OFDM system with the proposed I/Q modulation outperforms the direct modulation case in terms of BER as well as Q factor. The proposed system with an optical I/Q modulator shows a worst-case BER of about 1.8616×10^{-3} and a Q factor of about 10.9949 dB. The direct modulator, on the other hand, provides worst-case BER of about 2.9750×10^{-2} and Q factor of about 6.0206 dB, over a 5 km MMF link. Table 3 shows a comparative analysis of performance of different systems for MMF links.

With the proposed optical I/Q modulation technique, it thus becomes possible to extend the reach of OFDM-based MMF link up to 5 Km. The reach of the MMF link can further be extended up to 90 km using multispan configuration. However, it is practically impossible in case of direct

Table 3: Comparative performance analysis of the proposed work.

Parameter	Proposed DDO-OFDM system	Lowery and Armstrong [8]	Deng et al. [9]
Data rate (Gb/s)	10	10	**16.375**
MMF	Yes	Yes	Yes
Transmission reach (km)	**5**	0.3	0.2
I/Q modulation	Yes	No	No
BER (worst case)	$\mathbf{6.19635 \times 10^{-3}}$	$\sim 1 \times 10^{-3}$	2.3×10^{-3}
Q factor (worst case)	**11.106108**	--	--

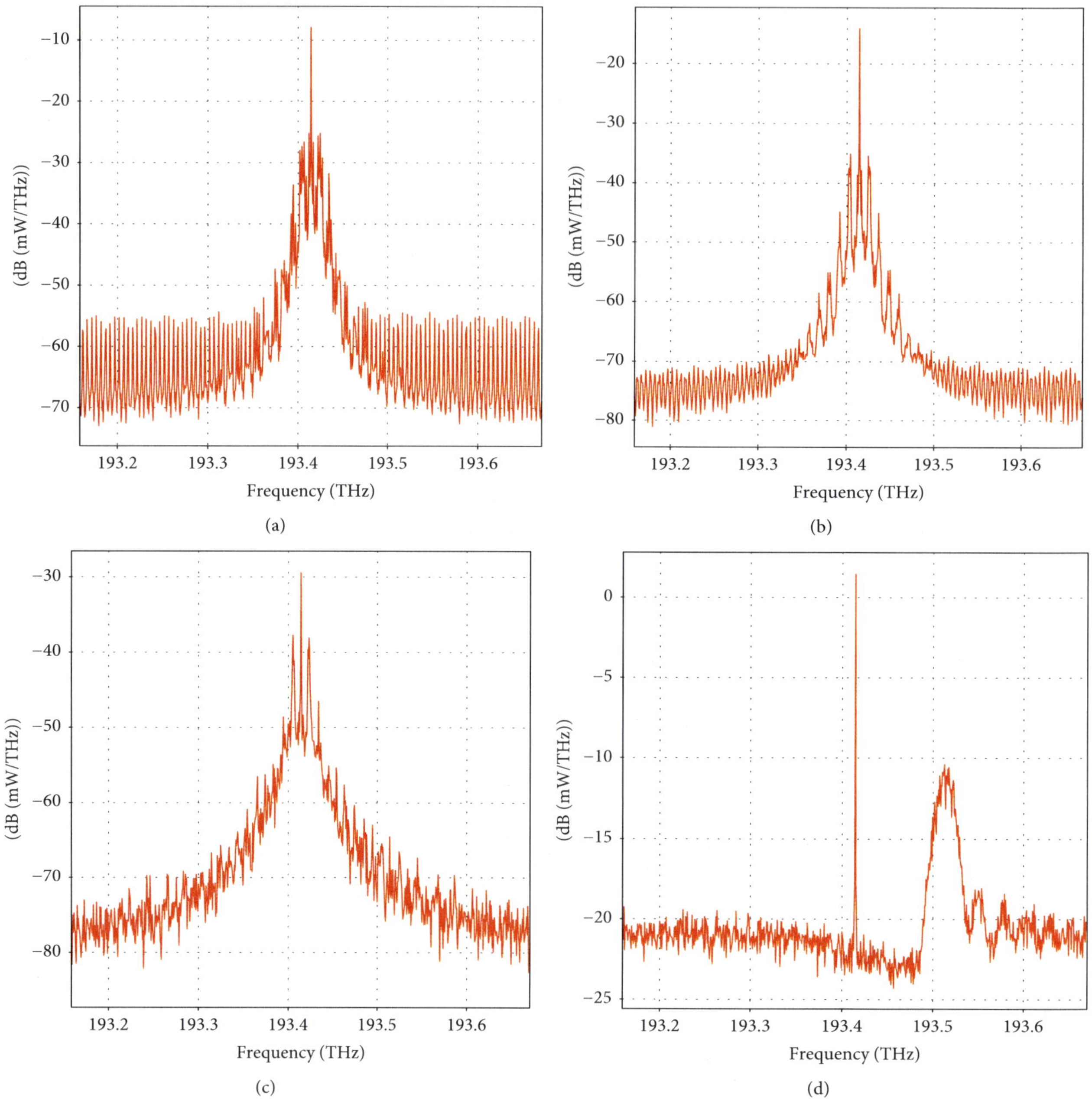

Figure 2: Optical spectra of the optical OFDM signals: (a) launch spectra in case of an optical I/Q modulator; (b) launch spectra in case of a direct modulator; (c) received spectra from the MMF link in case of an optical I/Q modulator; and (d) received spectra from the MMF link in case of a direct modulator.

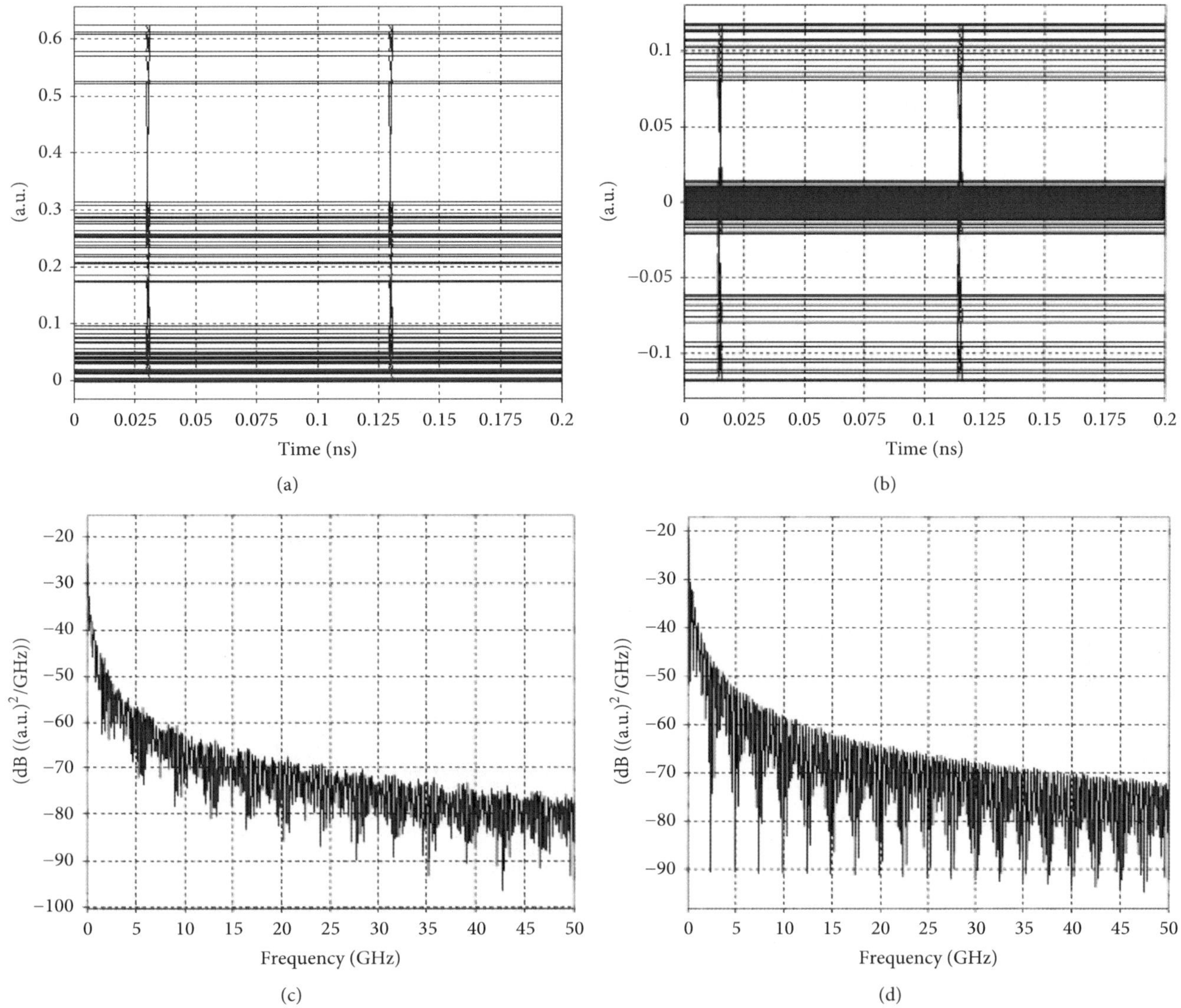

FIGURE 3: (a) Eye spectra in case of an optical I/Q modulator, (b) eye spectra in case of a direct modulator, (c) frequency spectra in case of an optical I/Q modulator, and (d) frequency spectra in case of a direct modulator.

modulation. Therefore, the performance of the DDO-OFDM system with an optical I/Q modulator has been evaluated for multispan MMF and EDFA, spaced at 10 km apart to achieve a transmission reach of 90 km. The block schematic of the system is shown in Figure 5. The gain of EDFA is adjusted to compensate for the span loss between two amplifiers. The span loss is represented by $\exp(-\alpha l)$, where α is the fiber attenuation coefficient.

The noise figure of EDFA is set to 5 dB and the dispersion coefficient of MMF is set to 17 ps/km-nm. Figures 6(a) and 6(b) show the respective BER and Q factor performance of the system with respect to launch power at 10 Gb/s data rate. It is clear from the figures that the proposed DDO-OFDM system provides excellent performance in terms of BER and Q factor in case of multispan configuration extended up to 90 km.

Thus, it is clear that the proposed optical I/Q modulation technique using VCSELs helps to achieve the extension of short reach in case of MMF links and also opens the possibilities for higher data rates, supporting the efficient bandwidth utilization of the DDO-OFDM system.

4. Conclusion

In this paper, we have proposed an optical I/Q modulation technique using VCSELs for the DDO-OFDM system over 10 Gb/s MMF link. The simulation results prove that improvement in the overall performance of the DDO-OFDM system takes place by the use of the proposed I/Q modulation technique over direct modulation. The effect of modal dispersion is significantly lower in case of I/Q modulated signal

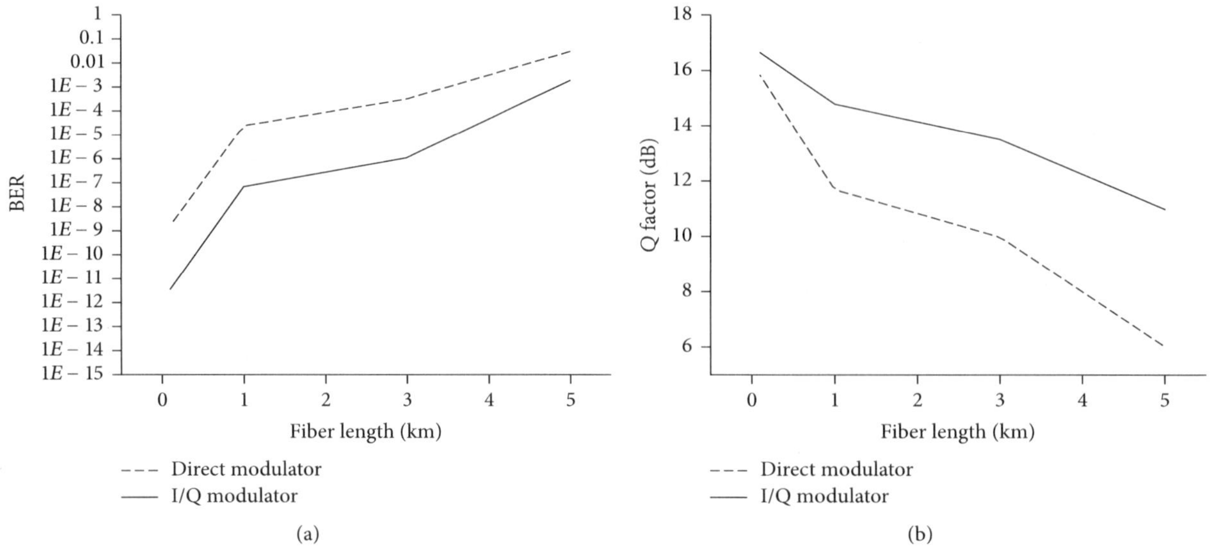

FIGURE 4: Performance analysis of the proposed DDO-OFDM system using an MMF link in terms of (a) BER and (b) Q factor.

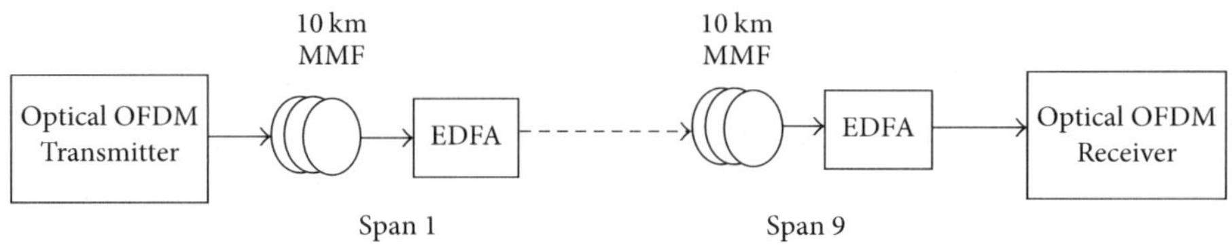

FIGURE 5: DDO-OFDM system in multispan configuration.

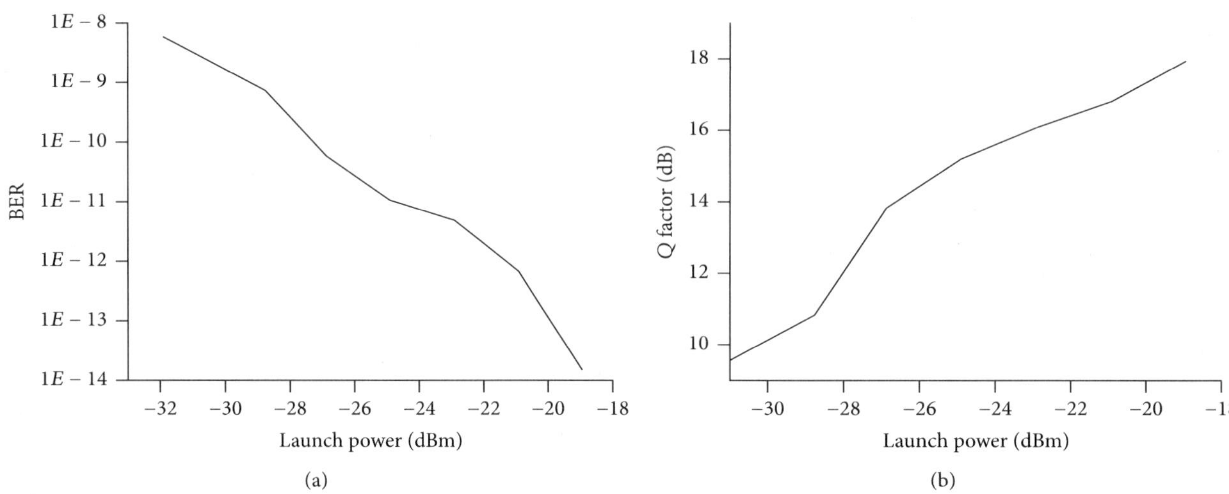

FIGURE 6: Performance analysis of the proposed DDO-OFDM system with respect to the launch power in terms of (a) BER and (b) Q factor.

in comparison to the direct modulated one. The proposed system shows worst-case BER of about 1.8616×10^{-3} and Q factor of about 10.9949 dB, whereas the direct modulator provides worst-case BER of about 2.9750×10^{-2} and Q factor of about 6.0206 dB, over a 5 km MMF link. Therefore, it becomes possible to extend the reach of OFDM-based MMF link up to 5 km using the proposed I/Q modulation. The reach of MMF link has further been extended up to 90 km using multispan configuration with excellent performance.

Conflicts of Interest

The authors declare that there are no conflicts of interest regarding the publication of this paper.

Acknowledgments

Monika Nehra would like to thank the UGC, India, for providing financial assistance in the form of JRF (Award no. 3608, dated 29-02-2016).

References

[1] Y. Sun, R. Lingle, R. Shubochkin et al., "SWDM PAM4 transmission over next generation wide-band multimode optical fiber," *Journal of Lightwave Technology*, vol. 35, no. 4, Article ID 7600413, pp. 690–697, 2017.

[2] T. Cseh and T. Berceli, "Dispersion compensation in millimeter wave radio over multimode fiber systems," *Microwave and Optical Technology Letters*, vol. 57, no. 1, pp. 204–207, 2015.

[3] X. Li, W.-D. Zhong, A. Alphones, C. Yu, and Z. Xu, "Channel equalization in optical OFDM systems using independent component analysis," *Journal of Lightwave Technology*, vol. 32, no. 18, Article ID 6860225, pp. 3206–3214, 2014.

[4] R. C. J. Hsu, A. Tarighat, A. Shah, A. H. Sayed, and B. Jalali, "Capacity enhancement in coherent optical MIMO (COMIMO) multimode fiber links," *IEEE Communications Letters*, vol. 10, no. 3, pp. 195–197, 2006.

[5] A. Boletti, P. Boffi, P. Martelli, M. Ferrario, and M. Martinelli, "Performance analysis of communication links based on VCSEL and silicon photonics technology for high-capacity data-intensive scenario," *Optics Express*, vol. 23, no. 2, pp. 1806–1815, 2015.

[6] K. Szczerba, P. Westbergh, J. Karout et al., "30 Gbps 4-PAM transmission over 200 m of MMF using an 850 nm VCSEL," *Optics Express*, vol. 19, no. 26, pp. B203–B208, 2011.

[7] C. Carlsson, A. Larsson, and A. Alping, "RF transmission over multimode fibers using VCSELs - Comparing standard and high-bandwidth multimode fibers," *Journal of Lightwave Technology*, vol. 22, no. 7, pp. 1694–1700, 2004.

[8] A. J. Lowery and J. Armstrong, "10 Gbit/s multimode fiber link using power-efficient orthogonal-frequency- division multiplexing," *Optics Express*, vol. 13, no. 25, pp. 10003–10009, 2005.

[9] M. L. Deng, N. Jiang, X. Duan et al., "Robust and tunable 16.375 Gb/s dual-band optical OFDM transmissions over directly modulated VCSEL-based 200 m OM2 MMFs," *Optics Express*, vol. 23, no. 1, pp. 373–383, 2015.

[10] C. Caspar, J. R. Kropp, V. A. Shchkin, and N. N. Ledentsov, "High speed transmission over multimode fiber with direct modulated single-mode VCSEL," in *Proceedings of the 9th ITG Symposium*, 2015.

[11] Y. Tang, W. Shieh, X. Yi, and R. Evans, "Optimum design for RF-to-optical up-converter in coherent optical OFDM systems," *IEEE Photonics Technology Letters*, vol. 19, no. 7, pp. 483–485, 2007.

[12] Z. Liu, J. Kakande, B. Kelly et al., "Modulator-free quadrature amplitude modulation signal synthesis," *Nature Communications*, vol. 5, article 5911, 2014.

[13] J. Estaran, R. Rodes, T. T. Pham et al., "Quad 14 Gbps L-band VCSEL-based system for WDM migration of 4-lanes 56 Gbps optical data links," *Optics Express*, vol. 20, no. 27, pp. 28524–28531, 2012.

[14] J. Wang and J. M. Kahn, "Impact of chromatic and polarization-mode dispersions on DPSK systems using interferometric demodulation and direct detection," *Journal of Lightwave Technology*, vol. 22, no. 2, pp. 362–371, 2004.

[15] M. Seimetz, M. Noelle, and E. Patzak, "Optical systems with high-order DPSK and star QAM modulation based on interferometric direct detection," *Journal of Lightwave Technology*, vol. 25, no. 6, pp. 1515–1530, 2007.

[16] J. Armstrong, "OFDM for optical communications," *Journal of Lightwave Technology*, vol. 27, no. 3, pp. 189–204, 2009.

17

Two-Step Phase Retrieval Algorithm using Single-Intensity Measurement

Cheng Zhang,[1,2] Meiqin Wang,[1] Qianwen Chen,[1] Dong Wang,[1] and Sui Wei[1]

[1] *Key Laboratory of Intelligent Computing and Signal Processing, Ministry of Education, Anhui University, Hefei, Anhui Province 230039, China*
[2] *Key Laboratory of Modern Imaging and Displaying Technology, Anhui University, Hefei, Anhui Province 230039, China*

Correspondence should be addressed to Meiqin Wang; 2417724433@qq.com

Academic Editor: Xiaowei Li

Aiming at the problem that the single-intensity phase retrieval method has poor reconstruction quality and low probability of successful recovery, an improved method is proposed in this paper. Our method divides the phase retrieval into two steps: firstly, the GS algorithm is used to recover the amplitude in the spatial domain from the single-spread Fourier spectrum, and then the classical GS algorithm using two intensity measurements (one is recorded and the other is estimated from the first step) measurements is used to recover the phase. Finally, the effectiveness of the proposed method is verified by numerical experiments.

1. Introduction

Most of the information in the optical field is contained in the phase, such as the depth and shape of the object and so on. The oscillation frequency of the optical field is as high as 10^{15} Hz. However, existing detection devices can only record intensity, directly [1]. Phase retrieval method recovers lost phase from recorded intensity measurements with some prior knowledge, which plays an important role in the imaging field. Phase retrieval problem arises in numerous of areas, such as crystallography [2–4], optics [5–9], astronomical imaging [10], microscopy [11, 12], biomedical [13], and holographic imaging[14, 15].

In 1972, Gerchberg and Saxton first proposed an alternate projection-based phase retrieval algorithm, the Gerchberg-Saxton (GS) algorithm [16]. The main idea of GS algorithm is to use the intensity of the spatial domain and the Fourier domain to recover the phase of the optical field. Subsequently, Fienup proved that the GS algorithm had obvious error-decreasing properties, Error Reduction (ER) algorithm, and Hybrid Input-Output (HIO) algorithm were proposed [17]. At present, ER algorithm and HIO algorithm are considered to be the most effective methods in the field of phase retrieval [18]. Since the above algorithms are only suitable for a positive linear transformation system, nobody cares about any linear transformation system. Therefore, Yang and Gu proposed the amplitude-phase detection theory in arbitrary linear transformation systems, namely, the Yang-Gu (Y-G) algorithm [19]. In 2015, Guo et al. optimized the iterative algorithm and proposed two improved GS iterative phase retrieval algorithms, which were the spatial phase perturbation GS algorithm and the combined GS HIO algorithm. For the two improved algorithms, the squared value of the squared error value rapidly drops to an acceptable value, and the lost phase can be successfully recovered in the spatial domain and Fourier domain, which means that both algorithms can jump out of the local minimum value and convergence to the global minimum [9].

The GS algorithm was originally proposed in connection with the problem of reconstructing phase given only two-intensity measurements in the spatial domain and Fourier domain, i.e., given two intensity measurements (one is recorded in the spatial domain and the other is estimated in the Fourier domain) measurements. Unfortunately, such two intensity measurements at two different planes cannot be measured in some cases. Therefore, a GS algorithm using single-intensity measurement with some prior knowledge combined was proposed to recover the phase. However the

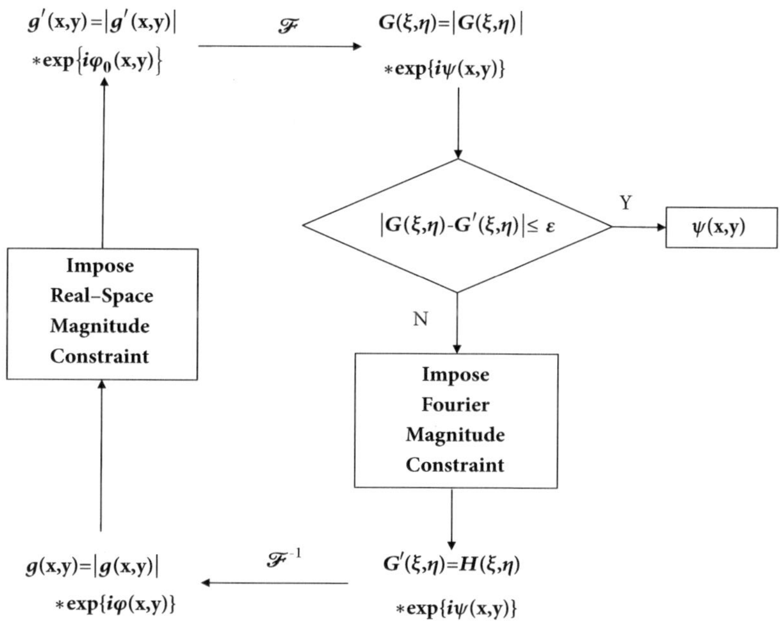

FIGURE 1: Flow chart of classical GS algorithms using two intensity measurements, i.e., TIPR.

single-intensity phase retrieval method has the drawbacks of poor reconstruction quality and low probability of successful recovery. This paper proposes an improved method based on single-intensity measurement, Two-Step Phase Retrieval (TSPR) from single-intensity measurement algorithm. The main idea of the TSPR algorithm is to first recover the amplitude (i.e., the square root of the intensity) in the spatial domain using the Single-Intensity Phase Retrieval (SIPR) algorithm with some prior knowledge and then recover the phase in the spatial domain using the Two-Intensity Phase Retrieval (TIPR) algorithm. Experimental results demonstrate that our proposed TSPR algorithm can effectively enhance the quality of phase recovered and the probability of successful recovery.

2. Basic Principles

The complex amplitude $g(x, y)$ is

$$g(x, y) = |g(x, y)| \exp(i\varphi(x, y)) \tag{1}$$

Its Fourier transform form is

$$G(\xi, \eta) = |G(\xi, \eta)| \exp(i\psi(\xi, \eta)) \tag{2}$$

where $g(x, y)$ and $G(\xi, \eta)$ denote the complex amplitude in the spatial domain and Fourier domain, respectively, $|g(x, y)|$ and $|G(\xi, \eta)|$ denote the amplitude in the spatial domain and Fourier domain, respectively, $\varphi(x, y)$ and $\psi(\xi, \eta)$ denote the phase in the spatial domain and Fourier domain, respectively. Phase retrieval refers to recovering the losing phase $\varphi(x, y)$ in the spatial domain from two amplitudes $|g(x, y)|$ and $|G(\xi, \eta)|$.

The flow chart of the classical GS algorithm is shown in Figure 1. This algorithm was first proposed and was easily used to solve the phase retrieval problem using two intensity measurements. $\exp(i\phi_0(x, y))$ denotes the initialized phase distribution function. The algorithm repeatedly iterates between the spatial domain and the Fourier domain until the error ε satisfies the termination condition [20].

3. Two-Step Phase Retrieval Using Single-Intensity Measurement

The TIPR algorithm uses the two intensity measurements recorded in the spatial domain and the Fourier domain, respectively, to recover the amplitude and phase in the spatial domain with high quality. Although the TIPR algorithm can achieve good results, in some case, it is difficult to obtain two intensity measurements simultaneously in the both spatial and Fourier domains.

Therefore, the SIPR algorithm is proposed, and it can recover the phase from single-intensity measurement in the Fourier domain with some prior knowledge. However, the SIPR method has low probability of successful recovery and poor quality. As shown in Figure 2, the purpose of the experiment is to test the SIPR method's performance to recover the amplitude and phase. The coded aperture is generated by a uniform distribution. The sampling rate of the coded aperture is used as a variable parameter in the experiment. The sampling rate gradually increased from 0.1 to 0.7; the step length is 0.05. The experiment runs 500 times independently to compute the probability of successful recovery. If the SNR of the phase reconstructed is greater than 25 dB, the reconstruction is deemed as successful. The probability of successful recovery versus sampling rate for both amplitude and phase is shown in Figure 2.

As depicted in Figure 2, it can be found that the SIPR method can achieve high probability of successful recovery

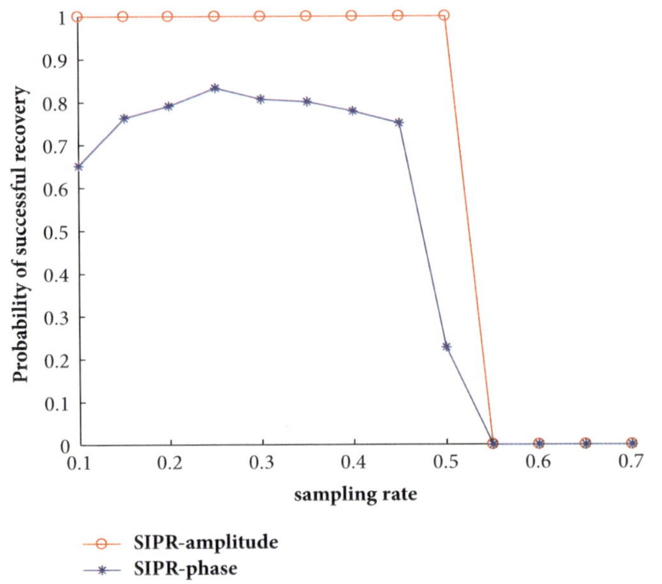

FIGURE 2: Probability of successful recovery of amplitude and phase versus sampling rate.

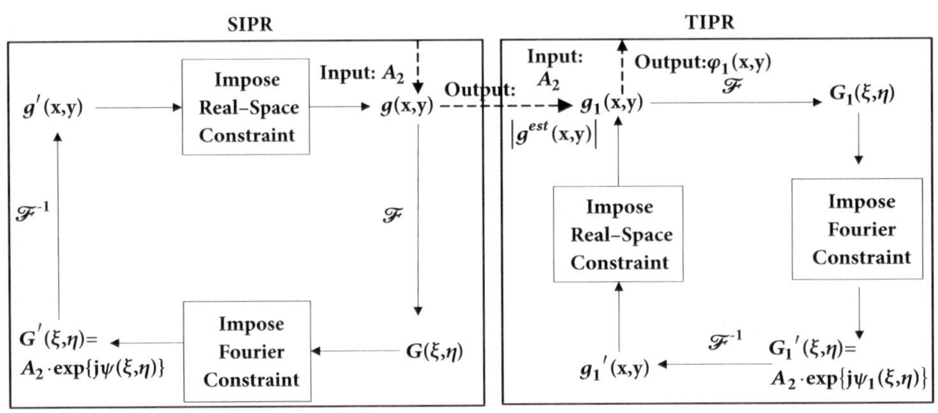

FIGURE 3: Flow diagram of the TSPR algorithm.

of the amplitude, but the probability of successful recovery of the phase is very low and unstable. Hence, our TSPR method is proposed to solve the phase retrieval problem from single-intensity measurement. Firstly, SIPR method is used to recover the amplitude in the spatial domain from the intensity in the Fourier domain and the support set in the spatial domain. Then, TIPR method using two intensity measurements (one of the two intensity measurements is recorded in the Fourier domain and the other is estimated in the spatial domain from the first step) is used to recover the phase in the spatial domain.

The flow chart of TSPR method is shown in Figure 3, which includes the following steps: (1) initialize the amplitude A_1 and phase $\varphi_0(x, y)$ in the spatial domain, i.e., use all-one amplitude and randomized phase. (2) Perform Fourier transform on the complex amplitudes and replace the amplitude after the Fourier transform in the Fourier domain with the recorded Fourier amplitudes A_2. (3) Perform the inverse Fourier transform on the synthesized complex amplitude in the Fourier domain and multiply it with the support in the spatial domain. (4) Repeat steps (2)-(3) until the amplitude converges in the spatial domain. (5) The classical TIPR algorithm using the recorded Fourier amplitude A_2 and the estimated amplitude $|g^{est}(x, y)|$ from the first step is used to iteratively recover the phase in the spatial domain.

The complex amplitude of the initialization is as follows:

$$g(x, y) = A_1 .* \exp\left[\varphi_0(x, y)\right] \quad (3)$$

The optical setup of imaging process of the TSPR method is shown in Figure 4. The coded aperture and CCD are placed on the front and back focal planes of the lens, respectively.

Firstly, the complex optical field $u(x, y)$ is filtered through the coded aperture M to obtain a complex optical field $u_M(x, y)$ and then imaged by a Fourier lens; finally the intensity is recorded in the CCD plane. The intensity recorded at the CCD plane is expressed as follows:

$$I_{CCD} = \left|\mathcal{F}\left[u_M(x, y)\right]\right|^2 \quad (4)$$

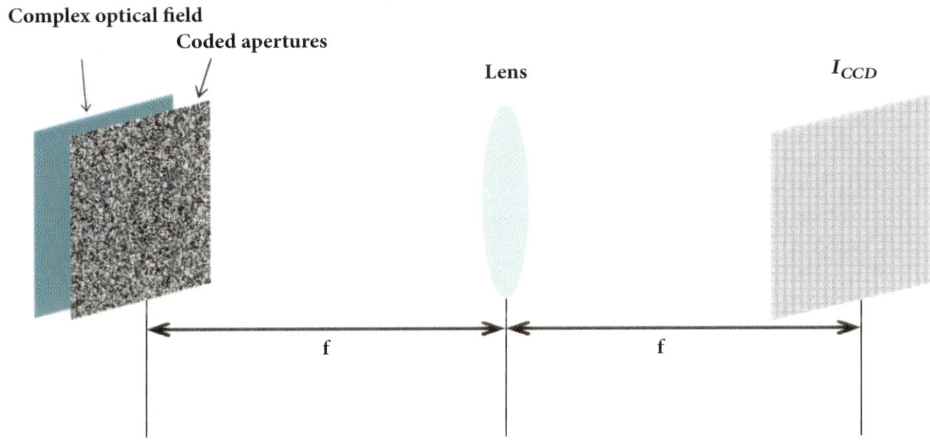

FIGURE 4: The optical setup of imaging process.

FIGURE 5: Comparison of SIPR algorithm and TSPR algorithm. (a) A 0/1 randomly distributed coded aperture. (b) and (c) are the amplitude and phase of the complex optical field behind the coded aperture. (d) The intensity recorded in the Fourier domain. (e) and (f) are the amplitude and phase recovered using SIPR algorithm. (g) and (h) are the amplitude and phase recovered using TSPR algorithm.

The main purpose of this paper is to recover the phase of the complex optical field $u_M(x, y)$ from the recorded intensity measurement I_{CCD}.

4. Numerical Experiments

In order to verify the effectiveness and superiority of our proposed method, three experiments are presented in this section. The first simulation experiment is the single reconstruction experiment. The second simulation experiment is to test the performance of TSPR algorithm under different coded apertures. The final simulation experiment is to compare the performance of the different phase retrieval algorithms.

4.1. Experiment 1: Single Reconstruction Experiment. The purpose of the experiment is to verify the feasibility of our proposed TSPR method. Two grayscale images ("Lena" and "Cameraman" with 256 × 256 pixels) are chosen in the experiment. The two images are multiplied by the coded aperture (256 × 256 pixels) to obtain the amplitude and phase, respectively. Then the complex amplitude is synthesized by amplitude and phase. The coded aperture is 0/1 random distribution. The sampling rate in this experiment is 0.4, and the phase retrieval results are shown in Figure 5.

Figure 5(a) is the coded aperture with 0/1 randomly distributed, and the sampling rate is 0.4. Figures 5(b) and 5(c) are the amplitude and phase of the complex optical field behind

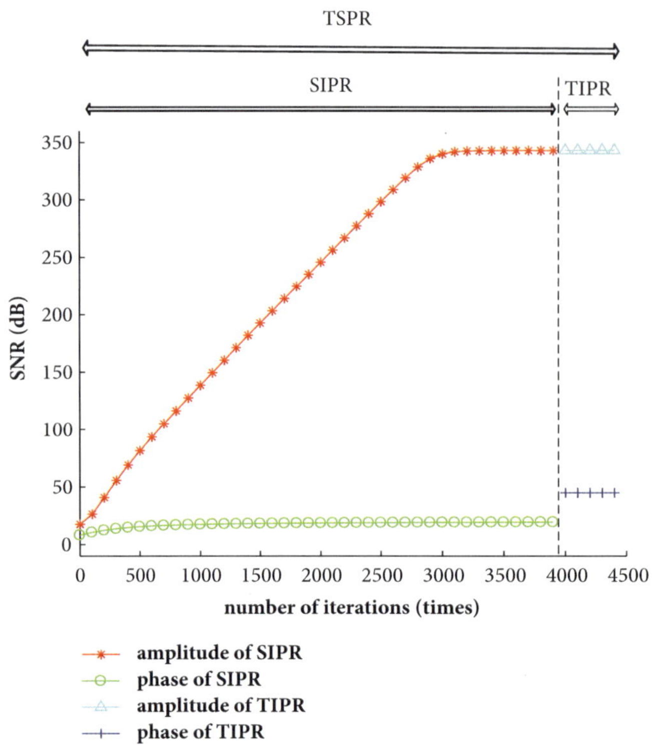

FIGURE 6: Amplitude and phase versus the number of iterations.

(a) (b) (c)

FIGURE 7: Three types of coded apertures with sampling rate of 0.3. (a) Uniform random sampling pattern. (b) Radial line sampling pattern. (c) Variable density sampling pattern.

the coded aperture to be retrieved. Figure 5(d) is the intensity recorded in the Fourier domain. Figures 5(e) and 5(f) are the amplitude and phase recovered using the SIPR algorithm, and the Signal-to-Noise Ratio (SNR) is 343.04 dB and 19.85 dB, respectively. Figures 5(g) and 5(h) are the amplitude and phase recovered using the TSPR algorithm, and their SNR is 343.04dB and 45.58dB, respectively. Comparing Figures 5(f)–5(h), it is easy to find that the quality of phase recovered is greatly improved by our TSPR method.

The trend of the amplitude and phase with the number of iterations in our TSPR method is shown in Figure 6. In the first step (i.e., SIPR) of TSPR, the amplitude can be well recovered, but the phase recovered is poor. As the number of iterations increases, the SNR of the phase recovered can only reach about 19dB. However, in the second step (i.e., TIPR) of the TSPR, the SNR of the phase recovered increases as high as 45 dB. Compared with SIPR method, TSPR method can greatly improve the SNR of phase reconstructed.

4.2. Experiment 2: Reconstruction Experiments under Different Coded Apertures. The coded aperture in Experiment 1 is a uniform random sampling pattern. The purpose of Experiment 2 is to test the performance of the TSPR method under different coded apertures.

In Figure 7, three different types of the coded apertures are selected, such as Uniform Random (UR) sampling, Radial Line (RL) sampling, and Variable Density (VD) sampling. The sampling rate of the coded aperture is used as a variable parameter in the experiment, and the sampling rate is set from 0.1 to 0.7. Each set of experiments with the

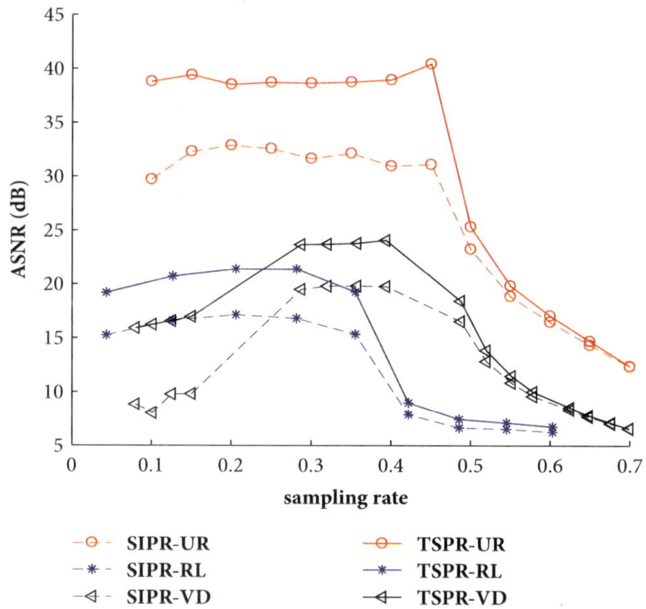

FIGURE 8: ASNR versus sampling rate.

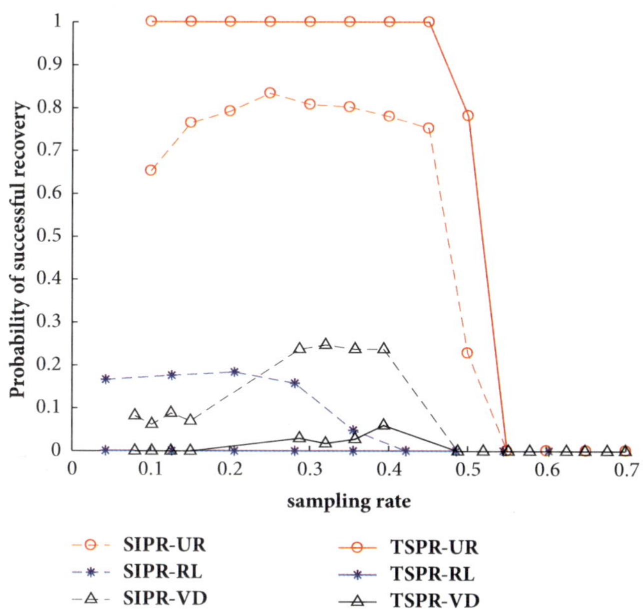

FIGURE 9: Probability of successful recovery versus sampling rate.

same parameters runs 500 times independently. Compute the recovered Average Signal-to-Noise Ratio (ASNR) and probability of successful recovery. The ASNR and probability of successful recovery versus sampling rate under different coded apertures are shown in Figures 8 and 9, respectively.

As shown in Figure 8, as the sampling rate increases, the TSPR algorithm achieves higher ASNR for phase recovered than the SIPR algorithm under three different coded apertures. In Figure 9, the best choice for coded aperture is uniform random sampling pattern, which achieve the optimal performance. The probability of successful recovery has also been significantly improved and has a good stability.

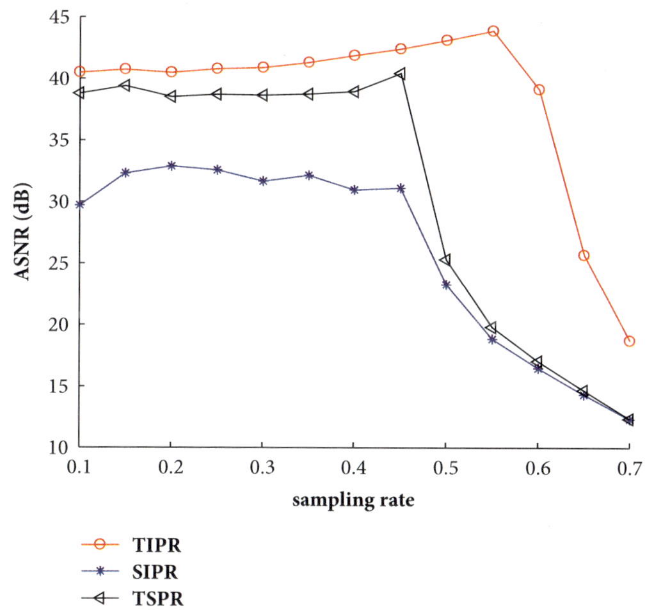

FIGURE 10: ASNR versus sampling rate.

The reason is that the support set of the uniform random sampling pattern is relatively uniform distributed in the spatial domain, while the support set of the other two coded apertures is more concentrated to the center, so the phase cannot be retrieved well.

4.3. Experiment 3: Reconstruction Experiment under Different Sampling Rates. The purpose of Experiment 3 is to compare the performance of phase retrieval between our TSPR method and the other two algorithms (TIPR algorithm and SIPR algorithm) under different sampling rates. The coded aperture uses a uniform random sampling pattern with a sampling rate varies from 0.1 to 0.7 and the step size is 0.05. Under the same parameters, experiments runs 500 times independently. The ASNR and probability of successful recovery versus sampling rate for different algorithms are presented in Figures 10 and 11, respectively.

As shown in Figures 10 and 11, the TIPR method is the best, the TSPR method is the second, and the SIPR method is the worst. The reason is that TIPR method has the most known information, i.e., the two intensity measurements at two different planes. The difference between the TSPR method and the SIPR method is that the TSPR method draws on the idea of the TIPR method. Firstly, use the SIPR method to estimate the amplitude in the spatial domain, then use the TIPR method to recover the phase. In summary, the TSPR method is superior to the SIPR method.

5. Conclusion

This paper proposes an improved TSPR algorithm to solve the problem of poor quality and low success rate of SIPR method. TSPR algorithm is completed in two steps: firstly, SIPR algorithm is used to recover the amplitude in the spatial domain, then TIPR algorithm is used to recover the phase in

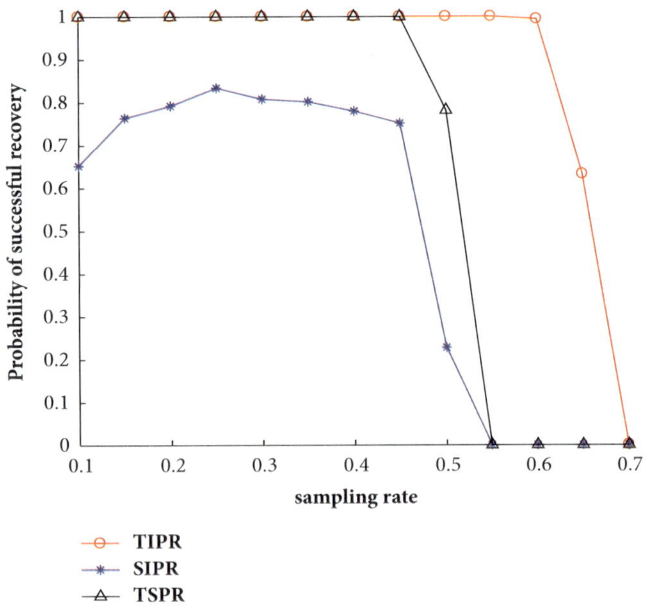

Figure 11: Probability of success versus sampling rate.

the spatial domain. Finally, the effectiveness and superiority of the TSPR method are verified. The experiment comparison results demonstrate that the proposed method can effectively improve the ASNR and probability of successful recovery under the same parameters. TSPR algorithm can recover the lost phase from known amplitude in the Fourier domain and support set in the spatial domain.

Conflicts of Interest

The authors declare that they have no conflicts of interest.

Acknowledgments

This work was supported by the National Natural Science Foundation of China (Grant Numbers 61501001, 61605002, 61301296, and 61377006) and the Natural Science Foundation of Anhui Province (Grant Numbers 1508085MF121, 1608085QF161).

References

[1] Y. Shechtman, Y. C. Eldar, O. Cohen, H. N. Chapman, J. Miao, and M. Segev, "Phase retrieval with application to optical imaging: a contemporary overview," *IEEE Signal Processing Magazine*, vol. 32, no. 3, pp. 87–109, 2015.

[2] R. W. Harrison, "Phase problem in crystallography," *Journal of the Optical Society of America A: Optics, Image Science & Vision*, vol. 10, no. 5, pp. 1046–1055, 1993.

[3] R. P. Millane, "Phase retrieval in crystallography and optics," *Journal of the Optical Society of America A: Optics, Image Science & Vision*, vol. 7, no. 3, pp. 394–411, 1990.

[4] J. A. Rodriguez, R. Xu, C.-C. Chen, Y. Zou, and J. Miao, "Oversampling smoothness: An effective algorithm for phase retrieval of noisy diffraction intensities," *Journal of Applied Crystallography*, vol. 46, no. 2, pp. 312–318, 2013.

[5] A. Walther, "The question of phase retrieval in optics," *Journal of Modern Optics*, vol. 10, pp. 41–49, 1963.

[6] M. C. Newton, "Compressed sensing for phase retrieval," *Physical Review E: Statistical, Nonlinear, and Soft Matter Physics*, vol. 85, no. 5, 2012.

[7] J. Miao, P. Charalambous, J. Kirz, and D. Sayre, "Extending the methodology of X-ray crystallography to allow imaging of micrometre-sized non-crystalline specimens," *Nature*, vol. 400, no. 6742, pp. 342–344, 1999.

[8] J. Hunt, J. Gollub, T. Driscoll et al., "Metamaterial microwave holographic imaging system," *Journal of the Optical Society of America A: Optics, Image Science & Vision*, vol. 31, no. 10, pp. 2109–2119, 2014.

[9] C. Guo, S. Liu, and J. T. Sheridan, "Iterative phase retrieval algorithms. I: optimization," *Applied Optics*, vol. 54, no. 15, pp. 4698–4708, 2015.

[10] C. Fienup and J. Dainty, "Phase retrieval and image reconstruction for astronomy," *Image Recovery: Theory and Application*, vol. 231, p. 275, 1987.

[11] J. Miao, T. Ishikawa, Q. Shen, and T. Earnest, "Extending X-ray crystallography to allow the imaging of noncrystalline materials, cells, and single protein complexes," *Annual Review of Physical Chemistry*, vol. 59, pp. 387–410, 2008.

[12] H. M. L. Faulkner and J. M. Rodenburg, "Movable aperture lensless transmission microscopy: A novel phase retrieval algorithm," *Physical Review Letters*, vol. 93, no. 2, Article ID 023903, 2004.

[13] M. Dierolf, A. Menzel, P. Thibault et al., "Ptychographic X-ray computed tomography at the nanoscale," *Nature*, vol. 467, no. 7314, pp. 436–439, 2010.

[14] G. Situ, J. P. Ryle, U. Gopinathan, and J. T. Sheridan, "Generalized in-line digital holographic technique based on intensity measurements at two different planes," *Applied Optics*, vol. 47, no. 5, pp. 711–717, 2008.

[15] D. Deng, W. Qu, W. He, Y. Wu, X. Liu, and X. Peng, "Phase retrieval for digital holographic microscopy with defocused holograms," *IEEE Photonics Journal*, vol. 10, no. 1, 2018.

[16] R. W. Gerchberg and W. O. Saxton, "A practical algorithm for the determination of the phase from image and diffraction plane pictures," *Optik - International Journal for Light and Electron Optics*, vol. 35, no. 2, pp. 237–250, 1972.

[17] J. R. Fienup, "Reconstruction of an object from the modulus of its Fourier transform," *Optics Expresss*, vol. 3, no. 1, pp. 27–29, 1978.

[18] J. R. Fienup, "Phase retrieval algorithms: a comparison," *Applied Optics*, vol. 21, no. 15, pp. 2758–2769, 1982.

[19] G. Yang, B. Gu, and B. Dong, "Theory of the amplitude-phase retrieval in any linear transform system and its applications," *International Journal of Modern Physics B*, vol. 07, no. 18, pp. 3153–3224, 1993.

[20] J. R. Fienup, "Phase retrieval algorithms: A personal tour," *Applied Optics*, vol. 52, no. 1, pp. 45–56, 2013.

18

Propagation Properties of Airy Beam through Periodic Slab System with Negative Index Materials

Long Jin and Xingqiang Zhang

Department of Basic Science, Hubei University of Automotive Technology, Shiyan 442002, China

Correspondence should be addressed to Long Jin; crazyjinlong@163.com

Academic Editor: Sulaiman Wadi Harun

Based on light transfer matrix and electric field vector equation, the evolution of Airy beam propagating in periodic slab system with three negative index materials (NIMs) and its transmission mechanism are investigated. The intensity profiles on emergent surface of periodic slab system and side view of Airy beam propagating in each right handed material (RHM) and double negative material (DNM) unit including lossless and losses DNMs are discussed. It is revealed that the self-recovery Airy beam can be achieved in long distance by using lossless periodic slab system as long as the negative refractive index $n_l = -n_r$ and each unit length $L = Z$. As to losses slab system contained DNMs, the smaller the collision frequencies are, the better the Airy beam quality is formed. It is expected that the proposed manner of beam transmission and corresponding conclusions can be useful for extension applications of optical control, especially for optical communication and optical encryption technique.

1. Introduction

Airy beam, which makes up one of the important nondiffracting waves, was first described by Berry and Balazs in 1979 by solving Airy solution of the paraxial wave equation [1]. Recently, Siviloglou et al. observed the 1D and 2D Airy phenomena through experiment in 2007 [2]; they utilized spatial light modulator combined with Fourier transform technique to generate the accelerating finite energy Airy beam. From then on, the particular properties of this beam, approximate nondiffraction, parabolic path of the central lobe, self-accelerating in the transverse, and self-recovery, have been an attractive topic for both theoretical and applied study in modern optics [3]. At the same time, a considerable number of frontier and interdisciplinary applications, such as optical microparticles manipulation [4], formation of Airy beam lasers [5], generation of optical bullets [6], routing of surface plasmon polaritons [7], generation of curved plasma channels [8], and creating curved electron beams without any external field [9], have been took advantage of the Airy function and Airy beam technology.

As far as the width and depth are concerned, optical film and related technology had been significantly developed with the progress of science and technology. The preparation process of optical film is using different refractive index materials to coat a periodic media structure [10]. These optical components possessing the properties of spectral transmittance, spectral reflection, spectral absorption, and altering the state of polarization and/or phase of light are widely used in manufacturing high reflectance film, antireflective film, interference filter, and so on [11]. Meanwhile, metamaterials so-called negative index materials (NIMs) have mushroomed in many aspects of material physics and chemistry science [12]. Double negative materials (DNMs) with both permittivity ε and permeability μ are known as left handed materials (LHMs), whose extraordinary electromagnetic (EM) properties were first presented by Veselago [13]. Beyond that, there are single negative materials (SNMs) among which only one of the EM parameters is negative; these materials include the mu negative media (MNGM, $\varepsilon > 0$, $\mu < 0$) and the epsilon negative media (ENGM, $\varepsilon < 0$, $\mu > 0$) [14, 15]. There exist abnormal negative refraction effect [16], inverse Cerenkov radiation, reverse Doppler effect [17], and negative radiation pressure [18] in NIMs because the electromagnetic wave group velocity and phase velocity of these materials are oriented in the opposite

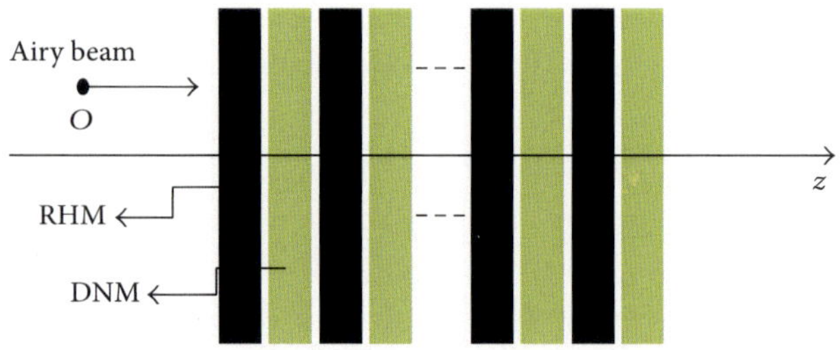

FIGURE 1: Light path schematic diagram of periodic slab system with DNMs.

direction; based on these principles, optical film contained NIMs has broad application prospects in microwave communication, large capacity storage medium, and superresolution imaging.

In the letter, we mainly research the evolution of Airy beam propagation in optical film with DNMs, but what we want to emphasize is that our optical film model is 1D periodic slab system, with its period number being very small and each unit is measured in metre in order to conveniently observe beam transmission mechanism. The rest of the paper is organized as follows: firstly, the general propagation formulae of the Airy beam propagating through the paraxial optical system are calculated in space domain by transfer matrix and the Collins integral. Furthermore, the intensity profiles on emergent surface of periodic slab system and side view of Airy beam transmission in each RHM and DNM unit including lossless and losses DNMs are discussed. Finally, some important conclusions are summarized. It is expected that the proposed manner of wave transmission and the corresponding conclusions can be useful for extension applications of optical control, especially for optical communication and optical encryption technique.

2. Physical Model and Formulae

The schematic diagram of 1D periodic slab system with NIMs is shown in Figure 1; the medium layer with black colour stands for RHM and the light green is NIM. Here we choose isotropic DNM with its relative permittivity and permeability given by the Drude model [19]:

$$\varepsilon_r(\omega) = 1 - \frac{\omega_{pe}^2}{\omega(\omega + i\gamma_e)}, \qquad \mu_r(\omega) = 1 - \frac{\omega_{pm}^2}{\omega(\omega + i\gamma_m)}, \tag{1}$$

where ω_{pe} and ω_{pm} are the electric plasma frequency and the magnetic plasma frequency. γ_e, γ_m are the electric and magnetic collision frequencies, respectively. The frequency $\nu = \omega/(2\pi)$ is measured in 10^2 THz, and ω is angular frequency. DNM's refractive index $n_l = -\sqrt{\varepsilon_r(\omega)\cdot\mu_r(\omega)}$; the length of each DNM unit is L. With regard to RHM, the refractive index and each unit length are n_r and Z, respectively.

Considering an Airy beam that propagates from O point along the direction of z, the electric field at $z = 0$ (O point) can be described as [20, 21]

$$E(x_1, y_1) = Ai\left(\frac{x_1}{x_0}\right)\exp\left(\frac{a_0 x_1}{x_0}\right) \times Ai\left(\frac{y_1}{y_0}\right)\exp\left(\frac{b_0 y_1}{y_0}\right), \tag{2}$$

where $Ai(\cdot)$ denotes the Airy function, x_0 and y_0 are beams horizontal size and vertical size, respectively, a_0 and b_0 represent exponential modulation factors, and $E(\cdot)$ is electric field intensity on incident surface. The propagation of beam through paraxial periodic slab system with NIMs can be described by the Collins integral [22]:

$$E_2(x, y, z) = \left(\frac{-i}{\lambda B}\right)e^{ikz}\iint_{s_1} E_1(x_1, y_1, 0) \cdot e^{((ik/2B)[A(x_1^2+y_1^2)+D(x^2+y^2)-2(x_1x+y_1y)])}dx_1dy_1, \tag{3}$$

where $\lambda = 1.55\,\mu$m is the wavelength of Airy beam, $k = 2\pi/\lambda$ is the wave number. A, B, C, and D are the transfer matrix elements of the total paraxial optical systems T:

$$T = \begin{pmatrix} A & B \\ C & D \end{pmatrix}. \tag{4}$$

Now, let us discuss some useful matrices of optical medium. When beam is traveling x distance in DNM or RHM, the transfer matrix can be written as

$$M(x) = \begin{pmatrix} 1 & x \\ 0 & 1 \end{pmatrix}. \tag{5}$$

While beam propagates from RHM into DNM slab, the transfer matrix form in RHM-DNM interface is given by

$$M(n_r, n_l) = \begin{pmatrix} 1 & 0 \\ 0 & \dfrac{n_r}{n_l} \end{pmatrix}. \tag{6}$$

Likewise, DNM-RHM interface transfer matrix is

$$M(n_l, n_r) = \begin{pmatrix} 1 & 0 \\ 0 & \dfrac{n_l}{n_r} \end{pmatrix}. \tag{7}$$

By cross-producing these matrices in accordance with the order of Airy beam passing through periodic slab system, we can obtain the general propagation formula of the Airy beam propagating through the paraxial optical systems in the space domain:

$$E(x) = \frac{1}{\sqrt{A}}Ai\left(\frac{x}{Ax_0} - \frac{B^2}{4A^2k^2x_0^4} + i\frac{a_0B}{Akx_0^2}\right) \cdot \exp\left(-i\frac{kCx^2}{2A}\right)\cdot\exp\left(\frac{a_0x}{Ax_0} - a_0\frac{B^2}{2A^2k^2x_0^4} - i\frac{B^3}{12A^3k^3x_0^6} + i\frac{a_0^2B}{2Akx_0^2} + i\frac{Bx}{2A^2kx_0^3}\right),$$

$$E(y) = \frac{1}{\sqrt{A}} Ai\left(\frac{y}{Ay_0} - \frac{B^2}{4A^2k^2y_0^4} + i\frac{b_0 B}{Aky_0^2}\right)$$
$$\cdot \exp\left(-i\frac{kCy^2}{2A}\right) \cdot \exp\left(\frac{b_0 y}{Ay_0} - b_0\frac{B^2}{2A^2k^2y_0^4}\right.$$
$$\left. - i\frac{B^3}{12A^3k^3y_0^6} + i\frac{b_0^2 B}{2Aky_0^2} + i\frac{By}{2A^2ky_0^3}\right),$$
$$E(x,y) = E(x) \times E(y). \quad (8)$$

And the light intensity is easily obtained by using complex conjugate of $E(x, y, z)$:

$$I = \frac{n}{2c\mu_0}|E(x,y,z)|^2 \propto |E(x,y,z)|^2. \quad (9)$$

3. Results and Discussion

To delve into the evolution of Airy beam passes through periodic slab system $(AB)_m$, where A and B are RHM and DNM, respectively, and $m = 2$ is period number of this system, we carry out Airy beam transmission numerical calculation and display some simulation findings. Here we assume $x_0 = y_0 = 0.2$ mm, $a_0 = b_0 = 0.1$, each DNM is lossless, and related parameters are set as $\gamma_e = \gamma_m = 0$, $\nu = 2 \times 10^2$ THz, $\omega_{pe} = \omega_{pm} = 2\pi\nu \times \sqrt{2}$, $2\pi\nu \times \sqrt{2.55}$, and $2\pi\nu \times \sqrt{3.1}$, resulting in $n_l = -1.00$, -1.55, and -2.10, respectively, $n_r = 1.55$ fixed, $L = Z = 0.25$ m. Firstly, we depict transverse normalized intensity distribution of incident Airy beam and three kinds of emerging beams on output surface. It is easily found that the incident beam distribution and emerging beam transverse profiles are completely the same when DNM's plasma frequencies $\omega_{pe} = \omega_{pm} = 2\pi\nu \times \sqrt{2.55}$ (i.e., abs(n_l) $= n_r = 1.55$) from Figures 2(a) and 2(b), and the normalized central main lobe intensity slightly decreases when $\omega_{pe} = \omega_{pm} = 2\pi\nu \times \sqrt{3.1}$, which was discerned from the colour bar in Figure 2(c); while $\omega_{pe} = \omega_{pm} = 2\pi\nu \times \sqrt{2}$ and emerging central maximum intensity is attenuated significantly, the quantity value is 0.57 compared to that of $\omega_{pe} = \omega_{pm} = 2\pi\nu \times \sqrt{2.55}$.

The intensity evolution of Airy beam propagating in periodic slab system $(AB)_2$ is deeply investigated in order to show how wave transfers from RHM to DNM; the simulation results are depicted in Figure 3; green arrow lines show the material boundaries; this reveals that the maximum intensity decreases and intensity distribution generates self-bending as the propagation distance increases while the Airy beam propagates in the RHM; when passing through the RHM-DNM interface, the beam undergoes negative refraction; in other words, both incident beam and refracted beam are on the same side of the interface normal; subsequently, the beam experiences inverse parabolic trajectory in DNM, which indicates the self-bending in RHM can be compensated and refocused again when wave passes through the DNM. In Figure 3(a), as DNM's $n_l = -n_r = -1.55$, the transverse pattern could go back to its original feature at 1st end of DNM and operate the stable and excellent beam quality; in consequence, we draw a conclusion that the intensity profiles of incident surface and emergent surface are completely the same on condition that $L = Z$ no matter what the value of period number m is; this case is called Airy beam which can just be compensated by each DNM, and related principle diagram is presented in Figure 3(b). The calculation result in Figure 3(c) reveals that the self-bending in RHM cannot be completely compensated by DNM when $n_l = -2.10 < -n_r$ and $L = Z = 2.5$ m; this effect accumulates accompanied by increasing of periodic slab system period number m and severely impacts the Airy beam's quality of imaging and transmission. We also depict corresponding principle diagram in Figure 3(d) to describe beam transmission characteristics under this circumstance. Now an interesting question naturally arises: How long is the DNM needed to just compensate self-bending when $n_l = -2.10$?

When Airy beam passes through the 1st RHM and DNM, the transfer matrix T can be described as

$$T = \begin{pmatrix} A & B \\ C & D \end{pmatrix}$$
$$= \begin{pmatrix} 1 & 0 \\ 0 & n_l \end{pmatrix} \times \begin{pmatrix} 1 & L \\ 0 & 1 \end{pmatrix} \times \begin{pmatrix} 1 & 0 \\ 0 & \frac{n_r}{n_l} \end{pmatrix} \times \begin{pmatrix} 1 & Z \\ 0 & 1 \end{pmatrix} \quad (10)$$
$$\times \begin{pmatrix} 1 & 0 \\ 0 & \frac{1}{n_r} \end{pmatrix}.$$

Substituting $Z = 2.5$ m, $n_r = 1.55$, and $n_l = -2.10$ into (10), the analytical transfer matrix element B can be formulated as

$$B = \frac{50}{31} - \frac{10}{21} \times L. \quad (11)$$

Hence, the necessary distance for just compensating self-bending is $L = 105/31$ m in DNM. The side intensity evolution of Airy beam propagating in quasiperiodic slab system with $Z = 2.5$ m and $L = 105/31$ m alternately is plotted in Figure 3(e); it is clearly found that the output Airy beam gets back to its original feature at the end of each DNM, which verifies our theoretical analysis. On the contrary, the length of DNM to just compensate self-bending when $n_l = -1.00$ is smaller than 2.5 m. By substituting $Z = 2.5$ m, $n_r = 1.55$, and $n_l = -1.00$ into (10), analytical transfer matrix element B can also be written as

$$B = \frac{50}{31} - L \quad (12)$$

and the quantity value is $L = 50/31$ m; therefore, the beam intensity self-bends again in the rest of the of 1st DNM because $L = 2.5$ m fixed; when passing through the RHM-DNM interface, the beam undergoes negative refraction as well and then experiences inverse parabolic trajectory even in the next RHM in Figure 3(f); Pendry once explained this overcompensation phenomenon in detail through photoelectromagnetic theory [23], and the principle diagram is depicted in Figure 3(g).

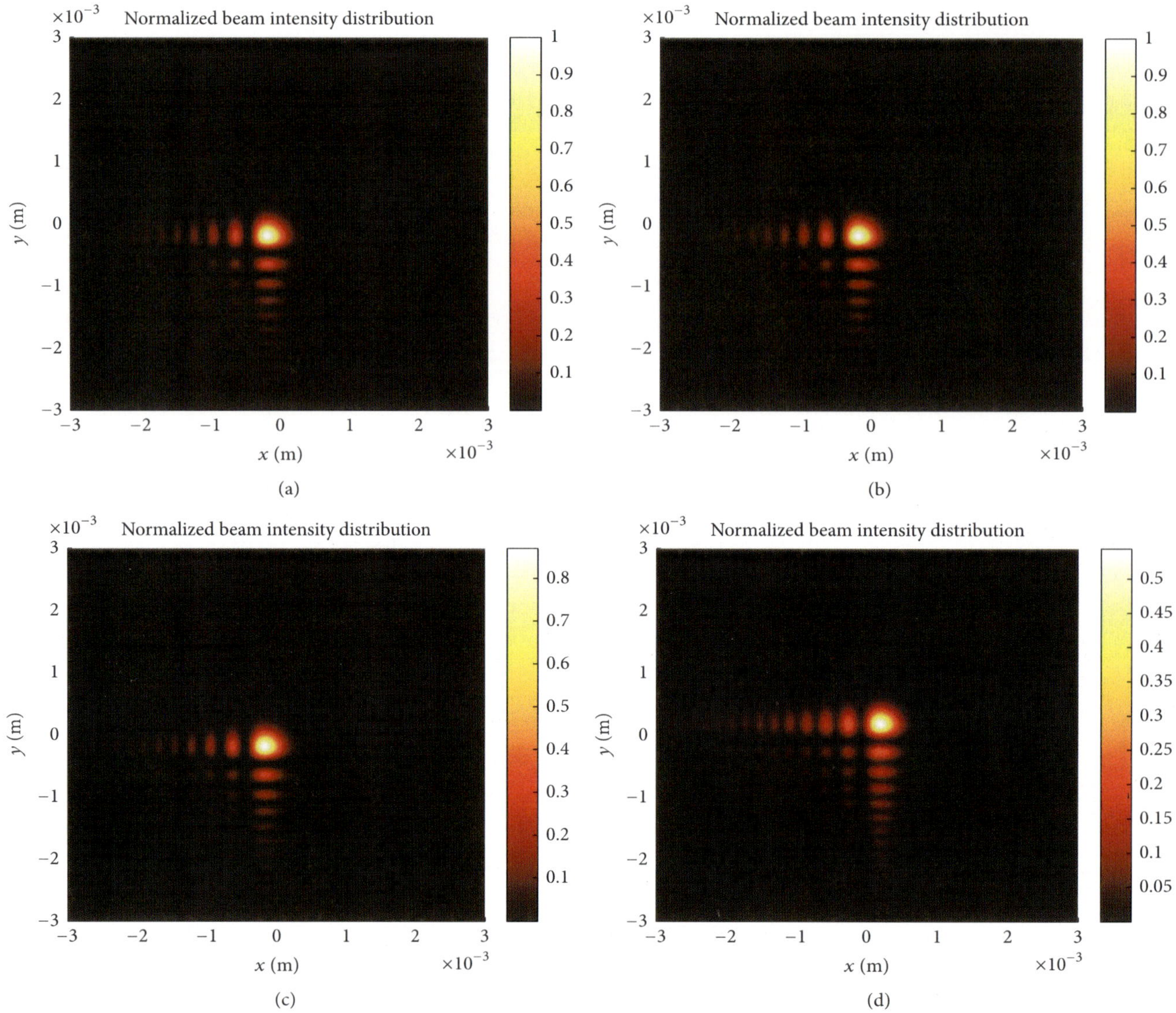

FIGURE 2: Transverse normalized intensity distribution of incident Airy beam and three kinds of emerging beams on output surface. (a) Incident Airy beam distribution, (b) emerging Airy beam distribution, DNM's $\omega_{pe} = \omega_{pm} = 2\pi\nu \times \sqrt{2.55}$, (c) emerging Airy beam distribution, DNM's $\omega_{pe} = \omega_{pm} = 2\pi\nu \times \sqrt{3.1}$, and (d) emerging Airy beam distribution, DNM's $\omega_{pe} = \omega_{pm} = 2\pi\nu \times \sqrt{2}$.

As we all known, the nondiffracting beams could be extended to infinite space and have infinite energy; they are generally unobtainable in the laboratory. However, from the above discussion, we can come to the conclusion that the self-recovery and self-reproducing Airy beam can be achieved in long distance by using periodic slab system with DNMs as long as $n_l = -n_r$, $L = Z$; the result is very important to form stable and excellent beam quality. When $n_l \neq -n_r$, the larger n_l requires smaller L to just compensate self-bending in RHMs; otherwise, the conclusion is also the opposite. $L \neq Z$ manifests that the slab system is quasiperiodic, which increases the difficulty of materials production process. Moreover, the quantitative relation between L and n_l of these quasiperiodic slab systems remains to explore and perfect further.

In practice, losses in DNMs are inevitable; in this subsection, emerging intensity profiles with four different collision frequencies γ_e are displayed in Figure 4. It is clearly acquired that the larger the collision frequencies γ_e are, the more serious the Airy beam deformation are formed. When $\gamma_e = 20$ THz, only main lobe of Airy beam existed; other side lobes are vanished or significantly weakened from Figure 4(d), which indicates that the beam is seriously distorted. Similarly, side views of Airy beam propagating in each RHM and DNM unit successively changing with two kinds of collision frequencies γ_e are exhibited in Figure 5. We can easily find that the beam propagation is rarely affected when $\gamma_e = 0.2$ THz from Figure 5(a); however, as γ_e gradually increases, the beam's attenuation and diffraction become more and more obvious; when $\gamma_e = 20$ THz, the interaction and diffraction between main lobe and side lobes lead to a distorted Airy beam; at this point, the bit error rate of communication system would greatly increase as to optical encryption. Gris-Sánchez et al. have designed and made a periodic optical fiber that guides an approximate Airy pattern as one of

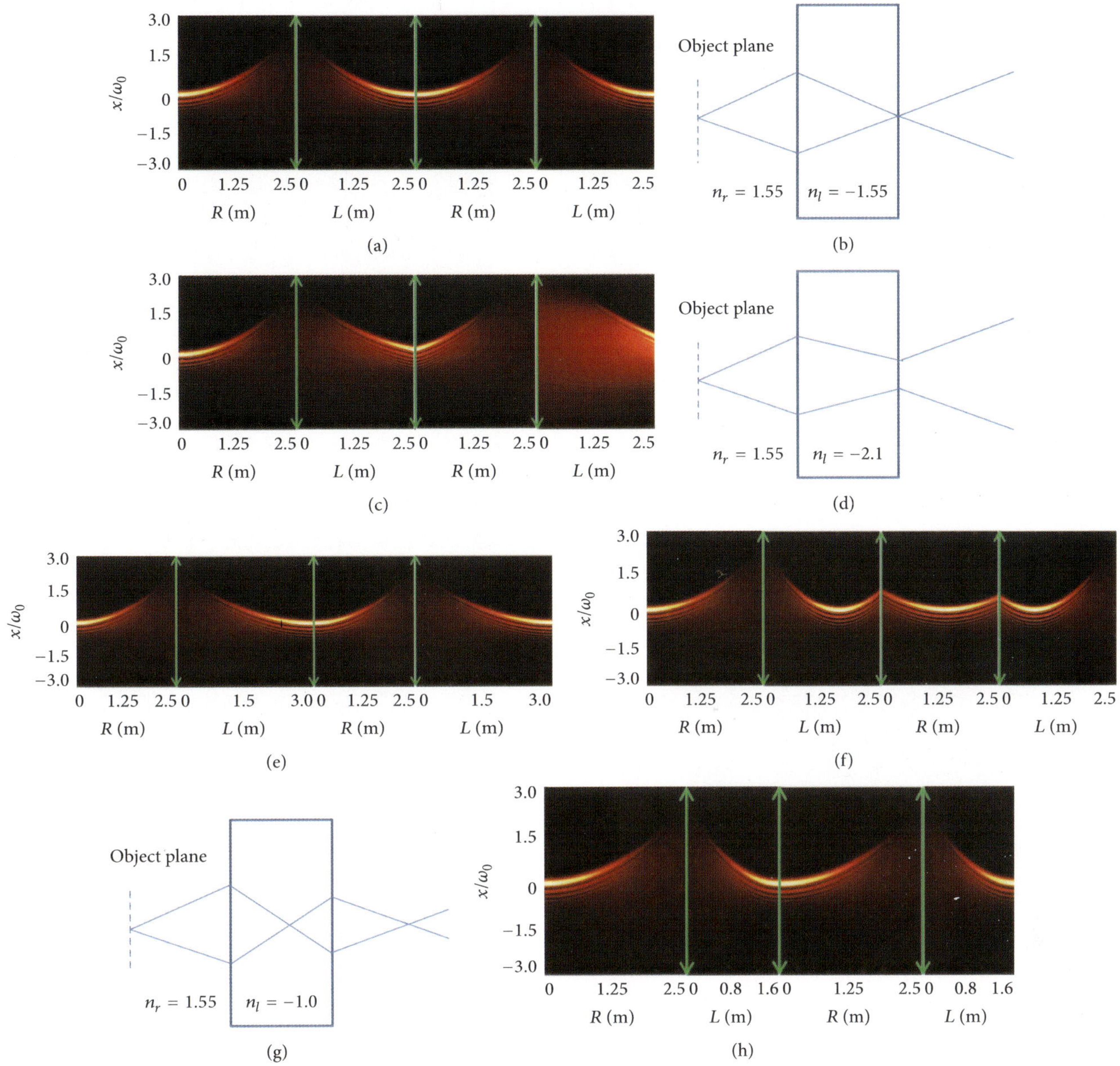

FIGURE 3: Side view of Airy beam propagating in each RHM and DNM unit successively with three kinds of DNMs and corresponding principle diagrams. (a) DNM's $\omega_{pe} = \omega_{pm} = 2\pi\nu \times \sqrt{2.55}$, (b) principle diagram where beam can just be compensated by each DNM, (c) DNM's $\omega_{pe} = \omega_{pm} = 2\pi\nu \times \sqrt{3.1}$, (d) principle diagram where beam cannot be completely compensated by each DNM, (e) Airy beam transmission in quasiperiodic slab system, $Z = 2.5$, $L = 105/31$, (f) DNM's $\omega_{pe} = \omega_{pm} = 2\pi\nu \times \sqrt{2}$, (g) principle diagram where beam is overcompensated by each DNM, and (h) Airy beam transmission in quasiperiodic slab system, $Z = 2.5$, $L = 50/31$.

the beam guided modes recently; the fiber's attenuation was 11.0 dB/km at 1.55 μm wavelength [24]. Here we estimate maximum collision frequencies value allowed by using their experimental result. Beam loss equation can be defined as [25]

$$\alpha(\lambda) = -\frac{10}{(Z+L)^2}\lg\frac{I_{out}}{I_{in}}, \quad (13)$$

where $\alpha(\lambda)$ is loss coefficient, I_{in}; I_{out} denote incident beam intensity and emerging beam intensity, respectively. By substituting $Z = L = 2.5$ m and $\alpha(\lambda) = 11.0$ dB/km into (13), we obtain $I_{out}/I_{in} = 0.975$. When DNM's collision frequencies γ_e and γ_m equal 0.2 THz, the corresponding $I_{out}/I_{in} = 0.980 \approx 0.975$; hence, we had better choose γ_e and γ_m no more than 0.2 THz (0.1% of the frequency ν) in order to achieve accurate transmission of optical communications.

4. Conclusion

In summary, we have studied the evolution of Airy beam propagating in periodic slab system with different negative

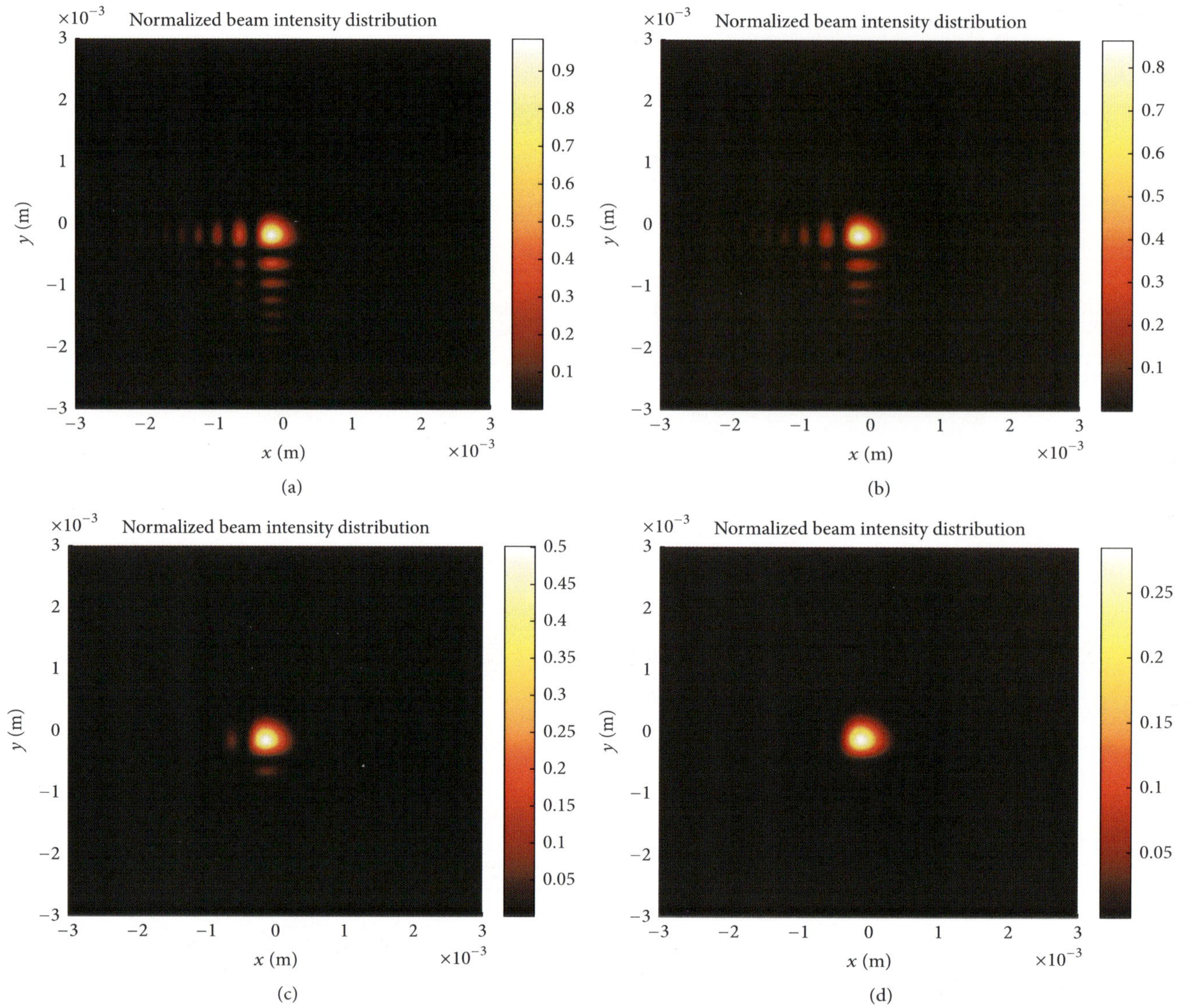

FIGURE 4: Emerging beam intensity profiles with four different collision frequencies γ_e. (a) $\gamma_e = 0.2$ THz, (b) $\gamma_e = 2.0$ THz, (c) $\gamma_e = 10$ THz, and (d) $\gamma_e = 20$ THz.

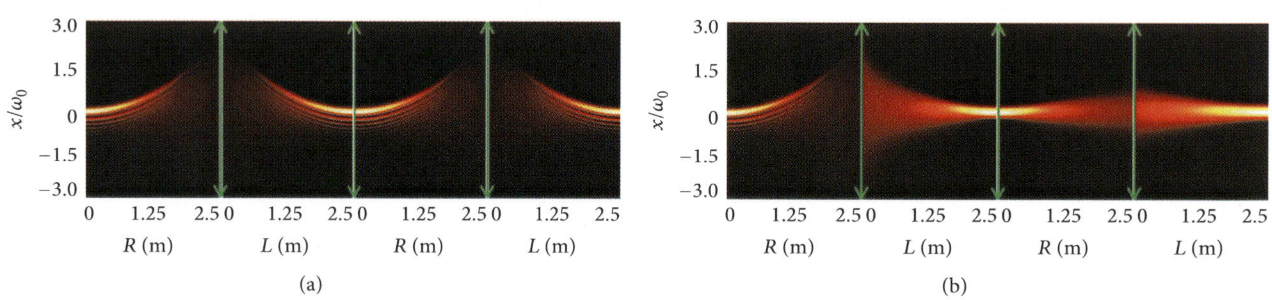

FIGURE 5: Side view of Airy beam propagating in each RHM and DNM unit changing with two different collision frequencies γ_e. (a) $\gamma_e = 0.2$ THz; (b) $\gamma_e = 20$ THz.

index materials and its transmission mechanism. The emerging intensity profiles on output surface of slab system and side view of Airy beam transmission in each RHM and DNM unit including lossless and losses DNMs have been discussed. It is revealed that the self-recovery Airy beam can be achieved in long distance by using lossless periodic slab system with negative index materials as long as $n_l = -n_r$, $L = Z$; the conclusion is very important to forming stable and excellent beam quality. As to losses periodic slab system contained DNMs, the larger the collision frequencies γ_e and γ_e are, the more serious the Airy beam deformation is formed; hence, we had better choose γ_e and γ_m no more than 0.1% of the frequency v in order to achieve accurate transmission of optical communications. It is expected that the proposed manner of wave transmission and the corresponding conclusions can be useful for the extension applications of optical control, especially for optical communication and optical encryption technique.

Conflicts of Interest

The authors declare that there are no conflicts of interest regarding the publication of this paper.

Acknowledgments

The authors gratefully acknowledge the financial support from Hubei Provincial Department of Education in China (no. B2017085).

References

[1] M. V. Berry and N. L. Balazs, "Nonspreading wave packets," *American Journal of Physics*, vol. 47, no. 3, pp. 264–267, 1979.

[2] G. A. Siviloglou, J. Broky, A. Dogariu, and D. N. Christodoulides, "Observation of accelerating airy beams," *Physical Review Letters*, vol. 99, no. 21, Article ID 213901, 2007.

[3] Y. Hu, G. A. Siviloglou, P. Zhang, N. K. Efremidis, D. N. Christodoulides, and Z. Chen, "Self-accelerating airy beams: Generation, control, and applications," *Springer Series in Optical Sciences*, vol. 170, pp. 1–46, 2012.

[4] J. Baumgartl, M. Mazilu, and K. Dholakia, "Optically mediated particle clearing using Airy wavepackets," *Nature Photonics*, vol. 2, no. 11, pp. 675–678, 2008.

[5] G. Porat, I. Dolev, and O. Barlev, "Airy beam laser," *Optics Letters*, vol. 36, no. 20, pp. 4119–4121, 2011.

[6] D. Abdollahpour, S. Suntsov, D. G. Papazoglou, and S. Tzortzakis, "Spatiotemporal Airy light bullets in the linear and nonlinear regimes," *Physical Review Letters*, vol. 105, no. 25, Article ID 253901, 2010.

[7] A. Minovich, A. E. Klein, N. Janunts, T. Pertsch, D. N. Neshev, and Y. S. Kivshar, "Generation and near-field imaging of airy surface plasmons," *Physical Review Letters*, vol. 107, no. 11, Article ID 116802, 2011.

[8] P. Polynkin, M. Kolesik, J. V. Moloney, G. A. Siviloglou, and D. N. Christodoulides, "Curved plasma channel generation using ultraintense airy beams," *Science*, vol. 324, no. 5924, pp. 229–232, 2009.

[9] N. Voloch-Bloch, Y. Lereah, Y. Lilach, A. Gover, and A. Arie, "Generation of electron Airy beams," *Nature*, vol. 494, no. 7437, pp. 331–335, 2013.

[10] H. A. Macleod, "Thin-Film Optical Filters," *CRC press*, 2001.

[11] E. Ritter, "Properties of optical film materials," *Applied Optics*, vol. 20, no. 1, pp. 21–25, 1981.

[12] D. R. Smith, J. B. Pendry, and M. C. K. Wiltshire, "Metamaterials and negative refractive index," *Science*, vol. 305, no. 5685, pp. 788–792, 2004.

[13] V. G. Veselago, "The electrodynamics of substances with simultaneously negative values of ε and μ," *Soviet Physics Uspekhi*, vol. 10, no. 4, pp. 509–514, 1968.

[14] A. Alù and N. Engheta, "Pairing an Epsilon-Negative Slab with a Mu-Negative Slab: Resonance, Tunneling and Transparency," *IEEE Transactions on Antennas and Propagation*, vol. 51, no. 10 I, pp. 2558–2571, 2003.

[15] Y.-T. Fang, J. Zhou, and E. Y. B. Pun, "High-Q filters based on one-dimensional photonic crystals using epsilon-negative materials," *Applied Physics B: Lasers and Optics*, vol. 86, no. 4, pp. 587–591, 2007.

[16] W. T. Lu, J. B. Sokoloff, and S. Sridhar, "Refraction of electromagnetic energy for wave packets incident on a negative-index medium is always negative," *Physical Review E*, vol. 69, no. 2, 26604 pages, 2004.

[17] J. Chen, Y. Wang, B. Jia et al., "Observation of the inverse Doppler effect in negative-index materials at optical frequencies," *Nature Photonics*, vol. 5, no. 4, pp. 239–242, 2011.

[18] A. Mizrahi and Y. Fainman, "Negative radiation pressure on gain medium structures," *Optics Letters*, vol. 35, no. 20, pp. 3405–3407, 2010.

[19] M. W. Feise and Y. S. Kivshar, "Sub-wavelength imaging with a left-handed material flat lens," *Physics Letters, Section A: General, Atomic and Solid State Physics*, vol. 334, no. 4, pp. 326–330, 2005.

[20] G. A. Siviloglou and D. N. Christodoulides, "Accelerating finite energy Airy beams," *Optics Letters*, vol. 32, no. 8, pp. 979–981, 2007.

[21] K. Cheng, J.-S. Xia, and X.-Q. Zhong, "Propagation dynamics and vortex trajectory of an Airy vortex beam in gradient-index media," *Acta Photonica Sinica*, vol. 43, no. 9, Article ID 0905002, 2014 (Chinese).

[22] H.-C. Lin and J.-X. Pu, "Propagation of Airy beams from right-handed material to left-handed material," *Chinese Physics B*, vol. 21, no. 5, Article ID 054201, 2012.

[23] J. B. Pendry, "Negative refraction makes a perfect lens," *Physical Review Letters*, vol. 85, no. 18, pp. 3966–3969, 2000.

[24] I. Gris-Sánchez, D. Van Ras, and T. A. Birks, "The airy fiber: An optical fiber that guides light diffracted by a circular aperture," *Optica*, vol. 3, no. 3, Article ID 030270, pp. 270–276, 2016.

[25] G. P. Agrawal, *Nonlinear Fiber Optics*, Academic Press, 2007.

19

Investigation and Suppression of Fiber Nonlinearities using Injection-Locking in OFDM-WDM System

Monika Nehra and Deepak Kedia

Department of Electronics and Communication Engineering, Guru Jambheshwar University of Science & Technology, Hisar, Haryana 125001, India

Correspondence should be addressed to Deepak Kedia; kedia29@gmail.com

Academic Editor: Sulaiman W. Harun

The fiber nonlinearities play a major role in optical communication with respect to the system performance and transmission capacity. Here, fiber nonlinear impairments are investigated through consideration of several parameters of optical fiber system, such as fiber length and core effective area. We also demonstrate the mitigation of fiber nonlinearities by using single-stage injection-locking in 3 × 10 Gb/s OFDM-WDM system. This paper focuses on selection of appropriate bias current for slave laser to suppress the four wave mixing (FWM) crosstalk effects. These findings resemble the importance of single-stage injection-locking for FWM suppression for high-speed data communication.

1. Introduction

Over the time, the requirements for bandwidth have been increased critically. The different possible solutions have come up which include architectural modifications in terms of modulator design, advance modulation formats, and integration of wireless communication with optical fiber technologies. However, there are several issues that need to be resolved such as fiber dispersion, noise, and fiber nonlinearities. With increased power levels, the different fiber nonlinearities come into picture such as four wave mixing (FWM), self-phase modulation (SPM), cross-phase modulation (CPM), stimulated Raman scattering (SRS), and stimulated Brillouin scattering (SBS). Out of these effects, the FWM and SBS processes have significant potential to alter the performance of transmission systems, even if the systems are operated using narrow linewidth single frequency lasers [1, 2]. These effects become more critical to handle in higher-channel WDM system for long distance transmission. The FWM products result throughout the changes in refractive index of fiber core with respect to optical intensity. If the FWM products fall in transmission window of baseband signals, these can give rise to crosstalk between significant channels propagating through the fiber [3]. These effects are analogous to intermodulation distortion in case of electrical systems.

The efficiency of FWM products generally depends on the fiber-chromatic dispersion, transmission length, and the channel separation. Therefore, diverse research efforts have been reported to mitigate the crosstalk due to FWM in appropriate transmission system design [4–7]. Guo and Shu [8] reported the suppression of FWM crosstalk using Raman assisted fiber optical parametric amplifier (FOPA). The Raman assisted FOPA system offered unique power evolution of signals that helped obtain 7 dB decrease in FWM crosstalk. Zhu et al. [9] reported microwave signals generation based on a novel technique utilizing optical sideband injection and FWM. This technique can produce microwave signals with frequency octupling, 12- and 14-tupling. However, the several limitations can be seen in terms of bandwidth limited photodetector that allows the signal generation up to 18 GHz only. Recently, injection-locking has emerged as one of the most promising techniques for suppression of fiber nonlinearities. Jignesh et al. [10] proposed the injection-locking technique to compensate the interchannel nonlinear phase noise that occurs mainly due to XPM. The injection-locking strongly follows the XPM-phase distortion that can be partially cancelled with the help of homodyne receiver. No doubt injection-locking technique is very effective for microwave signal generation or XPM-phase noise cancellation; it can outperform other techniques in the suppression of other fiber nonlinearities also.

Here, the fiber length and core effective area are investigated in terms of their effects on FWM power. The suppression of FWM products using master-slave injection-locking is a new concept and not much explored. The selection of bias current for slave laser has a significant impact over the system performance in terms of suppression of FWM products.

2. Concept of Fiber Nonlinearities

The terms linear and nonlinear optics basically relate to intensity-independent and intensity-depenedent phenomenon, respectively. The linear impairments usually originate due to chromatic dispersion (CD), fiber loss, and polarization mode dispersion (PMD). On the other hand, the nonlinear impairments are caused by two different mechanisms: (i) inelastic scattering of photons inside the fiber (e.g., SBS and SRS nonlinearities) and (ii) changes in the refractive index of fiber core via changes in optical power intensity (e.g., FWM, SPM, and XPM nonlinearities) [11]. Out of these nonlinear effects, FWM is a third-order nonlinearity that comes into picture when two or more than two signals are propagated through the fiber. The number of FWM crosstalks increases rapidly with the number of propagated signals (N) as $(1/2)N^3(N-1)$ [12]. The power level of FWM products is strongly dependent on channel spacing, fiber dispersion, and core effective area. The magnitude of FWM can be expressed [13, 14] as

$$P_{FWM} = \frac{\eta}{9} D^2 \gamma^2 P_i P_j P_k e^{(-aL)} \left\{ \frac{\left[1 - e^{(-aL)}\right]^2}{a^2} \right\}, \quad (1)$$

where η denotes the FWM efficiency, D denotes the CD of the fiber, P_i, P_j, and P_k denote input optical power with frequency f_i, f_j, and f_k, respectively (given that $k \neq i$), L denotes the fiber length, α denotes the attenuation coefficient, and γ is the nonlinear coefficient of the fiber, given as

$$\gamma = \frac{2\pi n_2}{\lambda A_{\mathrm{eff}}}, \quad (2)$$

where n_2 is the nonlinear refractive index of optical fiber, λ is the operating wavelength of the fiber, and A_{eff} is the core effective area. The FWM efficiency can be expressed [15] as

$$\eta = \frac{\alpha^2}{\alpha^2 + \beta^2} \left[1 + \frac{4 \exp(-\alpha L) \sin^2(\Delta\beta L/2)}{\left(1 - \exp(-\alpha L)\right)^2)} \right], \quad (3)$$

where $\Delta\beta$ is the phase-matching factor and can be expressed as

$$\begin{aligned} \Delta\beta &= \beta - \beta_o \\ &= 2\pi (f - f_o) \frac{d\beta}{d\omega} + \frac{1}{2} 2\pi (f - f_o)^2 \frac{d^2\beta}{d\omega^2} \\ &+ \frac{1}{6} 2\pi (f - f_o)^3 \frac{d^3\beta}{d\omega^3} + \frac{1}{24} 2\pi (f - f_o)^4 \frac{d^4\beta}{d\omega^4} \\ &+ \frac{1}{120} 2\pi (f - f_o)^5 \frac{d^5\beta}{d\omega^5}. \end{aligned} \quad (4)$$

Here, dispersion parameters are considered up to fifth order. The phase-matching factor varies along with the wave propagation through the fiber such as

$$\Delta\beta' = \Delta\beta - \gamma m (P_1 P_2 P_3) \left[\frac{1 - xp\left(-\alpha L_{eff}\right)}{\alpha L_{eff}} \right], \quad (5)$$

where m is an integer and L_{eff} is the effective fiber length, i.e., $L \gg L_{eff}$. Further, the effect of different parameters of optical fiber on FWM crosstalk is investigated in the next section.

3. Investigation of FWM Effects

In order to investigate the effect of FWM-induced crosstalk, the powers of all input channels are considered equal. The other parameters are considered as per the ITU:T recommendation G.653 as fiber CD (D) is 0.5 ps/km nm along with 0.25 dB/km attenuation factor (α). The dispersion effect is considered up to 5th order with $m = 0.63$. The effective core area is 67.43 μm^2 and fiber nonrefractive index (n_2) is chosen as 2.68×10^{-20} m^2/W. Figure 1 shows the FWM crosstalk induced in optical fiber system due to combined effect of dispersion up to fifth order. As the fiber length increases, the FWM power products become more critical to handle (refer to Figure 1(a)). Similarly, fiber effective area also puts limitation on its further decrement beyond 50 μm^2 that can give rise to higher FWM nonlinearities (refer to Figure 1(b)).

4. Suppression of FWM Effects using Injection-Locking

As already discussed in Section 3, the optical transmission power, channel length, and the core effective area put several limitations on the system in terms of their careful selection in order to suppress the fiber nonlinearities, i.e., FWM. The FWM effects occur due to confinement of optical field to the small effective area of fiber core over long distances. However, the core effective area as well as the optical transmission power and/or channel length cannot be increased and decreased respectively beyond certain limits. Therefore, we have put efforts over the suppression of FWM products by laser injection-locking technique. The injection-locking can suppress the FWM products by deviating the spacing between master and slave operating wavelengths [16]. The simulation setup for 3-channel OFDM-WDM system is shown in Figure 2. The generation of OFDM signals is presented elsewhere [17]. In brief, OFDM signal makes use of N closely spaced orthogonal subcarriers that are transmitted in parallel configuration. The advanced modulation technique (i.e., 16-quadrature amplitude modulation) is used to separately modulate the subcarriers. In the OFDM technique, guard band is also provided at the end of each OFDM stream in order to reduce the intersymbol interference (ISI). The three OFDM signals are generated and separately modulated with the help of Mach-Zehnder modulator (MZM) at different wavelengths. The MZM can generate optical combs with excellent properties in terms of spectral flatness. The MZM can also cause nonlinear effects depending on its static phase

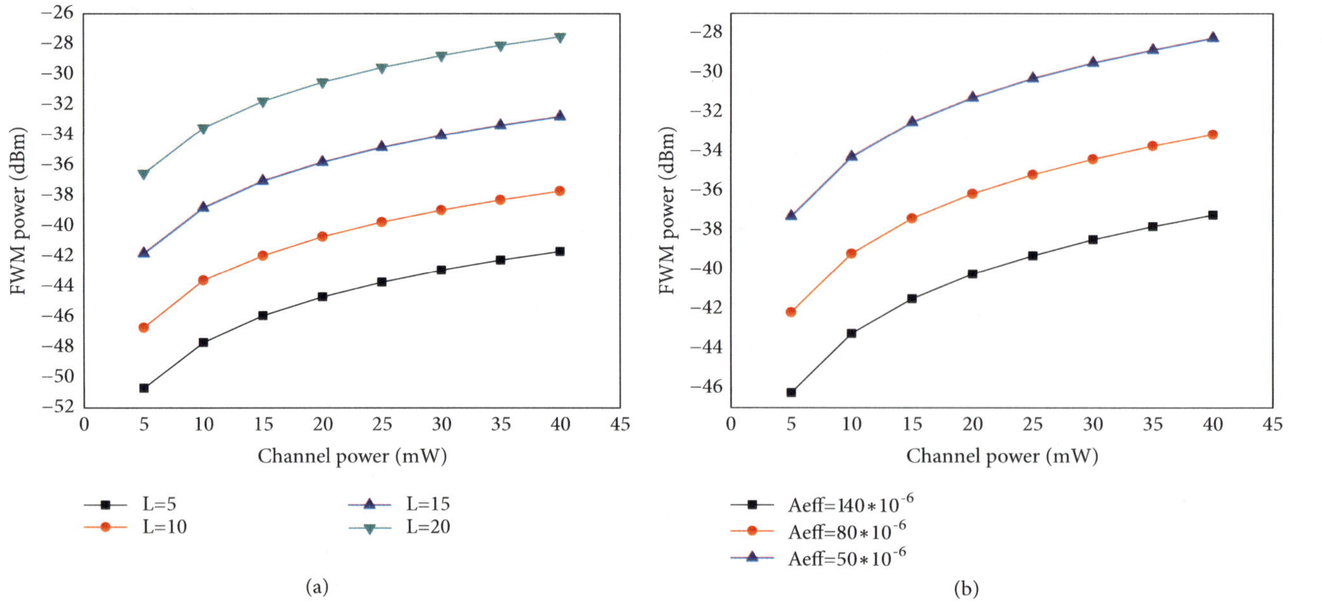

FIGURE 1: Dependence of FWM crosstalk products over channel power at different values: (a) fiber length (L) and (b) core effective area (A_{eff}).

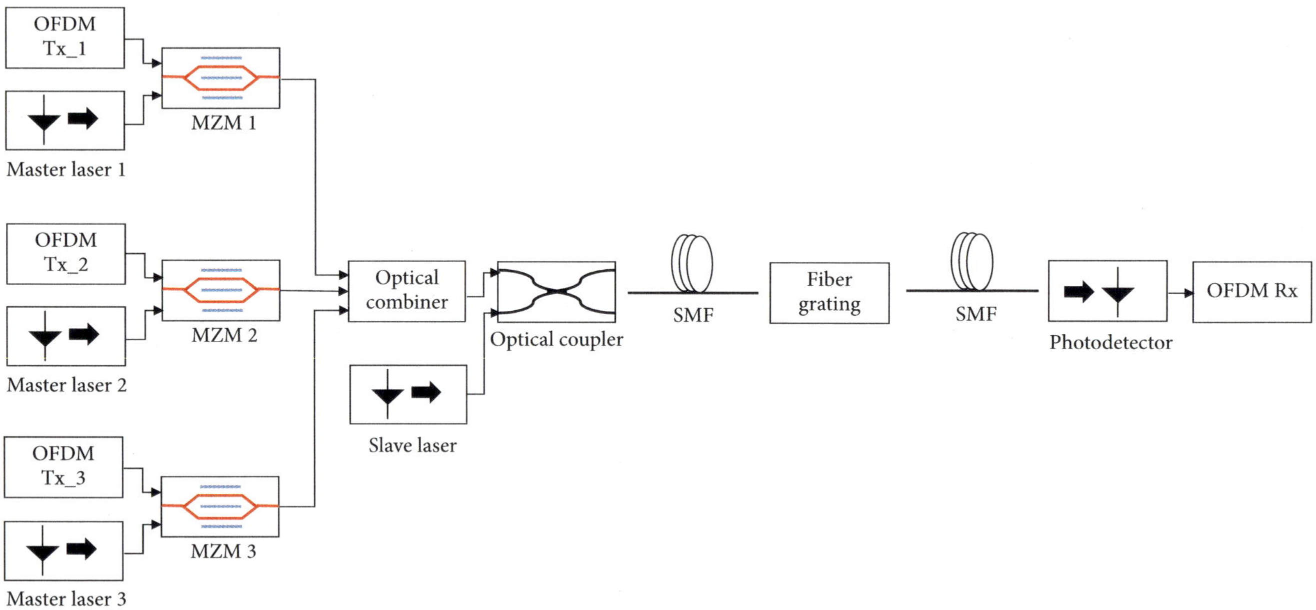

FIGURE 2: Configuration of 3-channel OFDM-WDM system using master-to-slave injection-locking technique.

shift or bias point. The power transfer characteristics of MZM are good at quadrature bias point for direct-detection systems [18].

The different parameters of OFDM-WDM system are listed in Table 1. Further, to suppress the effect of FWM crosstalk, injection-locking technique is employed. In the proposed technique, we have derived the slave laser at a nearby frequency of master lasers. This injection ratio can be defined as P_{master}/P_{slave}, where P_{master} and P_{slave} are the optical powers in case of master and slave laser, respectively. A comparative analysis of FWM crosstalk generation in optical system is presented in Figure 3. The injection-locking technique offers better performance by lowering the magnitude of FWM products, i.e. ~20 dB reduction in power level of FWM crosstalk products, than that of without injection-locking.

Further, the locking conditions of both (master and slave) lasers affect the system performance. In case of unstable injection-locking, the strong oscillations may appear in the output power caused due to beating in between the components of the optical field. The output power of injection-locking system depends upon the frequency detuning between the free-running slave laser and the injected

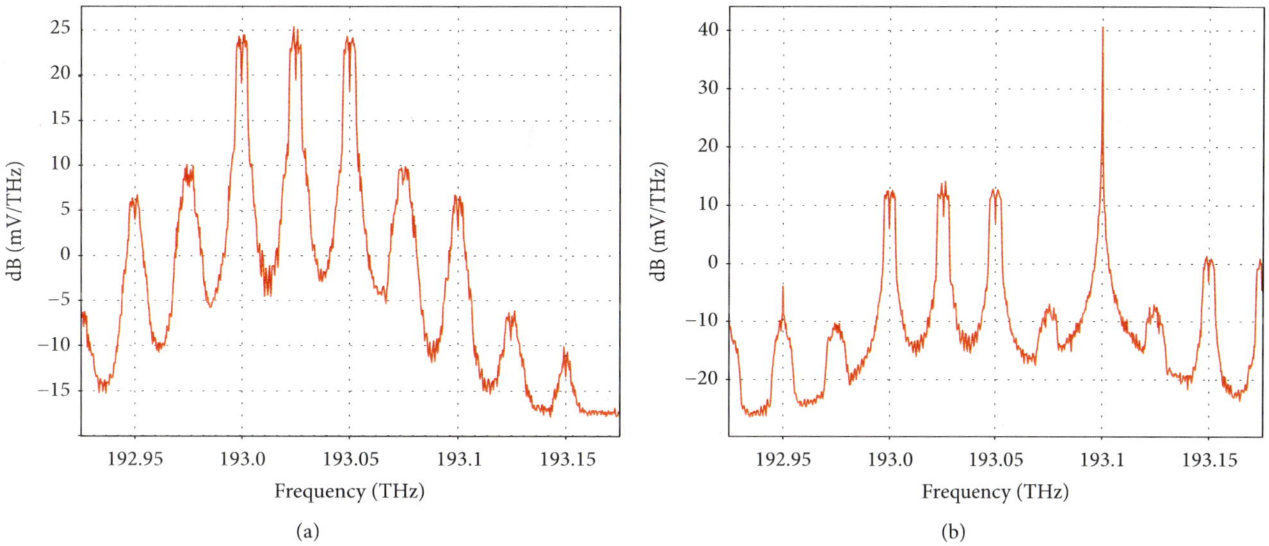

FIGURE 3: Received optical spectra of OFDM-WDM system (a) without injection-locking and (b) with injection-locking.

TABLE 1: Different parameters of 3-channel OFDM-WDM system.

Parameters	Value
Data rate	10 Gb/s
OFDM subcarriers	64
Operating wavelength for master laser 1	1553.33 nm
Operating wavelength for master laser 2	1553.13 nm
Operating wavelength for master laser 3	1552.93 nm
Operating wavelength for slave laser	1552.53 nm
Channel spacing	25 GHz
Fiber length	2×100 km
MZM bias point	$\pi/2$

optical beam. The biasing of semiconducting laser in such optical injection can help in locking of optical field components with respect to the current modulation and the optical injection [19]. The bias point should be carefully chosen in order to obtain perfect locking conditions. If the slave laser power is not strong enough to affect the locking conditions, the optical frequencies may operate in FWM regime. Figure 4 shows the variation in distribution of power among FWM cross-products as function of bias current for slave laser. The bias current of slave laser is varied from 8 mA to 40 mA. At high injection ratio, the FWM crosstalk occurs with high power levels and can degrade the system performance. Therefore, without increasing the optical power levels of master lasers, the strength of optical signal from slave laser can be tuned with the help of bias currents [20]. As biasing of slave laser increases from 8 mA to 40 mA, the power level of FWM products decreases significantly (refer to Figures 4(a)–4(f)). However, further increment in bias current of slave laser does not cause any significant reduction in power level of FWM products (refer to Figures 4(g)–4(h)). These variations in power of FWM products near the operating wavelengths of master(s) and slave lasers are plotted in Figure 5. The appropriate dc-biasing of slave laser offers improved modulation characteristics for better locking conditions in order to suppress the FWM products.

From Figure 5, it is clear that a threshold point has been achieved for appropriate biasing of slave laser in order to suppress the FWM products, i.e., 40 mA. The comparative analysis of injection-locking technique is also carried out with respect to existing literature (refer to Table 2).

Therefore, injection-locking technique offers effective suppression of FWM effects and can enhance the system performance by avoiding the complexity of the system.

5. Conclusion

In this paper, we have investigated the fiber nonlinearities, especially FWM. There are several parameters that may contribute to reducing the power level of FWM crosstalk products, i.e., appropriate selection of channel spacing and core effective area. However, significant decrement in FWM crosstalk power levels can be obtained by employing the injection-locking technique, i.e., ~20 dB reduction in power level of FWM crosstalk products. The slave laser should be biased appropriately in order to obtain stable-locking conditions. We have achieved excellent suppression of FWM products at a bias current of 40 mA for slave laser.

Conflicts of Interest

The authors declare that there are no conflicts of interest regarding the publication of this paper.

Table 2: Critical analysis of the proposed technique with respect to existing literature.

Research work(s)	Proposed mechanism	Outcomes
Bordonalli et al., 2015	Injection-locking technique	Optical comb generation with reduced phase noise spectral density (below -110 dBc/Hz)
Zhu et al., 2017	Combined effect of FWM and injection-locking	Photonic generation of microwave signals
Jignesh et al., 2018	Injection-locking technique	XPM-phase distortion cancellation
Hu et al., 2017	Multiple optical phase conjugation pairs	Fiber nonlinearity mitigation
Proposed work	**Single-stage injection-locking technique**	**Suppression (~20 dB) of power level of FWM crosstalk**

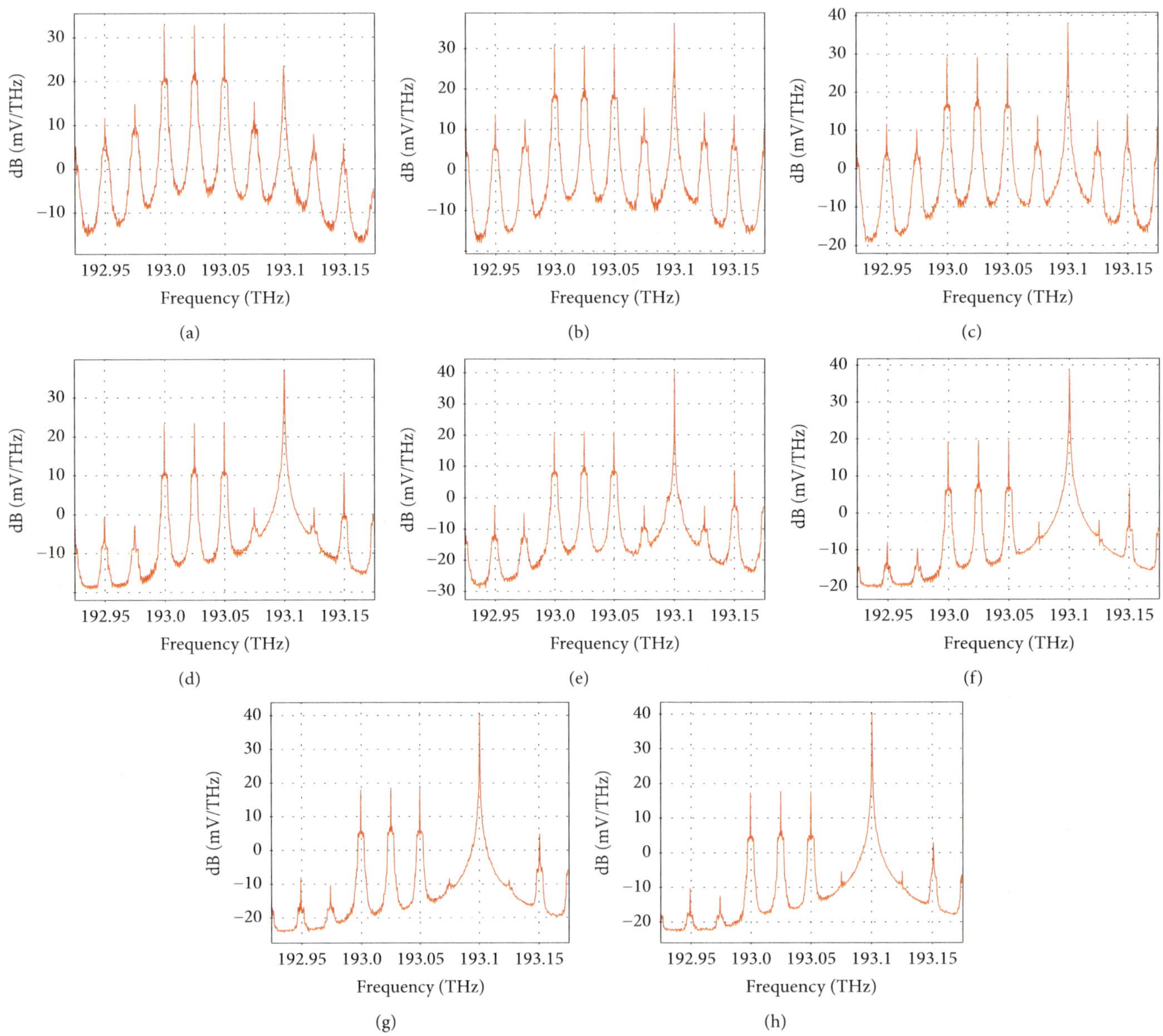

Figure 4: Received optical spectra with significant FWM crosstalk products as a function of bias current of slave laser: (a) 8 mA, (b) 9 mA, (c) 10 mA, (d) 20 mA, (e) 30 mA, (f) 40 mA, (g) 50 mA, and (h) 60 mA.

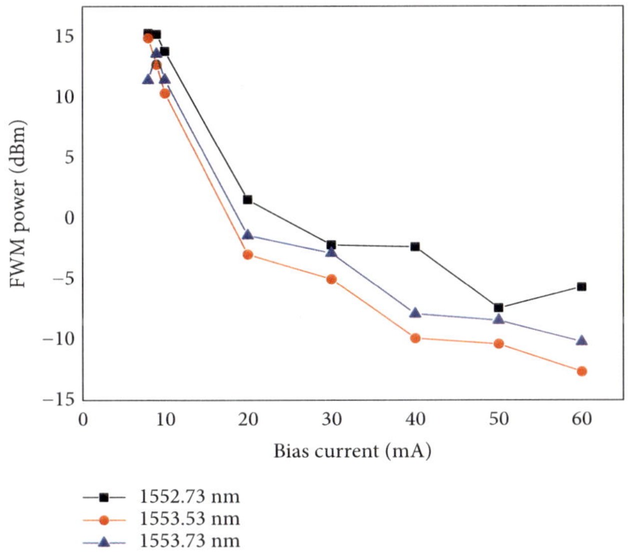

FIGURE 5: Optical power-to-bias current curves for FWM products near the operating wavelengths of master(s) and slave laser.

Acknowledgments

Monika Nehra thanks the UGC, India, for providing financial assistance in the form of JRF (Award no. 3608 dated 29-02-2016).

References

[1] S. Sugavanam, N. Tarasov, X. Shu, and D. V. Churkin, "Narrow-band generation in random distributed feedback fiber laser," *Optics Express*, vol. 21, no. 14, pp. 16466–16472, 2013.

[2] A. Sincore, N. Bodnar, J. Bradford, A. Abdulfattah, L. Shah, and M. C. Richardson, "SBS Threshold Dependence on Pulse Duration in a 2053 nm Single-Mode Fiber Amplifier," *Journal of Lightwave Technology*, vol. 35, no. 18, pp. 4000–4003, 2017.

[3] D. E. Ceballos-Herrera, R. Gutierrez-Castrejon, and J. A. Alvarez-Chavez, "Stimulated Raman scattering and four-wave mixing effects on crosstalk of multicore fibers," *IEEE Photonics Technology Letters*, vol. 30, pp. 63–66, 2018.

[4] E. Temprana, E. Myslivets, B. P.-P. Kuo et al., "Overcoming Kerr-induced capacity limit in optical fiber transmission," *Science*, vol. 348, no. 6242, pp. 1445–1448, 2015.

[5] J. Kaur and H. Singh, "Placement of optical phase conjugation in WDM systems to suppress FWM," *Microwave and Optical Technology Letters*, vol. 60, no. 3, pp. 704–708, 2018.

[6] M. Nehra and D. Kedia, "Design of Optical I/Q Modulator Using Dual-drive Mach-Zehnder Modulators in Coherent Optical-OFDM System," *Journal of Optical Communications*, vol. 39, no. 2, pp. 155–159, 2018.

[7] X. Guo and C. Shu, "Investigation of raman-assisted crosstalk reduction in multi-wavelength fiber optical parametric amplification," *Journal of Lightwave Technology*, vol. 33, no. 23, pp. 4746–4751, 2015.

[8] H. Hu, R. M. Jopson, A. H. Gnauck, S. Randel, and S. Chandrasekhar, "Fiber nonlinearity mitigation of WDM-PDM QPSK/16-QAM signals using fiber-optic parametric amplifiers based multiple optical phase conjugations," *Optics Express*, vol. 25, no. 3, pp. 1618–1628, 2017.

[9] H. Zhu, R. Wang, P. Xiang et al., "A novel approach to photonic generate microwave signals based on optical injection locking and four-wave mixing," *Optics Communications*, vol. 400, pp. 101–105, 2017.

[10] J. Jignesh, A. Lowery, and B. Corcoran, "Inter-channel nonlinear phase noise compensation using optical injection locking," *Optics Express*, vol. 26, no. 5, pp. 5733–5746, 2018.

[11] M. P. Yankov, K. J. Larsen, and S. Forchhammer, "Temporal probabilistic shaping for mitigation of nonlinearities in optical fiber systems," *Journal of Lightwave Technology*, vol. 35, no. 10, pp. 1803–1810, 2017.

[12] S. T. Le, K. Blow, and S. Turitsyn, "Power pre-emphasis for suppression of FWM in coherent optical OFDM transmission," *Optics Express*, vol. 22, no. 6, pp. 7238–7248, 2014.

[13] D. Uzunidis, C. Matrakidis, and A. Stavdas, "An improved model for estimating the impact of FWM in coherent optical systems," *Optics Communications*, vol. 378, pp. 22–27, 2016.

[14] K. Inoue and H. Toba, "Fiber Four-Wave Mixing in Multi-Amplifier Systems with Nonuniform Chromatic Dispersion," *Journal of Lightwave Technology*, vol. 13, no. 1, pp. 88–93, 1995.

[15] T. Yamamoto and M. Nakazawa, "Highly efficient four-wave mixing in an optical fiber with intensity dependent phase matching," *IEEE Photonics Technology Letters*, vol. 9, no. 3, pp. 327–329, 1997.

[16] H.-Y. Chen, Y.-C. Chi, C.-Y. Lin, C.-T. Tsai, and G.-R. Lin, "Four-wave-mixing suppression of master-to-slave injection-locked two-wavelength FPLD Pair for MMW-PON," *Journal of Lightwave Technology*, vol. 34, no. 20, pp. 4810–4818, 2016.

[17] D. Kedia and M. Nehra, "Extended Reach 10 Gb/s Transmission with an Optical I/Q Modulator Using VCSELs over OFDM-Based Multimode Fiber Link," *International Journal of Optics*, vol. 2018, Article ID 2395495, 7 pages, 2018.

[18] Y. Tang, W. Shieh, X. Yi, and R. Evans, "Optimum design for RF-to-optical up-converter in coherent optical OFDM systems," *IEEE Photonics Technology Letters*, vol. 19, no. 7, pp. 483–485, 2007.

[19] T. B. Simpson and F. Doft, "Double-locked laser diode for microwave photonics applications," *IEEE Photonics Technology Letters*, vol. 11, no. 11, pp. 1476–1478, 1999.

[20] J. Al Roumy, J. Perchoux, Y. L. Lim, T. Taimre, A. D. Rakic, and T. Bosch, "Effect of injection current and temperature on signal strength in a laser diode optical feedback interferometer," *Applied Optics*, vol. 54, no. 2, pp. 312–318, 2015.

20

Angularly Resolved Deep Brain Fluorescence Imaging using a Single Optical Fiber

Israel J. De La Rosa,[1] Mehdi Azimipour,[2] Patrick K. Cullen,[3] Fred J. Helmstetter,[3] and Ramin Pashaie[1]

[1]Electrical and Computer Engineering Department, University of Wisconsin-Milwaukee, Milwaukee, Wisconsin, 53201, USA
[2]Vision Science and Advanced Retinal Imaging Laboratory (VSRI), UC Davis Eye Centre, Sacramento, California, 95817, USA
[3]Department of Psychology, University of Wisconsin-Milwaukee, Milwaukee, Wisconsin 53201, USA

Correspondence should be addressed to Ramin Pashaie; pashaie@uwm.edu

Academic Editor: Stefan Wabnitz

In this article, we report the development of a minimally invasive fiber optic based fluorescence probe which can reach deep brain objects and measure the intensity and spatial distribution of fluorescence signals in the tissue. In this design, the brain is scanned by a single penetrating side-firing optical fiber which delivers excitation light pulses to the tissue at different depths and orientations and simultaneously collects samples of fluorescence emission signals. Signal-to-noise ratio of the measurements is improved by adapting the pulse compression technique and the theory of optimal filters. Effects of each design parameter on the overall performance of the scanner, including the spatial resolution and speed of scanning, are analyzed and experimentally measured. In vivo experiments show that the new device, despite the simplicity of the design, provides valuable information particularly useful in optogenetic stimulation experiments where the exact position of the fiber tip and the radiation orientation can change the outcome of a test.

1. Introduction

Recent advances in the discovery or synthesis of new fluorescent indicators have been followed by the development of sophisticated technologies for in vitro and in vivo fluorescence imaging offering spatial resolution close to the diffraction limit of adapted optical wavelengths [1]. Nonetheless, in vivo fluorescence imaging has remained limited only to the very immediate superficial areas of the brain tissue. A good example of such imaging modalities is the two-photon excitation microscopy, which allows imaging of living tissue up to one millimeter in depth and the system can image a large portion of the cortex in small rodents [1]. Since most high-level information processing operations take place in the cortex, invention of two-photon microscopes opened new horizons in the study of the cortical dynamics. Fluorescence tomography scanners are also made that employ mathematical (e.g., fluorescence diffuse optical tomography (FDOT) [2]) or statistical algorithms (e.g., fluorescence laminar optical tomography (FLOT) [3, 4]) that model light tissue interaction and use this information to translate the measured data to three-dimensional images that reveal the distribution of fluorescent molecules. Usually such tomography scanners can image slightly deeper areas; however, resolution is significantly sacrificed and spurious objects appear in generated images caused by the instability of algorithms that solve ill-posed inverse problems. As a result, none of these technologies can reliably image deep brain objects, such as the thalamus or hippocampus, which play crucial roles in vital brain functions including the early processing of sensory inputs or memory consolidation. For such applications, the main option is to conduct the experiments in brain slices which is an invasive process and certainly nonreversible. Therefore, there is high demand for the invention of new instrumentation that can reach deep brain objects to perform fluorescence imaging.

A reasonable approach to achieve optical access to deep brain regions is the use of thin optical fibers. An optical fiber can function as a reliable tool to guide the power of a coherent or incoherent light source to any target area

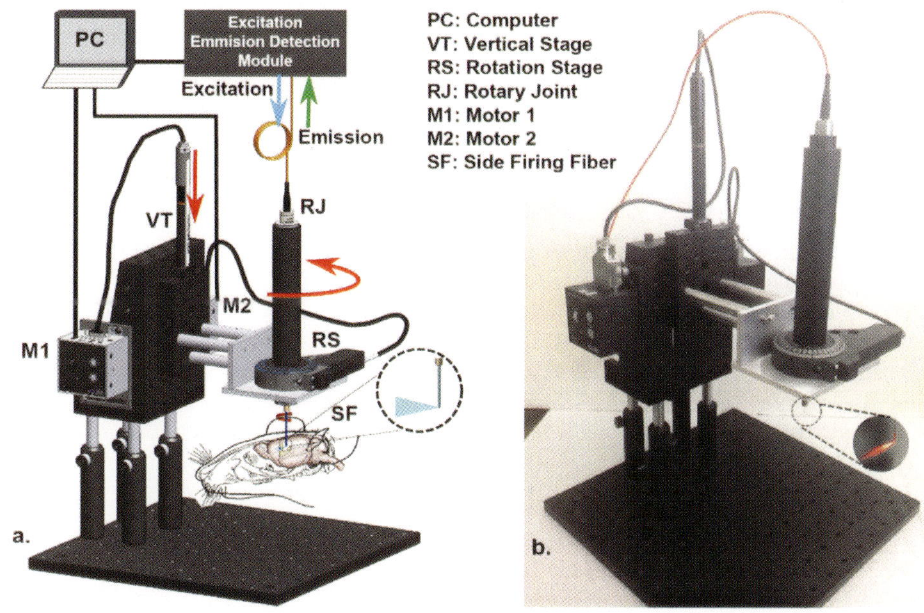

FIGURE 1: (a) Schematic of the setup: two computer controlled actuators adjust the position and orientation of a side-firing optical fiber inside the tissue. A fiber rotary joint is used to help rotate the fiber with no mechanical stress. At each position, the system sends an excitation laser pulse and detects a sample of the corresponding emission signal. Next, the collected data is processed by a computer to generate images that show the distribution of fluorescence signal in a cylinder surrounding the fiber; (b) picture of the developed system.

with minimum loss. Such waveguides are used in many challenging biomedical applications, for example, to perform fluorescent detection (e.g., see [5, 6]) spectroscopy (e.g., see [7–9]) or the assessment of tissue optical property (e.g., see [8, 10, 11]) for optical biopsy applications. The performance of such mechanisms is also theoretically and empirically analyzed in recent years (e.g., see [12, 13]).

Here, we introduce a new approach in which the brain tissue is scanned by advancing and simultaneously rotating a single side-firing optical fiber inside the tissue. This fiber delivers the excitation light pulses at each site and guides a sample of the emission signal back to the photodetectors. In the brain research applications, we prefer to reduce the size of the optical fiber to minimize the damage to the tissue under test. Reducing the fiber diameter translates to some major reduction in the number of collected photons and as a result, the signal-to-noise ratio dramatically drops in such measurements. We recently introduced a single optical fiber probe for fluorescence measurement and optogenetic stimulation in which the excitation laser pulses were frequency modulated in the form of linear or nonlinear chirped pulses [5, 14]. Next, the system detects the corresponding chirped emission signals and passes them through a correlation filter to compress the energy of the pulse and momentarily maximize signal-to-noise ratio. Since the noise of the system was estimated to be white Gaussian, the correlation filter functions as the optimal filter and helps to detect the most fainted fluorescence signals. The balanced photodetection mechanism adapted in the platform also helps to remove the effect of systematic distortions including the instability of the excitation laser or noise of the detectors [5]. Here, this fluorescence detection module (FDM) is connected to the new head in which the tissue is scanned by the rotating side-firing optical fiber. Details of the design and results from computer simulations and experimental work are discussed in the article.

2. Materials and Methods: Design and Components

The main component of the new scanner is the side-firing fiber. A side-firing fiber is an optical fiber in which the delivery end is angled to deflect the fiber's radiation profile. When designed correctly, the main radiation lobe of the side-firing fiber is almost orthogonal to the fiber's optical axis which is an ideal factor in the design of the probe. Nonetheless, electromagnetic analysis of the structure shows that the divergence of the beam increases as the radiation angle gets closer to 90° with respect to the fiber's axis. Therefore, there is a trade-off between the direction and directivity of radiation. The appearance of sidelobes in the radiation profile of side-firing fibers, which causes considerable light power leaking out in unintended directions, is another design concern.

We have proved mathematically and experimentally that by selecting the right value for the tip angle, based on the refractive index and numerical aperture of the fiber, we can design an optimal side-firing fiber for which (1) total internal reflection occurs for all incident rays hitting the angled tip and as a result no leakage occurs, (2) the radiation angle is close to normal, and (3) beam divergence is minimized to help improve the spatial resolution of the scanner [5].

The proposed system is shown in Figure 1 and includes a custom made high hydroxyl ion (high OH) silica core side-firing fiber (Polymicro Technologies, IL, USA), rotation arm (rotary joint, lens tube, and fiber chuck), two servo motor actuators with their corresponding servo motor controllers, structural components needed to hold the system together, a laptop computer, and the FDM system. The FDM system is not shown here but the schematic of the design is available in Figure 1 of [5]. Motor controllers, M1 and M2, drive the vertical and rotational translation stages, VT and RS, to precisely position the optical fiber at all target coordinations while the FDM delivers excitation light pulses and detects fluorescence emission through the side-firing fiber, *SF*. For the side-firing optical fiber that we used in this research, we measured the numerical aperture experimentally and we found NA = 0.34 in air at room temperature. Our measurements also showed that the fiber radiates the optical power at 73.81° angle with respect to the optical axis of the fiber. As a result, the main radiation lobe is not completely orthogonal to the fiber and this effect needs to be compensated during image reconstruction process.

3. Materials and Methods: Design Considerations

The rotary joint, *RJ*, allows the fiber to rotate without twisting that can affect or even damage the waveguiding property of the fiber. The stationary side of the rotary joint is connected to the optical fiber that guides light pulses from the FDM to the scanner's head. The rotating side of the rotary joint holds the side-firing fiber that is mounted in a fiber chuck (HFR001, Thorlabs, NJ, USA) and positioned within the mechanism of the rotary servo motor. We use FC/PC to FC/PC mating sleeves to connect the fibers to the joints. It is important to make the fiber rotate completely along its axis. Aligning the optical fiber is a vital step and it needs to be adjusted sequentially three set screws, one revolution at a time, until the chuck that holds the fiber is firmly secured. If this is not done properly, the tip of the fiber will follow an elliptical trajectory that is quite destructive inside the soft tissue of the brain; see Figure 2(a).

Another important consideration for the rotary joint is the fluctuations in the coupling of the light from the stationary fiber to the side-firing fiber as a function of the angular position of the tip. To estimate this effect, we connected the stationary fiber to a stable source (M47051, 470nm fiber coupled LED, Thorlabs, NJ, USA) while we were measuring the light power radiating from the rotating fiber by an optical power sensor (S130C, Thorlabs, NJ, USA). We repeated the experiment six times and averaged the data. Results are shown in Figure 2(b). Left panel displays the normalized coupling factor as a function of the position and orientation of the tip. From these results, we noticed that the rotary joint had a tendency to increase the detection intensity at the angle of approximately 150°. The maximum amplitude of these fluctuations compared to the base line of the measured power is small; nonetheless, such tiny distortions can still affect the outcome particularly when we are measuring faint fluorescence signals. To compensate this effect, we generated a complementary map, shown in Figure 2(b) right panel, and used that to normalize the collected data and eliminate this effect.

4. Results and Discussion

4.1. Simulation Results. To better understand and analyze the performance of this scanner, we used the stochastic method of Monte Carlo (MC) [15, 16] to simulate light propagation in a scattering and absorbing medium. Based on our previous experience of running these simulations for light brain tissue interaction [17], we chose to launch and trace 5×10^6 photons in the tissue which seems to be enough to generate a good estimation of how many photons find their way back to the fiber tip for different parameters of the medium, optical fiber, or wavelengths. Scattering coefficient of the brain tissue was modelled by $\mu'_s = a(\lambda/500nm)^{-b}$, where λ is the excitation or emission wavelength and a is a scaling factor which represents the reduced scattering coefficient at $\lambda = 500nm$. The parameter b in this equation is the scattering power. Values for a and b are 2.37mm^{-1} and 1.15, respectively [15]. This calculation gives $\mu'_s = 23.2cm^{-1}$. The absorption coefficient, μ_a, was calculated by $\mu_a = BS\mu_{HbO_2} + B(1-S)\mu_{Hb} + W\mu_{water}$. Here, B is the average blood volume fraction, S is the hemoglobin oxygen saturation, and W is the water content. The values for B, S, and W, are 0.028, 0.62, 0.62, respectively [15]. The absorption coefficients for hemoglobin and water were taken from the Oregon Medical Laser Center website [18]. This calculation gives $\mu_a = 3.36cm^{-1}$. The value of g was set to 0.9 and the value of $\mu_s = \mu'_s/(1-g) = 232.0cm^{-1}$.

We used these values to define a scattering and absorbing medium in the MC simulation software (https://omlc.org/software/mc/) and we modelled the performance of the scanner when a 250μm diameter fluorescent particle, emitting 510nm wavelength, was placed at the radial distance of 0.5mm from the fiber tip. First, we looked at the effect of the fiber's *NA* on the spatial resolution of the scanner. Figure 3 shows the simulation results when the *NA* varies from 0.2 to 0.6. In each test, we computed the full-width half-maximum (FWHM) of the image of the particle produced by the scanner as a measure of the resolution. The left panel in this figure shows the FWHM along the vertical axis and the right panel displays the radial FWHM. From these plots, we observe that increasing the *NA* translates to some noticeable increase in the FWHM or reduction of the spatial resolution. Therefore, we prefer side-firing fibers that offer smaller numerical apertures; however, reducing the divergence of the radiation cone for a side-firing fiber usually comes with the main lobe of radiation being less orthogonal to the optical axis of the fiber. As a result, there is a trade-off in this design. The dashed line in these graphs represent *NA* = 0.34 which is the numerical aperture of the side-firing fiber we used for our experiments and as mentioned before the beam makes the angle of 73.81° with respect to the fiber's optical axis.

4.2. Phantom Results. To test the precision of our simulations, we built eight tissue phantoms. The phantoms were made by suspending a 250μm fluorescence microsphere (using an ultra-thin wire and optically clear epoxy (Loctite)) in liquid mixtures of milk and India ink in different concentrations to achieve any combination of scattering and absorption

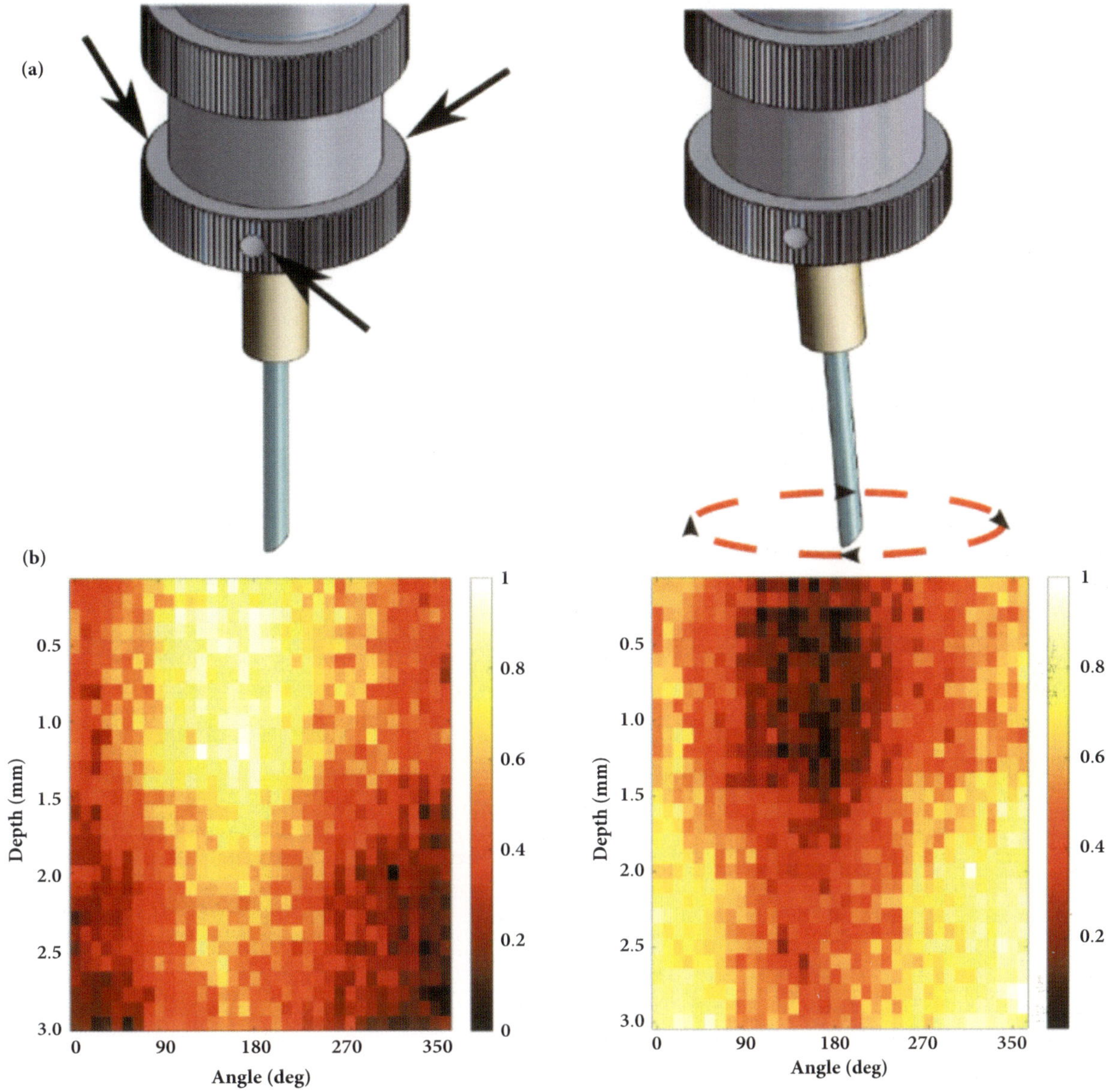

FIGURE 2: (a) Arrows point to the position of the set screws that need to be adjusted to keep the chuck aligned along the vertical axis (Left). Undesired trajectory of optical fiber resulting from improper set screw adjustment (Right). (b) Fluctuations caused by the rotary joint (left) introduce some nonuniformity in fluorescence intensity recording which in this case shows a spike at 150 degrees. To eliminate the effect, this nonuniformity was measured in a uniform fluorescence solution. Next, the complement of this data was created as a distortion compensation matrix (Right) and this matrix was used to cancel the effect.

coefficients. We scanned these phantoms one by one and we compared the measured resolution with the estimated values from simulations; see Figure 4. Based on this comparison, the measured FWHMs are within 20% of the simulation results when the scattering coefficient was fixed but the absorption was changing over a reasonable range. On the other hand, when we were changing the scattering coefficients for a medium with fixed absorption, the results were within 25% of the simulated results which are still reasonable. The resolution of optical imaging systems is limited because of scattering in the tissue. The impact of absorption coefficient is not more than reducing the absolute signal. Figure 4 shows the higher impact of reduced scattering on the resolution compared to the effect of absorption on the resolution.

Simulations provide an intuition to the response of the system when making measurements in the tissue. One important parameter for in vivo experiments is the trade-off between the speed of scanning and the resolution of the reconstructed images. Reducing the vertical or radial distance between the measurement points potentially leads to better resolution but makes the scanning process highly time consuming. For instance, scanning with axial and radial resolutions of $100\mu m$ and 10° and taking two measurements for each point can take up to $45 min$, which is too long. We

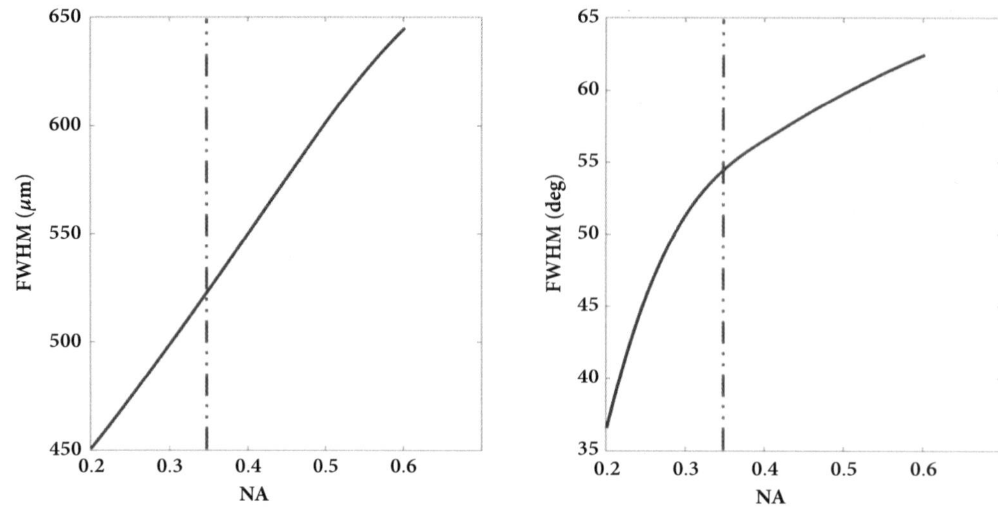

FIGURE 3: Plots of the effects from a varying *NA* on the FWHM for the fluorescent of a 250μm particle. Left plot shows the behavior of the FWHM vertically while right plot shows the behavior radially. The red lines represent the *NA*=0.34 which is the *NA* of the optical fiber in our system.

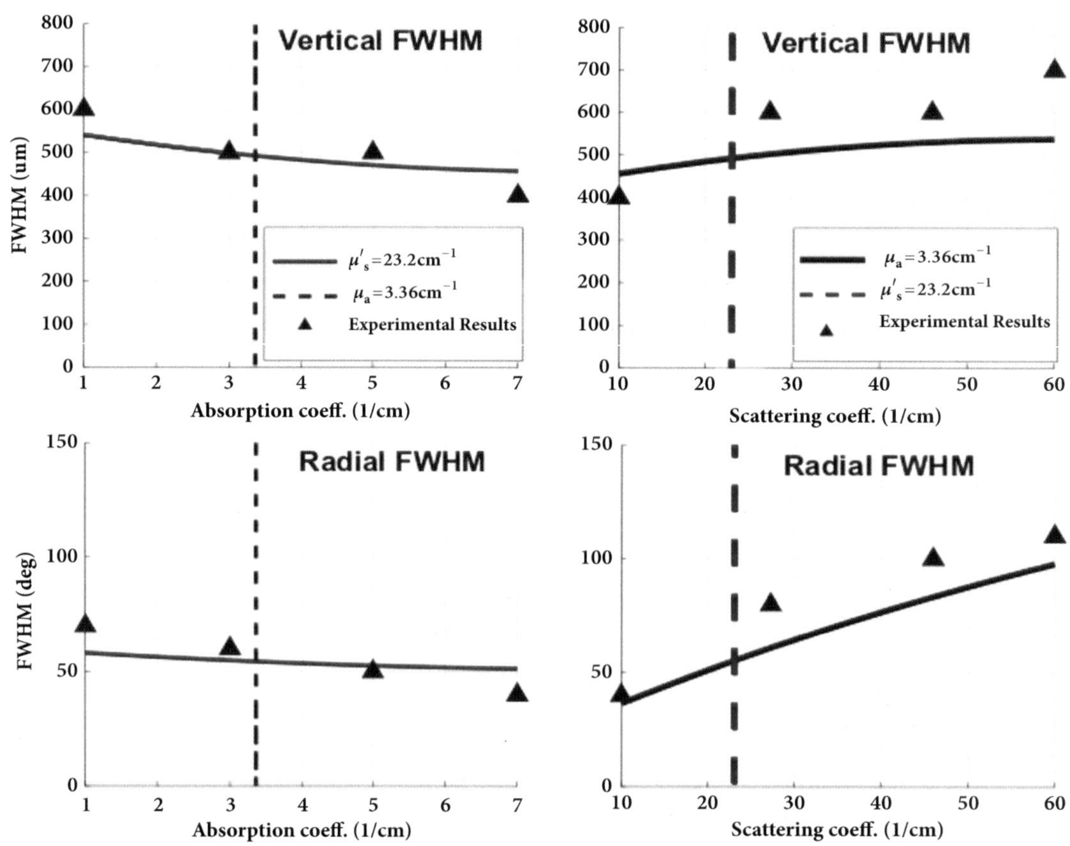

FIGURE 4: Results from phantom experiments compared to simulations measuring resolution in vertical and radial direction. For the absorption coefficient, μ_a, four data points were collected at 1.0, 3.0, 5.0, 7.0cm^{-1}. Our estimation for the value of μ_a in brain tissue is 3.36 cm^{-1}. Phantom results are within 20% of our simulations (Left). For the reduced scattering, μ'_s, four data points at 10.0, 26.0, 43.0, 60.0 cm^{-1} were collected. The value of μ'_s we used for simulations was 23.2 cm^{-1}. These results are within 25% of the simulated values for the FWHM.

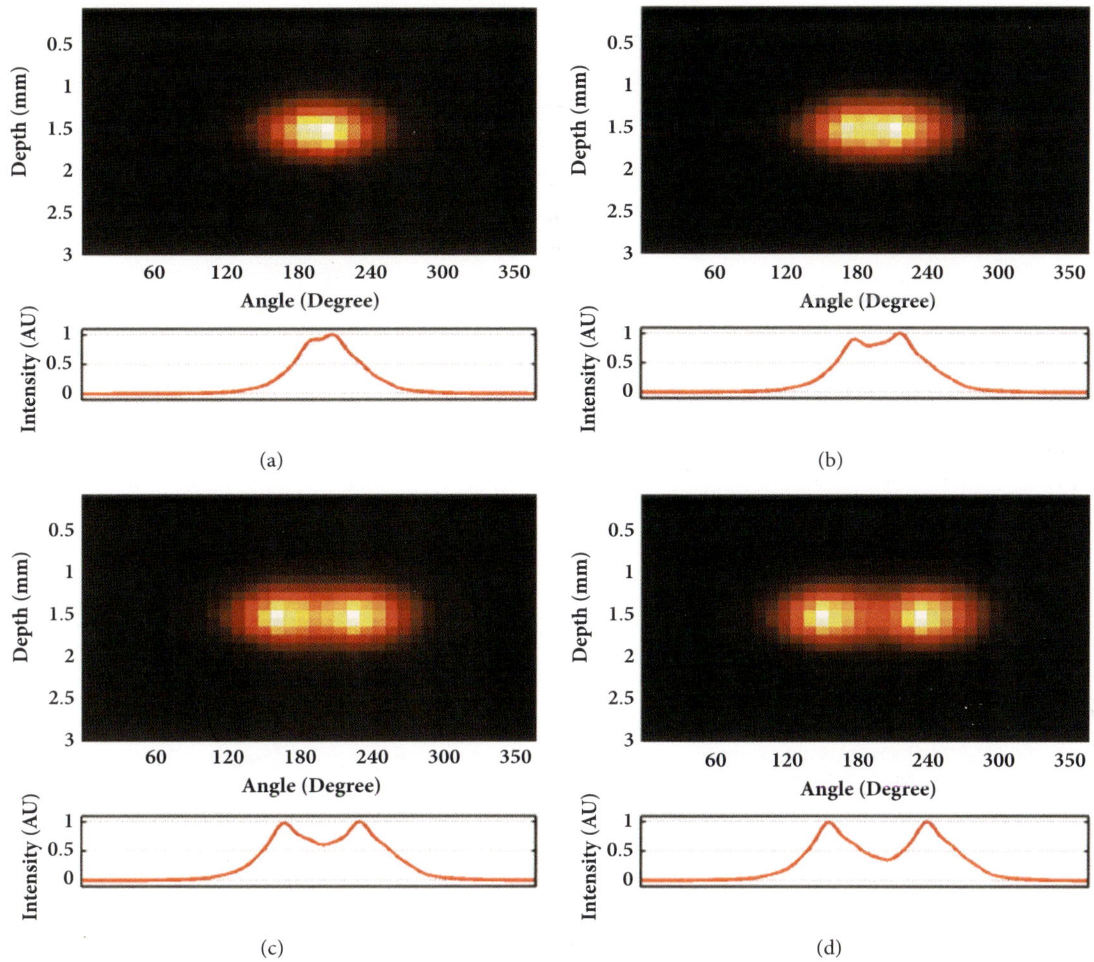

FIGURE 5: Results from a simulation to find the resolution of our proposed system. Based on this result, when the particles are positioned at the distance of 0.5mm from the fiber tip, the particles are still detectable as two separate particles and the angular distance between them is larger than 50°.

also know that, for a fiber with a given NA, reducing the distance between the points of measurements will not even provide more information and the collected data will become redundant. Figures 5 and 6 show the results of our simulation where we placed two 250μm diameter fluorescence particles 0.5*mm* apart from the tip of the side-firing fiber and we imaged them while changing the vertical or horizontal distance between the particles. In these simulations, photons were launched uniformly within the diameter of the optical fiber used in our system (200μm) and the corresponding NA (0.34). The optical properties used were μ's=23.2cm-1, μa = 3.36cm-1, and g=0.9. In these figures we can see the reconstructed images when the particles are placed at the horizontal and vertical distances of 0.3*mm* to 0.6*mm* from each other.

It is clear from the figures that the particles are not distinguishable when the distance between them is less than 0.5mm. This means that the rotation angle needs to be less than 50°. Nonetheless, 50° is still too large since, for the 50° angular scanning step size, one of these particles can be missed. Therefore, the limit for detecting particle of this size at this distance is half of 50°. In Figure 6 we can see that the two particles are not distinguishable as separate particles when the vertical distance between them is less than *0.49mm*. Consequently, the vertical steps need to be at least half of this value if the goal is to detect particles of this size (250μm) at this distance (0.5*mm*). In Figure 7 axial and angular resolutions for the same particle size are plotted as a function distance from the fiber tip. In most in vivo experiments, we use the scanning angle of 20° and the vertical scanning steps of 0.2*mm*.

Another consideration is changing the wavelength of the system. By simulating multiple wavelengths, Figure 8, we see that changing the wavelength affects the resolution of the system. As the wavelength increases, the scattering coefficient for the brain tissue decreases but so does the absorption coefficient. Since the tissue does not absorb many photons for longer wavelengths, the smaller number of photons that do scatter affects the FWHM of an imaged particle and the resolution of the system.

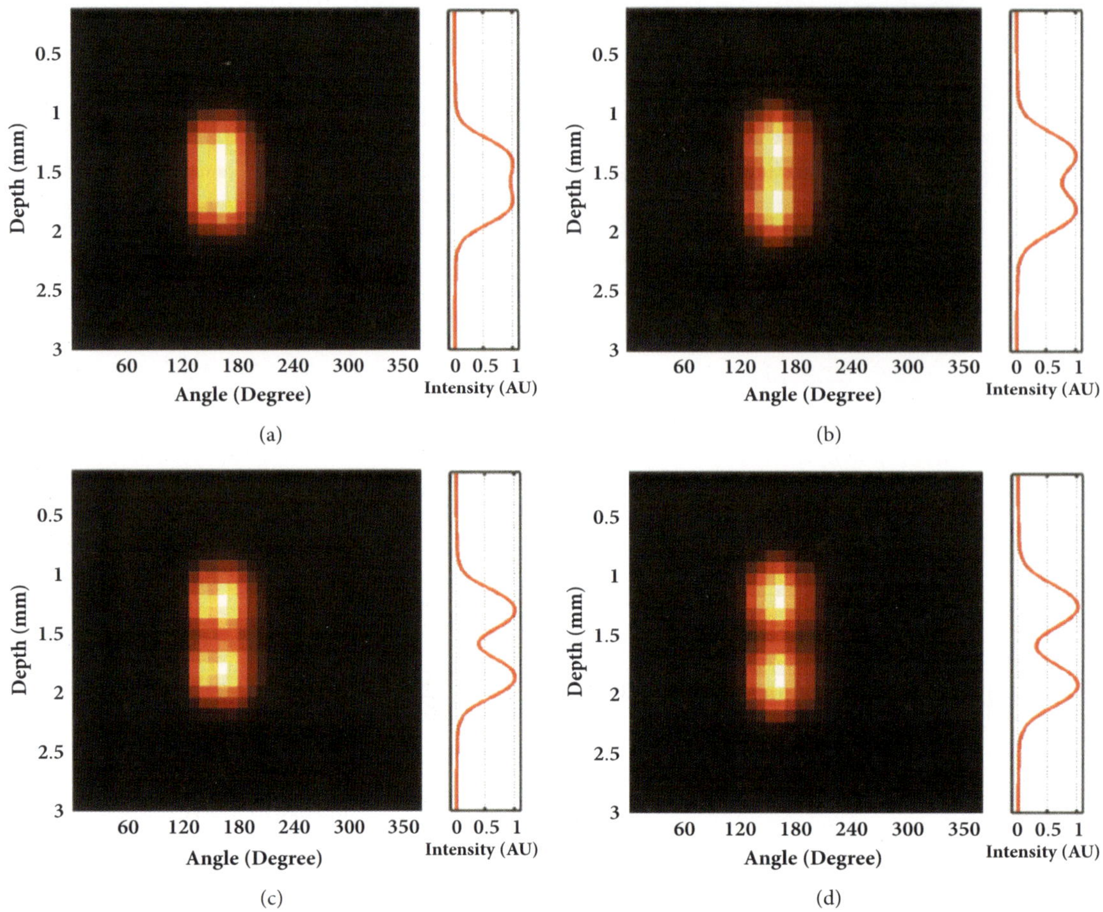

FIGURE 6: Results from a simulation to find the resolution of our proposed system. Based on this result, when the particles are positioned at the distance of $0.5mm$ from the fiber tip, the particles are still detectable as two separate particles and the vertical distance between them is larger than $0.49mm$.

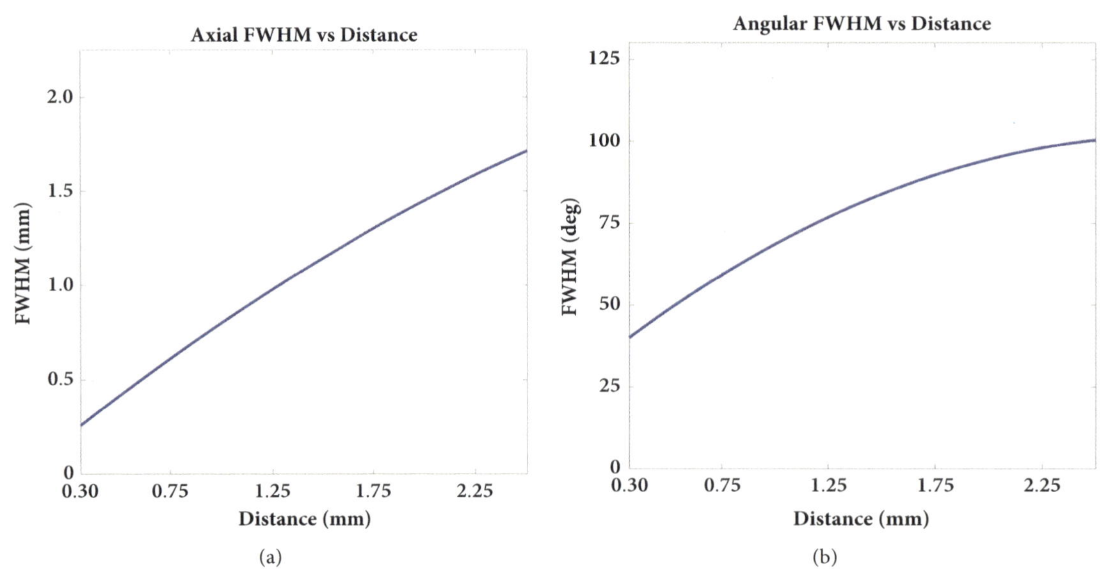

FIGURE 7: Axial (a) and angular (b) resolutions displayed as a function of distance from the fiber tip. Resolution is measured as the FWHM of a $250\mu m$ fluorescence particle.

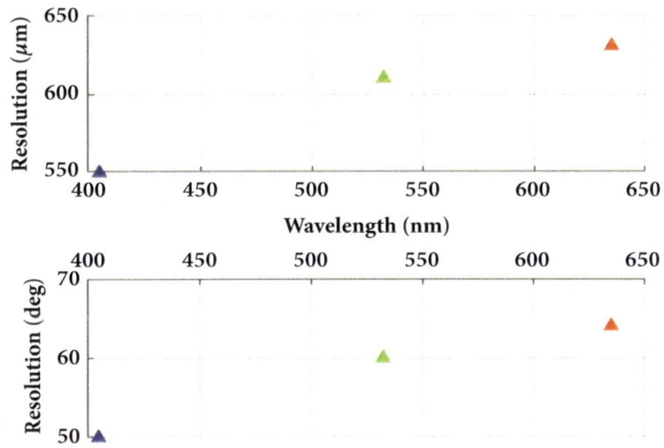

FIGURE 8: Results from spectral response simulations. Increasing the wavelength has a negative effect on the resolution of the proposed system. Resolution is measurement in terms of the FWHM of the imaged fluorescence particle.

4.3. In Vivo Results. To test the performance of the system when imaging the brain of a live animal, we used a male Long-Evans rat (300*g*-400*g*, Envigo, WI). All procedures were carried out in a facility accredited by the Association for Assessment and Accreditation and Laboratory Animal Care (AAALAC) and approved by the University of Wisconsin-Milwaukee Institutional Animal Care and Use Committee (IACUC) and conducted within the ethical guidelines of the National Institutes of Health (NIH). Two weeks prior to the experiments, the animal was anesthetized with isoflurane in 100% oxygen (induction occurred with 4% isoflurane and maintained with 2%). The rat was mounted in a stereotaxic apparatus (Kopf Instruments, CA, USA). Purified AAV9-CAG-GFP virus (UNC Vector Core) was infused into ventral hippocampus (A/P -5.28mm, M/L + 5.4mm, D/V 5.7mm) to express green fluorescence protein (GFP) in the tissue. The injection volume was 0.3ml. The virus was infused using a 10μl nanofil syringe and 34-gauge needle (World Precision Instruments) mounted to a stereotaxic automated injector (Micro4 MicroSyringe Injector; World Precision Instruments) at a rate of 0.05ml/min. The injector was left in place for 10min following the injection to allow diffusion of the virus away from the injector.

The experiment was conducted by scanning the brain tissue starting at the depth of 4.4mm inside the tissue and exactly at the coordination of the viral injection. Then, we scanned the brain tissue with the side-firing optical fiber using the 20° angular and 200μm vertical step sizes. A 473nm blue laser was used which is suitable for excitation since its wavelength is close to the peak of the GFP excitation curve which is the fluorescence target in this experiment. The data was collected for all measurement points and stored in the computer. Immediately following the completion of the experiment, the animal was anesthetized with isoflurane, transcardially perfused with 0.2M PBS followed by 10% buffered formalin. The brain was removed and post-fixed in 10% formalin for 24 hours before being transferred to 30% sucrose/PB. Next, the brain was frozen, sliced into coronal sections of 200μm thickness which were mounted on glass microscope slides, and cover slipped with UltraCruz mounting medium (Santa Cruz, CA). Next, we imaged all slices via confocal fluorescence microscope (Olympus Fluoview FV 1200).

Figure 9 shows the mapped measurements for each depth and for each angle with respect to the position of the fiber's main lobe of radiation. The data is presented after interpolating and processing using gamma correction with a gamma coefficient of 6.0. Starting from a 3D reconstruction of a section of the brain shown in panel (a), we can find the exact location of the optical fiber entry point to the brain. Two cross-sections were of specific interest to validate our results. The first cross-section is a slice imaged by the confocal microscopy shown in panel (b). The path of the optical fiber is marked in the image. Maximum fluorescence appears at approximately 800μm away from the fiber and 6.6mm deep inside the tissue, based on confocal microscopy images. The data collected by the optical fiber probe, displayed in panel (d), shows that the maximum intensity is occurring at the depth of 6.3mm. This difference in measuring the coordination of maximum fluorescence protein expression is the result of the angle of the main lobe of radiation which is not completely orthogonal to the fiber's optical axis. Taking the firing angle of 73.81° into account, we can use simple geometry to compensate for this discrepancy. The second cross-section of interest is the slice of the brain along the depth of maximum expression which is shown in panel (c). The lobe of radiation is initially at the 0° and as the fiber rotates clockwise collecting measurements every 20°, it detects the nearest distribution of fluorescence at 135°. This agrees with our collected data which shows maximum expression at exactly 135°. While the distribution expands well into 180°, with respect to the initial position at 0°, the area of expression that is of nearest proximity to the optical fiber has the largest contribution to the measurement.

5. Conclusions

It is important to consider the position of an optical fiber as well as the direction of light delivery during optogenetic

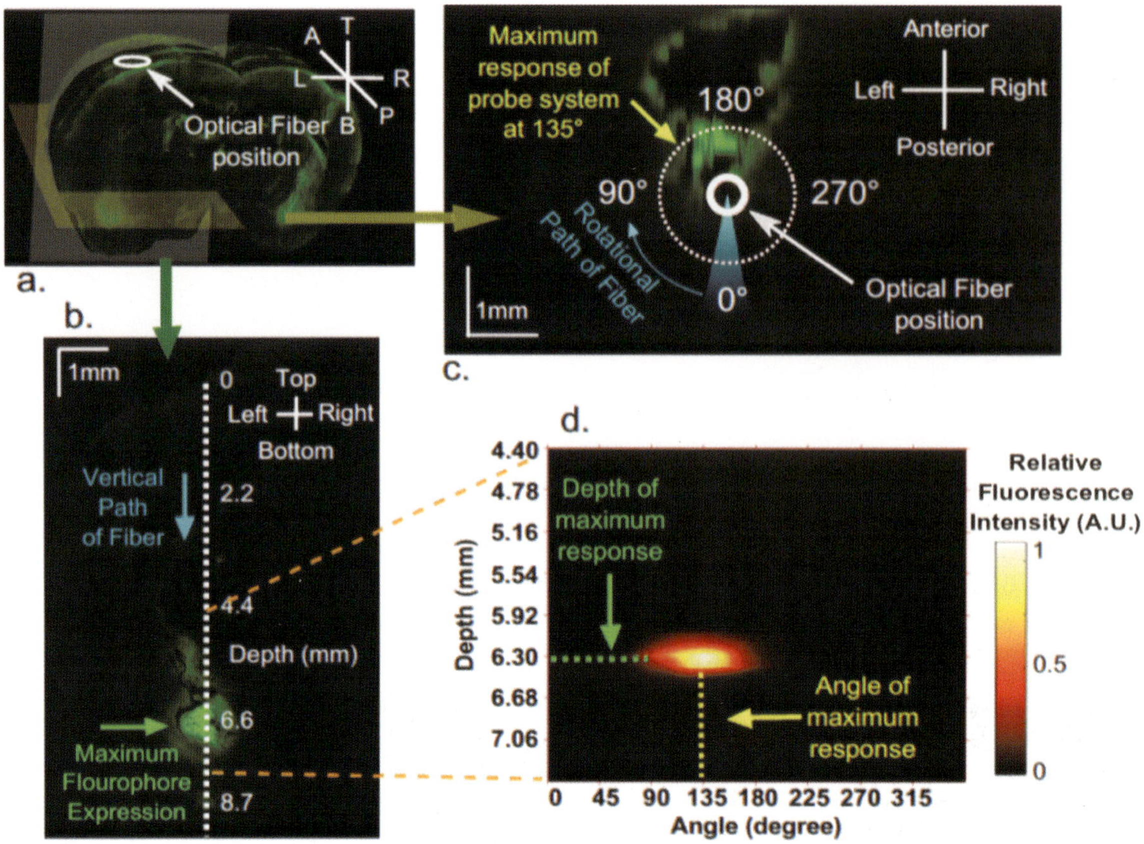

FIGURE 9: (a) In vivo results. 3D image reconstruction of a section of the brain under test. (b) Following the path of the fiber, the maximum expression is found at approximate depth of 6.6*mm* from the surface. (c) Following the rotational path of the fiber, we found the maximum expression closest to the fiber tip at approximately 135°. (d) Data collected with our system and presented using gamma correction factor of 6.

stimulation [19, 20] or other neuroscience experiments. The measured data clearly showed that the expression of fluorescence protein can diffuse in different directions with respect to the fibers main lobe of radiation. With the proposed system, we could accurately detect the depth and the orientation at which the fluorescence protein expression occurs. This information can be used to optimize the light delivery process, for example, in optogenetic experiment to make the optical stimulation more effective or to reduce the amount of optical power necessary to activate targeted neurons and avoid phototoxicity or generating excessive amount of heat in the tissue. While optogenetics is only one application, the proposed system can be used in applications where the distribution of fluorescence needs to be determined in a variety of different soft tissue.

Conflicts of Interest

The authors declare that there are no conflicts of interest regarding the publication of this paper.

Acknowledgments

This project was supported by the Army Research Office (ARO) grant# 68750-LS-II, the National Science Foundation (NSF) Career Award grant# 1454300, the National Institute of Mental Health (NIMH) grant# R01MH11241, Brain and Behavior Research Foundation (NARSAD) Young Investigator Award grants# 20610 and 23620, and the University of Wisconsin research growth initiative (RGI) grants# 101X213 and 101X254.

References

[1] T. Vo-Dinh, *Biomedical Photonics Handbook*, CRC Press, 2003.

[2] D. A. Boas, D. H. Brooks, E. L. Miller et al., "Imaging the body with diffuse optical tomography," *IEEE Signal Processing Magazine*, vol. 18, no. 6, pp. 57–75, 2001.

[3] E. M. C. Hillman, D. A. Boas, A. M. Dale, and A. K. Dunn, "Laminar optical tomography: Demonstration of millimeter-scale depth-resolved imaging in turbid media," *Optics Expresss*, vol. 29, no. 14, pp. 1650–1652, 2004.

[4] M. Azimipour, M. Sheikhzadeh, R. Baumgartner et al., "Fluorescence laminar optical tomography for brain imaging: System implementation and performance evaluation," *Journal of Biomedical Optics*, vol. 22, no. 1, 2017.

[5] R. Pashaie and R. Falk, "Single optical fiber probe for fluorescence detection and optogenetic stimulation," *IEEE Transactions on Biomedical Engineering*, vol. 60, no. 2, pp. 268–280, 2013.

[6] K. R. Diamond, M. S. Patterson, and T. J. Farrell, "Quantification of fluorophore concentration in tissue-simulating media by fluorescence measurements with a single optical fiber," *Applied Optics*, vol. 42, no. 13, pp. 2436–2442, 2003.

[7] S. Brooks, C. L. Hoy, A. Amelink, D. J. Robinson, and T. E. Nijsten, "Sources of variability in the quantification of tissue optical properties by multidiameter single-fiber reflectance and fluorescence spectroscopy," *Journal of Biomedical Optics*, vol. 20, no. 5, Article ID 2290871, 2015.

[8] S. C. Kanick, U. A. Gamm, M. Schouten, H. J. C. M. Sterenborg, D. J. Robinson, and A. Amelink, "Measurement of the reduced scattering coefficient of turbid media using single fiber reflectance spectroscopy: Fiber diameter and phase function dependence," *Biomedical Optics Express*, vol. 2, no. 6, pp. 1687–1702, 2011.

[9] Q. Liu and N. Ramanujam, "Experimental proof of the feasibility of using an angled fiber-optic probe for depth-sensitive fluorescence spectroscopy of turbid media," *Optics Expresss*, vol. 29, no. 17, pp. 2034–2036, 2004.

[10] U. A. Gamm, S. C. Kanick, H. J. C. M. Sterenborg, D. J. Robinson, and A. Amelink, "Quantification of the reduced scattering coefficient and phase-function-dependent parameter γ of turbid media using multidiameter single fiber reflectance spectroscopy: Experimental validation," *Optics Expresss*, vol. 37, no. 11, pp. 1838–1840, 2012.

[11] U. A. Gamm, S. C. Kanick, H. J. C. M. Sterenborg, D. J. Robinson, and A. Amelink, "Measurement of tissue scattering properties using multi-diameter single fiber reflectance spectroscopy: In silico sensitivity analysis," *Biomedical Optics Express*, vol. 2, no. 11, pp. 3150–3166, 2011.

[12] P. R. Bargo, S. A. Prahl, and S. L. Jacques, "Collection efficiency of a single optical fiber in turbid media," *Applied Optics*, vol. 42, no. 16, pp. 3187–3197, 2003.

[13] T. J. Pfefer, K. T. Schomacker, M. N. Ediger, and N. S. Nishioka, "Multiple-fiber probe design for fluorescence spectroscopy in tissue," *Applied Optics*, vol. 41, no. 22, pp. 4712–4721, 2002.

[14] I. Diester, M. T. Kaufman, M. Mogri et al., "An optogenetic toolbox designed for primates," *Nature Neuroscience*, vol. 14, no. 3, pp. 387–397, 2011.

[15] C. Zhu and Q. Liu, "Review of Monte Carlo modeling of light transport in tissues," *Journal of Biomedical Optics*, vol. 18, no. 5, Article ID 050902, 2013.

[16] L. Wang, S. L. Jacques, and L. Zheng, "MCML-Monte Carlo modeling of light transport in multi-layered tissues," *Computer Methods and Programs in Biomedicine*, vol. 47, no. 2, pp. 131–146, 1995.

[17] M. Azimipour, R. Baumgartner, Y. Liu, S. L. Jacques, K. Eliceiri, and R. Pashaie, "Extraction of optical properties and prediction of light distribution in rat brain tissue," *Journal of Biomedical Optics*, vol. 19, no. 7, Article ID 075001, 2014.

[18] Oregon Medical Laser Center, http://omlc.org.

[19] R. Pashaie, P. Anikeeva, J. H. Lee et al., "Optogenetic brain interfaces," *IEEE Reviews in Biomedical Engineering*, vol. 7, pp. 3–30, 2014.

[20] O. Yizhar, L. E. Fenno, T. J. Davidson, M. Mogri, and K. Deisseroth, "Optogenetics in neural systems," *Neuron*, vol. 71, no. 1, pp. 9–34, 2011.

Efficient Tandem Organic Light Emitting Diode using Organic Photovoltaic Charge Generation Layer

Akanksha Jetly and Rajesh Mehra

Electronics and Communication Department, NITTTR, Chandigarh, India

Correspondence should be addressed to Akanksha Jetly; akankshajetly@yahoo.com

Academic Editor: Sulaiman W. Harun

Highly efficient tandem red OLED is proposed using planar organic heterojunction based charge generation layer (CGL) of C70 (Fullerene)/CoPc (Cobalt Phthalocyanine). The proposed charge generation layer generates charges and exhibits organic photovoltaic type (OPV) behavior in the red emitter zone which helps in slower efficiency roll-off at higher current densities. In addition an ultrathin layer of Al/LiF is used as electron injection layer (EIL) so as to enhance the injection of electrons from charge generation layer to emissive unit. It is found that the tandem red device can reach current efficiency and luminance of 26 cd/A and 26039 cd/m^2 at 20 mA/cm^2 and 100 mA/cm^2, respectively, which is 1.663 times and 1.665 times higher than the corresponding single emitter device. Moreover, it is known that the tandem device requires double turn-on voltage than conventional devices. In this work, the turn-on voltage of red tandem device (5.85 V) is found to be less than twice the voltage of conventional device (3 V). Further with the introduction of Distributed Bragg Reflector (DBR), enhancement in current efficiency and brightness of tandem red device are observed due to high reflectance property of DBR. Thus, the effective charge generation and OPV function of proposed charge generation layer as well as remarkably high injection property of EIL layer lead to the high performance tandem OLED.

1. Introduction

Developed in 1987 and based on the electroluminescent phenomenon, organic light emitting diodes have turned out to be a blessing for both display and lightning applications because of their ability to produce more vibrant colors and amazing features like low operating voltage, wide viewing angle, low power consumption, flexibility, thin, light weight, high resolution, and many more. The basic OLED structure comprises of electroluminescent unit that contains anode and cathode and emitting and transport layers between these electrodes. The light is generated in the emitting layer upon the application of external supply because of the radiative recombination of electrons and holes injected by the electrodes. To make the conventional OLEDs more efficient, the theory of tandem OLED was demonstrated in the year 2003. Tandem OLED is an advancement of basic OLED structure that comprises of two or more electroluminescent units that are placed one over the other and are separated by the interconnecting or charge generation (CGL) layer. They have the potential to achieve higher efficiency and luminance at minuscule current density with the increase in the number of electroluminescent units. They have longer lifespan in contrast to the conventional OLEDs as they have the tendency to reduce efficiency roll-off caused by the higher current density.

The enhanced performance of tandem OLED is attributed to the charge generation layer which serves exactly like the heart of tandem OLEDs. As a fundamental interconnector in tandem OLEDs, CGLs must have high transparency, low electrical resistance, matched HOMO LUMO levels with the adjacent transport layers, low power loss, etc. When the external voltage is supplied, charge carriers are generated in the CGL, which are then injected into the neighboring emitting unit via the transport layers, where they combine with the opposite charges injected by the electrodes to generate light.

TABLE 1: The detail device structures of red and blue OLEDs.

Device	Device Name	Layer Structure
Red conventional OLED	A	ITO/EL/LiF (2 nm)/Al (150 nm)
Red Tandem OLED	R	ITO/EL/CGL/EL /LiF (2 nm)/Al (150 nm)
Blue conventional OLED	B1	ITO/EL 1/LiF (2 nm)/Al (150 nm)
Blue Tandem OLED	B2	ITO/EL 1/CGL/EL 1/LiF (2 nm)/Al (150 nm)

EL: MoO_3 (3 nm)/NPB (40 nm)/α-NPD: $Ir(piq)_3$ (30 nm, 20% by volume)/Bphen (30 nm).
EL 1: MoO_3 (3 nm)/NPB (40 nm)/sDPVBi (30 nm) /Bphen (30 nm).
CGL: Al (3 nm)/LiF (2 nm)/C70 (5 nm)/CoPc (5 nm).

The CGL is typically a p-n junction that consists of metal/metal or metal oxide/metal oxide bilayer structure, organic doped n-type layer/p-type layer, or organic photovoltaic heterojunctions. Out of these, CGLs comprising metal layers or metal oxide layers are comparatively less transparent and electrically more resistant as they can easily react with dopants present in the adjacent transport layer to form complexes, thus enhancing the driving voltage and degrading the performance of the device. Further in OLEDs because of the waveguide modes, surface plasmon, and optical losses, the 75-80% of the total produced light is lost; only 20-25% of the persisting light can be efficiently drawn out of device. Thus, for enhancing the efficiency and performance of device, it becomes very necessary to utilize that lost part of light. To surpass all these limitations organic photovoltaic (OPV) type interconnectors formed of the organic materials with the low energy gap are mostly used [1, 2].

They have the tendency to generate the charges called electrically generated charges upon the application of external voltage and they have the additional advantage of absorbing a fraction of wasted photons (packets of light energy) from the emissive layer to form the excitons and then at the interface of heterojunction separates them into free electrons and holes. These free electrons and holes are known as photo generated charges. Thus, due to their unique feature of producing photo as well as electric generated charges, this OPV heterojunctions are contemplated as very efficient interconnectors in tandem OLEDs. In this work, the conjunction of organic fullerene derivative C70 and CoPc (Cobalt Phthalocyanine) heterojunction is proposed as planar charge generation layer of tandem OLEDs, where C70 behaves as n-type semiconductor material that has high thermal stability with exceptionally high electrons mobility and CoPc acts as p-type semiconductor material that has high holes' mobility. Further, to boost up the injections of electrons from the CGL to the neighboring emitting unit, the combination of thin layers of Al (aluminium) along with LiF (Lithium Fluoride) is used as electron injection material. The advantage of using aluminium is that it helps in decreasing the work function of C70, thus increasing the device performance.

2. Materials and Methods

In this work, the CGL comprises the multilayers of Al/LiF/C70/CoPc. The coalition of C70/CoPc acts as the planar organic heterojunction at which the charges are generated whereas the Al/LiF works as the effective electron injection layer that helps in enhancing the injection of electrons from CGL to the adjacent emitting units. To exhibit the efficacy of the proposed CGL, various devices are designed and simulated. The detail device structures are given in Table 1. The Schematic diagram of tandem device R is shown in Figure 1. In these devices ITO (Indium Tin Oxide) and Al/LiF act as anode and cathode, respectively, NPB (N,N0-bis-(3-naphthy)-N,N′-biphenyl-(1,1′-biphenyl)-4,4′-diamine) as the hole transport layer, and Bphen(4,7-diphenyl-1,10-phenanthroline) as electron transport layer in the two electroluminescent units. To prove the OPV behavior of the proposed CGL, red and blue devices are designed as given in Table 1 using α-NPD (N, N′-Bis(naphthalen-1-yl)-N,N′-bis(phenyl)-2,2-diMe) as host material for red emitter $Ir(piq)_3$(Tris [1-phenylisoquinoline-C^2,N] iridium(III)) and sDPVBi (4,4′-bis(2,2′-diphenylvinyl)-1,1′-spirobiphenyl as blue emitting material. The transport layers and electrodes optical constants are drawn from the [1]. The device emitting area is kept at 3 mm X 3 mm and the temperature is kept at 290K for this work. The Fluxim Setfos 4.6 is used to study the electrical and optical properties of the device.

To prove the efficacy of proposed OPV planar heterojunction and the ultrathin electron injection layer (Al/LiF), the conventional single emitter device structure is kept the same as used in [1] except that of the emitter that was CBP: $Ir(piq)_2acac$ (8% by volume). Due to the simulations environment limitation, the emitter used in this work is α-NPD: $Ir(piq)_3$ (20% by volume). However, the emitting zone and the peak wavelength of α-NPD: $Ir(piq)_3$ are the same as that of CBP: $Ir(piq)_2acac$. Both the emitters emit in the red wavelength zone of 600-800 nm and have peak wavelength around 625 nm. The charge generation layer used in [1] comprised the multilayers of LiF/C60 (Fullerene Derivative 60) /CuPc (copper phthalocyanine) /MoO_3, where C60/CuPc exhibits the organic photovoltaic type behavior and LiF and MoO_3 are the interface layers that increase the injection of charge carriers from CGL to the emissive layer.

3. Results and Discussion

The capacitance versus voltage plot as known from the former studies is the most suitable method to designate that the free positive and negative charge pairs are produced at the interface of the proposed organic heterojunction [3–5]. For this purpose, three devices D1-D3 is designed and simulated.

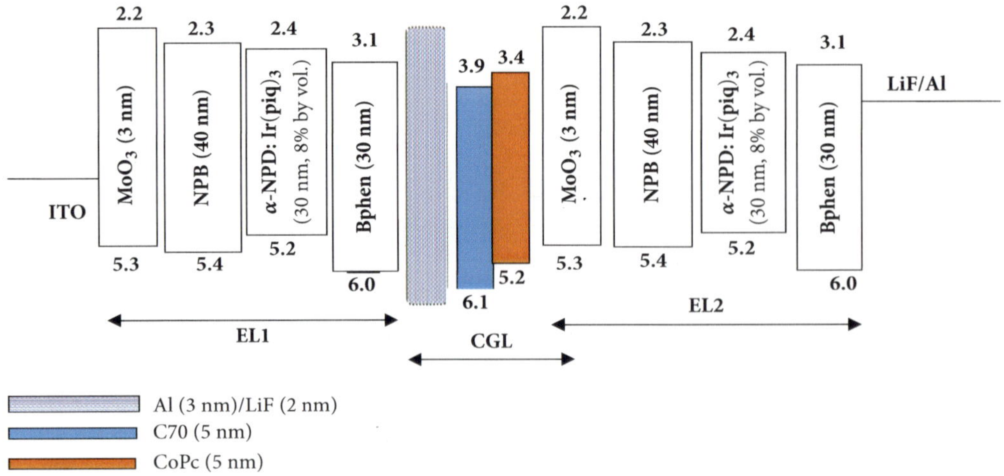

Figure 1: Structure of device R.

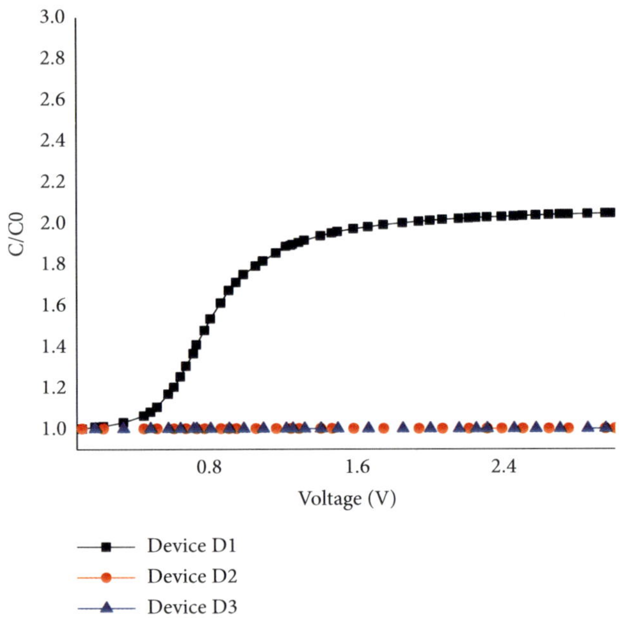

Figure 2: CV characteristics of devices D1, D2, and D3.

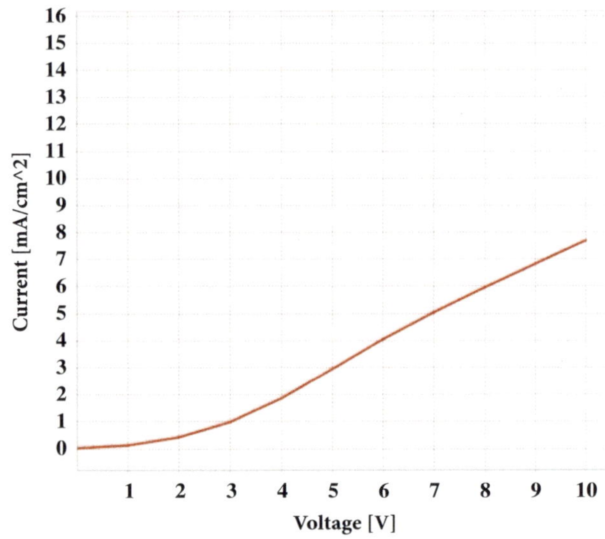

Figure 3: JV plot of device C.

The detail structure of the all the three devices is summarized in Table 2. To block the injection of holes and electrons from electrodes thick layers of LiF (50 nm) and MoO_3(50 nm) are used alongside anode and cathode, respectively. As can be seen from Figure 2 that, with the increase in the external voltage, the capacitance also increases for device D1 whereas for the devices D2 and D3, the capacitance remains invariable with the voltage. The rise in capacitance with voltage of device D1 should be attributed to the production of free charges at the interface of the proposed organic heterojunction of C70/CoPc because charge carriers from the electrodes are blocked by the thick insulating layers of LiF and MoO_3.

To show further the charge generation capability of C70/CoPc based planar heterojunction, a current limited device is fabricated. The detail structure of device C is summarized as follows: ITO (100 nm)/LiF (50 nm)/C70 (5 nm)/CoPc (5 nm)/MoO_3(50 nm)/Al (150 nm). Here the 50 nm thick layers of LiF and MoO_3 act as the holes and electrons blocking layers, respectively, that blocks the injection of holes and electrons from the two electrodes. It means that, as shown in Figure 3, the current in the current limited device is only because of the production of charge carriers at the C70/CoPc interface because the injection of charge carriers from the two electrodes are blocked by the thick blocking layers. The whole operating process of the device is shown in Figure 4. On application of voltage, the large amount of charges are produced at the C70/CoPc interface, which moves towards the electrodes from the interface, which are then collected by the electrodes, exhibiting unique diode behavior in the device. So, JV characteristics is one of the most appropriate methods to depict the charge generation ability of the charge generation layer.

TABLE 2: C-V characteristics of three devices D1, D2, and D3.

Device	Layer Structure
D1	ITO/LiF (50 nm)/C70(5 nm)/CoPc (5 nm)/MoO_3(50 nm)/Al(150 nm)
D2	ITO/LiF (50 nm)/C70(5 nm)/NPB(5 nm)/MoO_3(50 nm)/Al(150 nm)
D3	ITO/LiF (50 nm)/Bphen(5 nm)/CoPc (5 nm)/MoO_3(50 nm)/Al(150 nm)

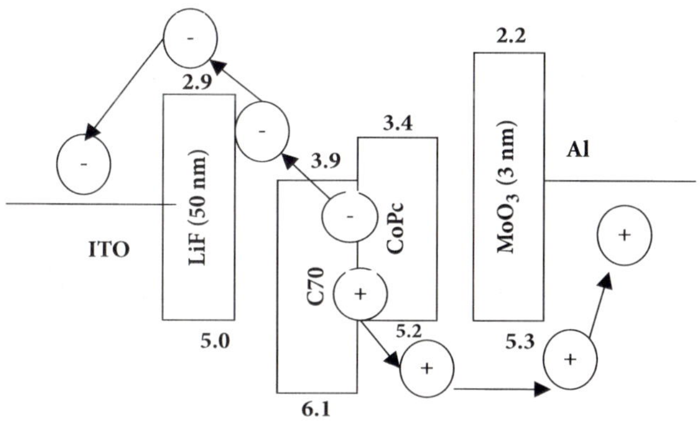

FIGURE 4: Operating process of device C.

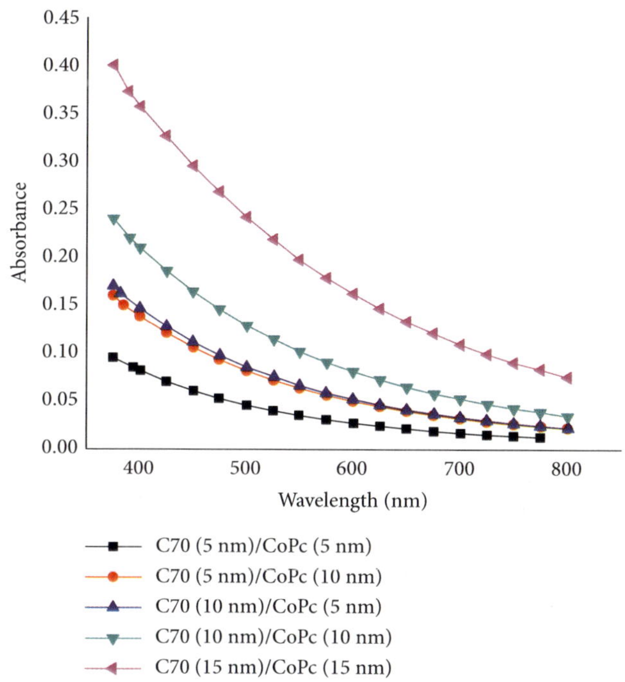

FIGURE 5: Absorbance plot of CGL at various thicknesses.

The energy level alignment of CGL with the adjacent transport layers is depicted in Figure 1. It can be seen that the HOMO (highest occupied molecular orbit) level difference between NPB (5.4 eV) and CoPc (5.2 eV) is only 0.2 eV and LUMO (lowest unoccupied molecular orbit) level difference between C70 (3.9 eV) and Bphen (3.1 eV) is 0.8 eV. Due to the small energy level difference between CGL and the neighboring transport layers, free charge carriers are produced at the interface of C70/CoPc planar heterojunction and transfer to the adjacent emitting layers i.e., holes can easily transfer from HOMO of CoPc to that of NPB while electrons can easily jump from LUMO of C70 to that of Bphen. Then from HOMO of NPB and LUMO of Bphen, the holes and electrons can move easily to the adjacent emitting units where they combine with the opposite charge carriers injected by the electrodes to form photons.

The optical and electrical properties are used to study the effect of thickness variation of CGL on the behavior of the device. The effect of thickness variation of CGL on the optical properties is studied by calculating the transmittance for different thickness of C70/CoPc from the absorption spectra as shown in Figure 5 by using the formula Transmittance= antilog (2-absorption). Thus, transmittance is calculated to be 94.5 %, 84.62%, and 71.56 %, respectively, for 5 nm, 10 nm, and 15 nm thickness of both C70 and CoPc while for 5 nm thickness of C70 and 10 nm thickness of CoPc, transmittance is 90.28 % and for 10 nm thickness of C70 and 5 nm thickness of CoPc, transmittance is 84.62 % at 625 nm. Thus, the planar organic heterojunction displays the highest transmittance for 5 nm thickness of both C70 and CoPc.

The effect of thickness variation of CGL on the electrical properties is studied by observing the current density versus voltage plot of different thickness of C70/CoPc. A test device T with the structure of ITO/Bphen (30 nm)/Al (3 nm)/LiF (2 nm)/C70 (x nm)/CoPc (y nm)/NPB (40 nm)/Al is designed and simulated for this purpose. The voltages at 200 mA/cm^2 are noted for different thickness sets of x and y (x, y). From Figure 6, it is found that voltages are 12.45 V, 13.3 V, 13.4 V, 14.4 V, and 16.46 V, respectively, for thickness sets (5nm, 5 nm), (5nm, 10 nm), (10nm, 5 nm), (10 nm, 10 nm), and (15nm, 15 nm). Thus the minimum voltage is found for 5 nm thickness of both C70 and CoPc for the same current density. Thus, it can be noted that increasing the thickness of CGL causes the difficulty in electron extraction from CGL. That is why, more voltage is required to extract the electrons at same current density with the increasing thickness [6, 7]. Therefore, the optimized thickness of both C70 and CoPc is found to be 5 nm from both the electrical and optical properties.

Further, tandem devices T1-T4 made with distinct interconnectors are designed and simulated to prove the effectiveness of proposed CGL consisting of multilayers of Al/LiF/C70/CoPc and also the advantage of using C70 over C60. The detail design structure of devices T1-T4 is given in Table 3. It can be shown from the current efficiency and luminance plots of devices T1-T4 and device A and R as given in Figures 7 and 8, respectively, that, out of all the devices given in Table 3, the red tandem device R with CGL layer consisting of Al/LiF/C70/CoPc shows the best performance in terms of current efficiency and luminance with respect to

TABLE 3: Detailed layer structures of red tandem OLED with different interconnectors.

Device	Layer Structure
T1	ITO/EL/IC 1/EL/LiF(2 nm)/Al(150 nm)
T2	ITO/EL/IC 2/EL/LiF(2 nm)/Al(150 nm)
T3	ITO/EL/IC 3/EL/LiF(2 nm)/Al(150 nm)
T4	ITO/EL/IC 4/EL/LiF(2 nm)/Al(150 nm)

EL: MoO_3 (3 nm)/NPB (40 nm)/α-NPD: $Ir(piq)_3$ (30 nm, 20% by volume)/Bphen (30 nm).
IC 1: C70 (5nm)/CoPc (5 nm).
IC 2: LiF (2 nm)/C70 (5 nm)/CoPc (5 nm).
IC 3: Al (3 nm)/C70 (5nm)/Pentacene (5 nm).
IC 4: Al(3 nm)/LiF (2 nm)/C60 (5 nm)/CoPc (5 nm).

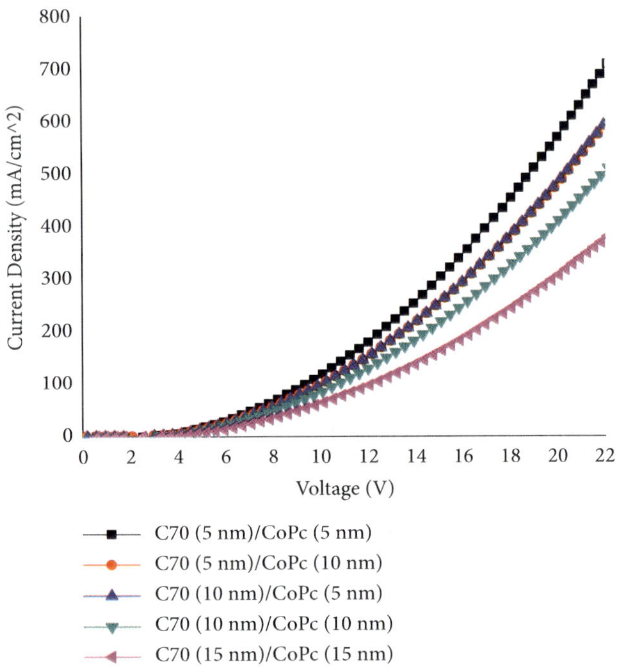

FIGURE 6: JV plot of CGL at various thicknesses.

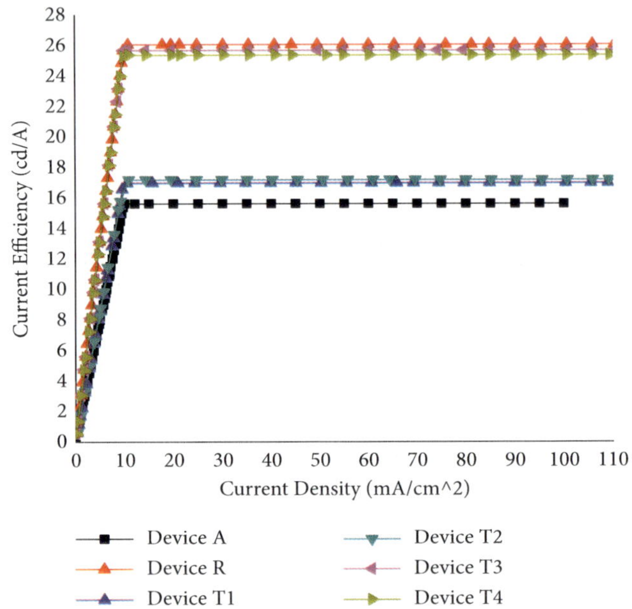

FIGURE 7: Current efficiency plot of various devices.

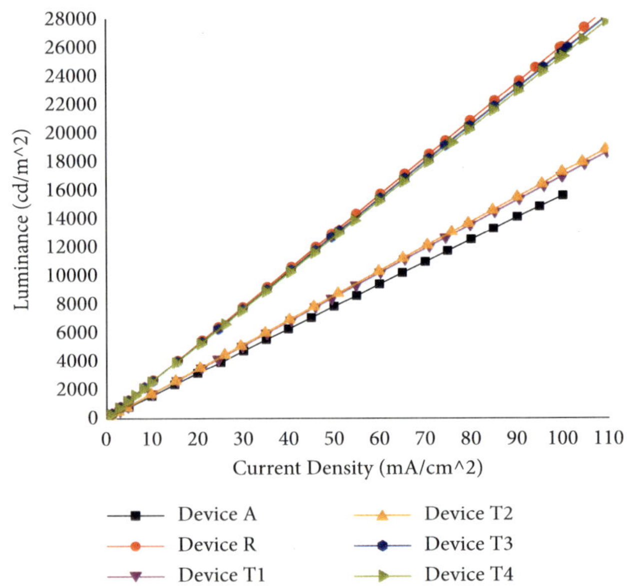

FIGURE 8: Luminance plot of various devices.

single emitter device A. The current efficiencies of device T1-T4 and device R, respectively, are found to be 16.97, 17.21, 25.65, 25.37, and 26 cd/A at 20 mA/cm^2 and luminance is 16970, 17217, 25660, 25378, and 26039 cd/m^2 at 100 mA/cm^2. It can be observed that the device R exhibits almost two times enhanced current efficiency and luminance in contrast to device A (15.63 cd/A and 15,630 cd/m^2).

This enhanced performance can be attributed to the interface layer or injection layer used with CGL which are crucial because of their functional efficacy and excellent charge injection property which eventually causes carrier recombination and balance in the emitting layer and also to the large energy level alignment between CoPc and C70 as well as good transport ability of C70 than that of C60. The HOMO of CoPc is 5.2 eV and LUMO of C60 is 3.7 eV whereas that of C70 is 3.9 eV. Therefore electrons have to cross a large energy barrier of 1.5 eV to jump to LUMO of C60 from the HOMO of CoPc in contrast to C70/CoPc where

TABLE 4: Comparison of proposed devices with the corresponding devices used in [1].

Parameters	Red Single Emitter Device A	Red Single Emitter Device [1]	Red Tandem Device R	Red Tandem Device [1]
Current Efficiency (cd/A) @ 20 mA/cm^2	15.63	8	26	17
Luminance (cd/m^2) @ 100 mA/cm^2	15630	7593	26039	16061
EQE @ 1000 cd/m^2	17.9	8.8	29.3	14.1

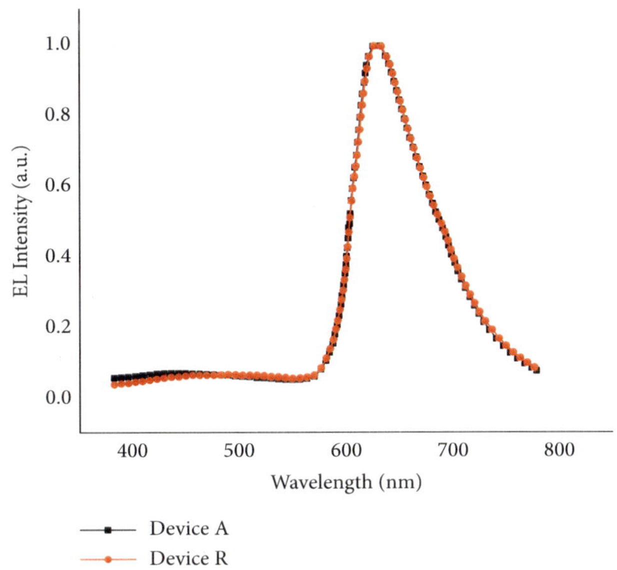

FIGURE 9: EL spectra of devices A and R.

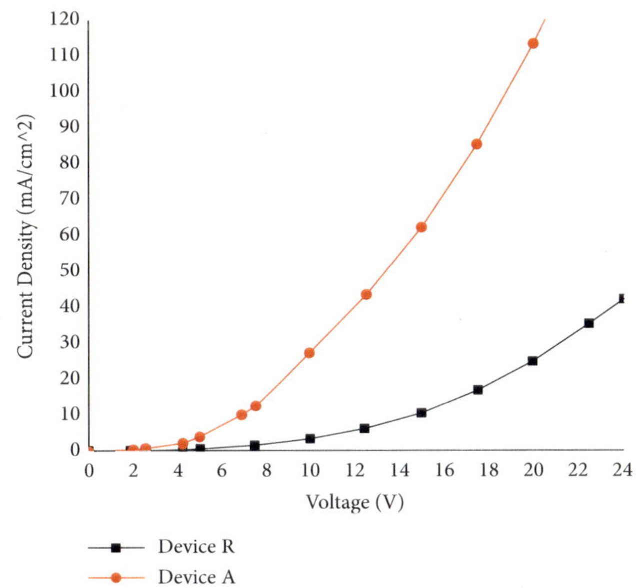

FIGURE 10: JV plot of devices A and R.

the energy barrier is 1.3 eV. Thus C70/CoPc heterojunction is highly efficient than the C60/CoPc heterojunction [8].

The EL spectra of both the structures A and R are shown in Figure 9. It can be noticed that the two spectra almost coincide over each other; i.e., there is very minuscule difference between the two spectra. Both the devices show peak wavelength located in red emitter zone at 628 nm. Thus, the proposed CGL has negligible impact on EL spectra relative to red single emitter device. From Luminance versus current density plot studied at 1000 cd/m^2 (Figure 8) it is found that current density of structure A is 6.39 mA/cm^2 which is 1.6 times higher than device R having current density of 3.84 mA/cm^2. Thus device R requires less current density than the device A to reach the same luminance. Therefore there is slow efficiency roll-off in device R than device A which increases with the current density. Thus, tandem devices are more efficient than conventional devices.

Normally it has been known that the turn-on voltage of tandem OLED is over two times greater in contrast to conventional OLED, as in tandem device two EL units are used [9–11]. But from the current density versus voltage graph as shown in Figure 10, the turn-on voltage of device A is found to be 3 V and device R is found to be 5.85 V which is less than two times the turn-on voltage of device A. It should be because of less voltage drop across the proposed CGL, thus showing its effectiveness. From Figure 8, the luminance of the tandem device R is 26,039 cd/m^2 that is 1.66 times the luminance of device A (15,630 cd/m^2) at 100 mA/cm^2. In addition, the current efficiency of device R is 26 cd/A which is almost twice as compared to device A (15.63 cd/A). Further enhancement in EQE of device R is also observed in contrast to device A. The detail EL performances of the device R and A are concluded in Table 4.

The difference in the performance of tandem device and single emitter unit device is because of the microcavity effect, interference from light, and many more. The enhancement in performance of tandem devices is because of the CGL that produced almost the same number of free charges as are inoculated by the anode and cathode, thus causing charge carriers balance which leads to reduction in quenching and the plasmonic losses and also increases the microcavity effect. Also, the emission quenching caused by relaxed CoPc molecules is reduced because of the large distance between emitting unit and CGL. Further, the electric field intensity of tandem device is found to be less in contrast to the single emitter device, thereby reducing the quenching effect caused by electric field in tandem device in contrast to single emitter device. In addition, in tandem devices both the emitting layers are located at the two antinodes so as to take the benefit of increased light output from the microcavity effect [12–15].

To observe the OPV behavior of proposed heterojunction, the UV-vis absorption spectra of C70/CoPc and the

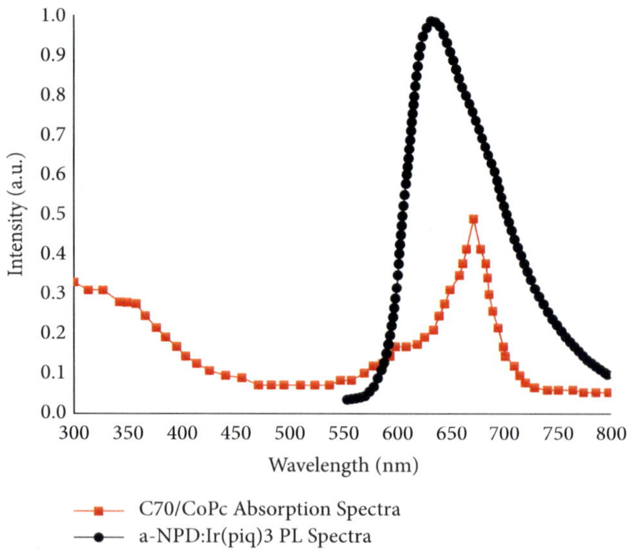

FIGURE 11: Absorption, PL spectra of CGL, and red emitter.

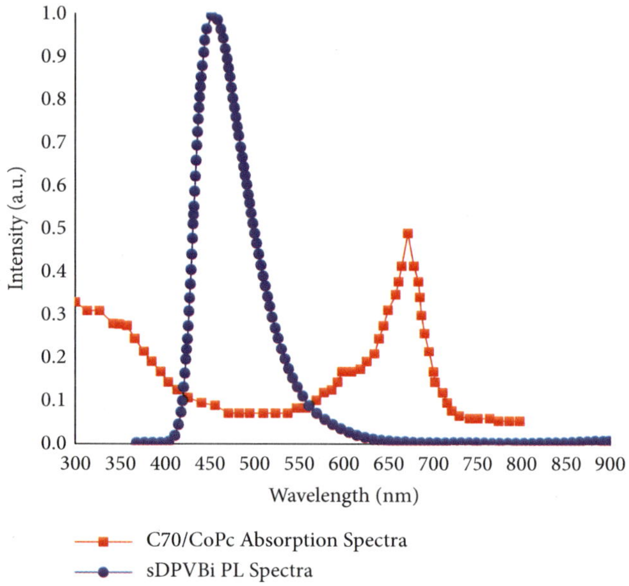

FIGURE 12: Absorption, PL spectra of CGL, and blue emitter.

photoluminescent spectra of α-NPD: $Ir(piq)_3$ are studied. As shown in Figure 11 both spectra overlap each other in the red emitter zone of 600-800 nm. So it can be presumed that the red wasted photons due to surface plasmon and waveguide modes in the emitter can be easily absorbed by C70/CoPc heterojunction to form excitons and further dissociate them into free charges, thus enhancing the efficiency. A blue tandem device B2 as shown in Table 1 is further designed and simulated to show the organic photovoltaic behavior of the C70/CoPc heterojunction. From the UV-vis absorption spectra of C70/CoPc and that of the blue emitter used as shown in Figure 12, it can be seen that two spectra do not coincide; i.e., the peak of the emitter does not fall in the main absorbing area of the C70/CoPc. So, it can be concluded that the proposed CGL is not able to observe the blue photons generated in the emitter.

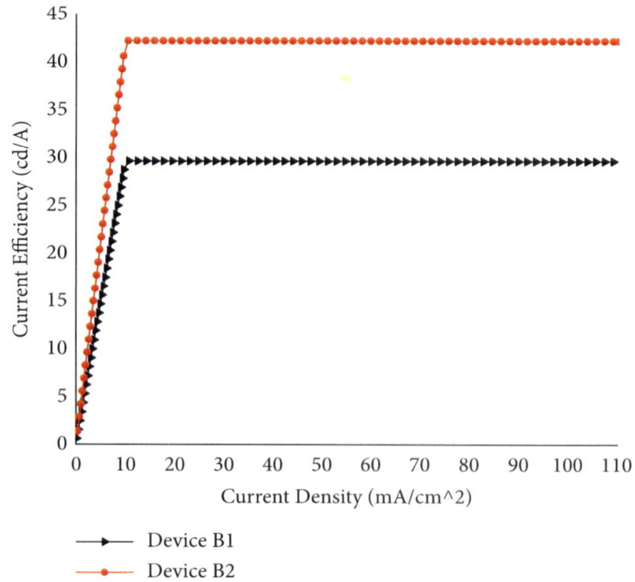

FIGURE 13: Current efficiency plot of devices B1 and B2.

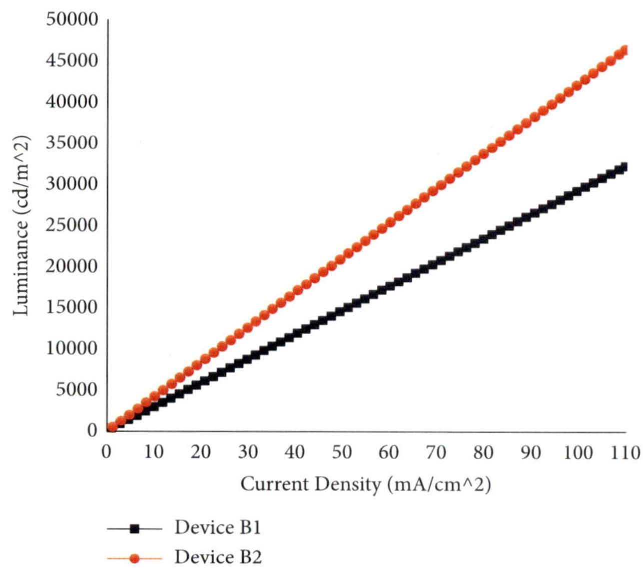

FIGURE 14: Luminance plot of device B1 and B2.

Also, from Figures 13 and 14, it can be seen that current efficiency and luminance of tandem device B2 (42.25 cd/A and 42250 cd/m^2) at 20 mA/cm^2 and 100 mA/cm^2, respectively, is only 1.42 times the current efficiency and luminance of device B1 (29.67 cd/A and 29670 cd/m^2) whereas because of the OPV type behavior of device R which is the absorption of red wasted photons by the proposed heterojunction, the current efficiency of device R at the same current density is 1.6 times the efficiency of device A. The JV characteristics of device B1 and B2 are shown in Figures 15 and 16, respectively. It can be observed that the turn-on voltage of device B2 is about 4.2 V which is 2.3 times greater than that of device B1 having turn-on voltage of 1.8 V. On the other hand, from

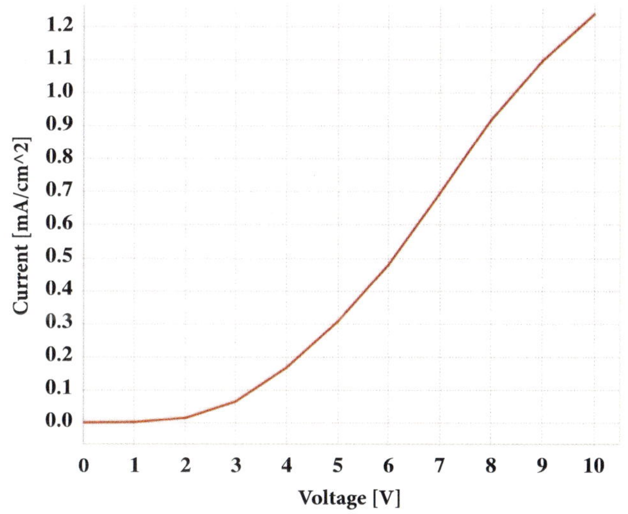

FIGURE 15: JV plot of of device B1.

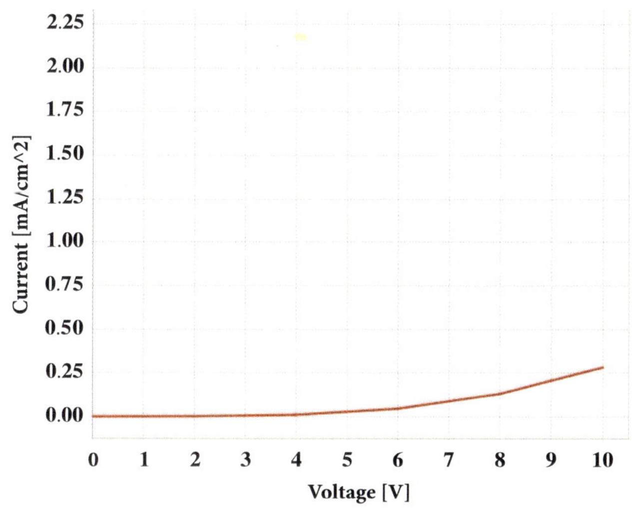

FIGURE 16: JV plot of device B2.

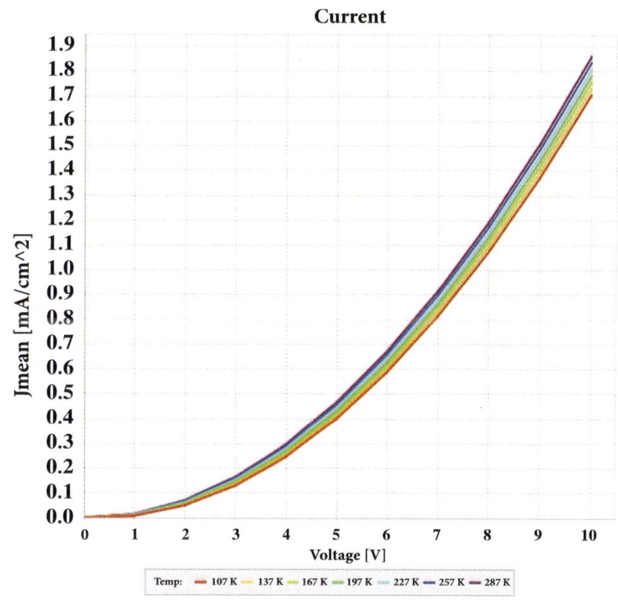

FIGURE 17: JV plot of device R at various temperatures.

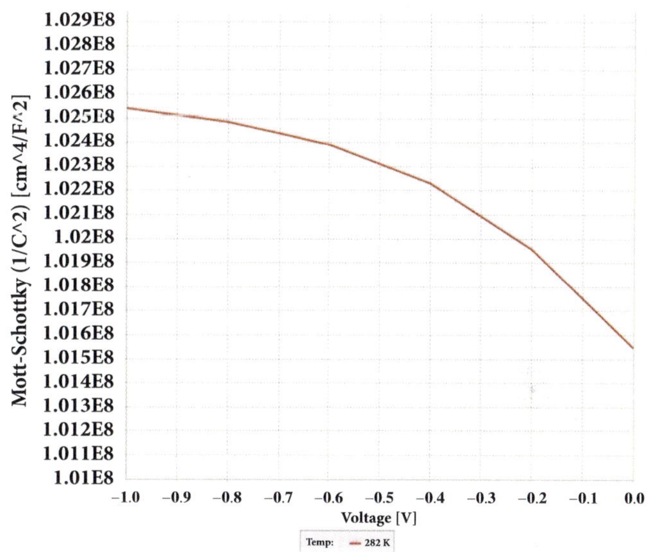

FIGURE 18: Mott-Schottky plot of device R at 282K.

Figure 10, it can be observed that the turn-on voltage of device R is only 1.9 times greater than that of device A. It is because in device B2, the CGL does not exhibit OPV type behavior, so the excitons are not present in the CoPc layer. So, the charge generation at the low current density is due to the transfer of electrons from HOMO of CoPc to the LUMO of C70 and because of the large energy barrier between the C70 and CoPc interface as depicted in Figure 1, greater value of electric field is needed for the charge carriers transfer, thereby reducing the efficiency and requiring high turn-on voltage.

The basic requirement in CGL is the generation of charges at the CGL interface. Many models have been proposed till date to show the charge generation capability of CGL like Kroger et al. who in [16] have suggested that the charge production capability of interconnector does not depend on temperature. Sun et al. in [17] have demonstrated through JV characteristics the charge production ability of interconnector under reverse bias. In this work, the charge generation capability of CGL is explained through J-V and C-V plot of tandem device R as shown in Figures 17 and 18, respectively. It can be seen that, at about 1.1 V forward voltage, the current begins to increase; on the other hand correspondingly there is a drastic decrease in the capacitance which can be attributed to the large injection and recombination of charge carriers. Also, at about 1V reverse voltage there is critical decrease in capacitance as well as current breakdown which can be attributed to large generation of charge carriers at C70/CoPc heterojunction [16]. This is because under the reverse voltage, hole and electron transport layers should stop the transport of electrons and holes, respectively. So, the large reverse current in the device R should be because of the charge carriers production rather than because of the injection by anode and cathode under the reverse bias. The entire working process of tandem device can be concluded as follows. At small current density, due to the generation of charges at heterojunction, current efficiency increases rapidly with the increase in luminance of device. Then, the current efficiency becomes invariable, after attaining the maximum height. This is because of the recombination of these generated charges at CGL with the charges inoculated by electrodes in the emissive layer to produce photons. Due to the OPV type behavior, CGL fetches a part of wasted photons from emissive layer to generate more charges and again the whole process repeats causing the slower efficiency roll-off. Thus, this OPV type CGL is very effective and results in longer lifespan of the device.

Further to improve the performance of device R in terms of efficiency, longer lifespan, and low power consumption and to elucidate microcavity effect for red emitter, DBR substrates consisting of quarter wavelength stack two layers of high refractive index Zirconium Dioxide (ZrO_2) and low refractive index Silicon Dioxide (SiO_2) is used. The reflectance of Distributed Bragg Reflector (DBR) structure depends upon variation in the refractive index of layers and number of layer pairs used in the structure. The more the gap and number of pairs are, the greater the reflectance will be. The DBR structures with high reflectance help in achieving highly effective cavity structures and thus help in improving brightness of the device. The optimized quarter wavelength thickness of ZrO_2 is found to be 64 nm by using the formula ($\lambda/4n_{ZrO_2}$ = 64nm, where n_{ZrO_2} = 2.45, λ = 628 nm) and that of SiO_2 is found to be 107 nm ($\lambda/4n_{SiO_2}$ = 107nm, where n_{SiO_2} = 1.4575, λ = 628 nm). Here λ = 628 nm is the peak wavelength of device R. The reflectance (Figure 19) of single pair of ZrO_2 and SiO_2 is found to be approximately 56% at 625-628 nm wavelength. Further enhancement in current efficiency and luminance (Figure 20) at 20 mA/cm^2 and 100 mA/cm^2, respectively, of device R is observed with the DBR structure which is found to be 31.80 cd/A and 31,800 cd/m^2 respectively. This enhancement is due to the constructive interference of light because of DBR structure.

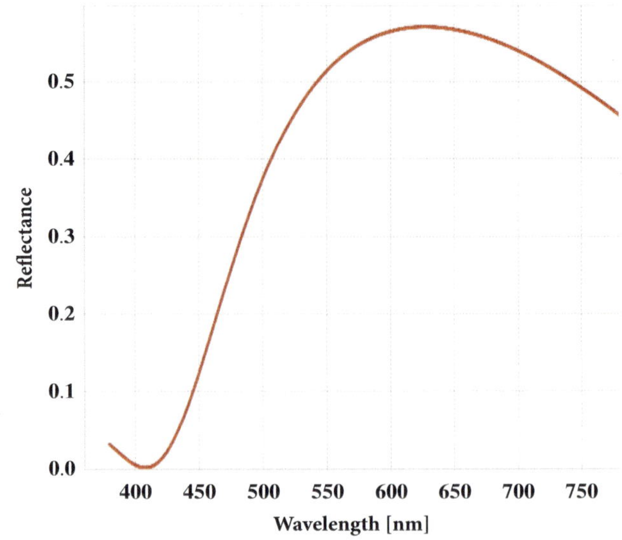

FIGURE 19: Reflectance of DBR structure.

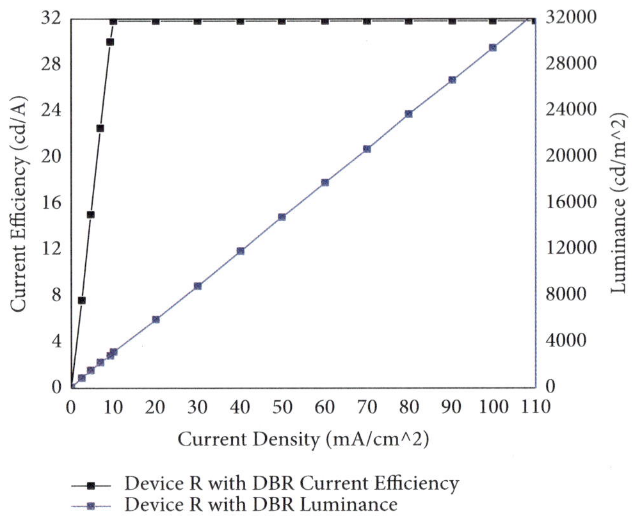

FIGURE 20: Current Efficiency and Luminance plot of DBR structure.

4. Conclusions

In this paper, the novel planar organic heterojunction of C70/CoPc is proposed as planar charge generation layer which shows organic photovoltaic type behavior in red wavelength range from 600 to 800 nm. The advantage of using organic photovoltaic CGL is that they can generate charges and can absorb the wasted photons from emitter to form excitons and dissociate them into free electrons and holes. As a result, there is rather slower efficiency roll-off as compared to other devices. The tandem device exhibits 1.63 and 1.65 times enhanced current efficiency and luminance in contrast to conventional device. Also, remarkable reduction in turn-on voltage of the proposed tandem device as compared to other tandem device is observed. The turn-on voltage of tandem device is found to be only 1.9 times greater than that of the single emitter unit device. This remarkable improvement in performance can be attributed to the excellent charge generation, separation ability of interconnector, and excellent transport capability of Al/LiF electron injection layer which can be seen from the great rise of capacitance and current density with voltage. Also, from -J-V and C-V curves it can be confirmed further that the produced carriers are due to the reverse bias current. Further, by inculcating DBR structure of high refractive index ZrO_2 and low refractive index SiO_2, the performance of tandem device in terms of current efficiency and luminance got enhanced by almost two times in contrast to conventional OLED. Thus, the doping free planar organic heterojunction yields high performance tandem OLED with the low fabrication cost. Also, the proposed work opens the path to explore the behavior of organic photovoltaic heterojunctions for fabricating highly efficient OLEDs.

Conflicts of Interest

The author declare that there are no conflicts of interest regarding the publication of this paper.

Acknowledgments

The author would like to thank Dr. S.S Pattnaik, Director National Institute of Technical Teachers' Training & Research, Chandigarh, India, and for constant motivation and support throughout this research work.

References

[1] D. Zhao, H. Liu, Y. Miao et al., "A red tandem organic light-emitting diode based on organic photovoltaic-type charge generation layer," *Organic Electronics*, vol. 32, pp. 1–6, 2016.

[2] Z. Lü, Y. Wang, X. Li, J. Xiao, and Z. Deng, "Role of MoO3-modified organic photovoltaic-type charge generation layer in tandem organic light-emitting diodes," *Synthetic Metals*, vol. 229, pp. 47–51, 2017.

[3] M. Ramar, P. Tyagi, C. K. Suman, and R. Srivastava, "Enhanced carrier transport in tris(8-hydroxyquinolinate) aluminum by titanyl phthalocyanine doping," *RSC Advances*, vol. 4, no. 93, pp. 51256–51261, 2014.

[4] S. Altazin, S. Züfle, E. Knapp et al., "Simulation of OLEDs with a polar electron transport layer," *Organic Electronics*, vol. 39, pp. 244–249, 2016.

[5] J. Yang, Y. Xiao, Y. Deng et al., "Electric-Field-Assisted Charge Generation and Separation Process in Transition Metal Oxide-Based Interconnectors for Tandem Organic Light-Emitting Diodes," *Advanced Functional Materials*, vol. 22, no. 3, pp. 600–608, 2012.

[6] Y. Chen, Q. Wang, J. Chen, D. Ma, D. Yan, and L. Wang, "Organic semiconductor heterojunction as charge generation layer in tandem organic light-emitting diodes for high power efficiency," *Organic Electronics*, vol. 13, no. 7, pp. 1121–1128, 2012.

[7] Y. Chen, D. Ma, H. Sun et al., "Organic semiconductor heterojunctions: electrode-independent charge injectors for high-performance organic light-emitting diodes," *Light: Science & Applications*, vol. 5, no. 3, pp. e16042–e16042, 2016.

[8] K. L. Mutolo, E. I. Mayo, B. P. Rand, S. R. Forrest, and M. E. Thompson, "Enhanced open-circuit voltage in subphthalocyanine/C_{60} organic photovoltaic cells," *Journal of the American Chemical Society*, vol. 128, no. 25, pp. 8108-8109, 2006.

[9] L. Duan, T. Tsuboi, Y. Qiu, Y. Li, and G. Zhang, "Tandem organic light-emitting diodes with KBH_4 doped 9,10-bis(3-(pyridin-3-yl)phenyl) anthracene connected to the charge generation layer," *Optics Express*, vol. 20, no. 3, pp. 64–72, 2012.

[10] Y. Deng, J. Wang, Q. Ou et al., "The electro-optic performance and photovoltaic effect of organic devices based on cesium carbonate/Al/molybdenum trioxide intermediate connector," *Organic Electronics*, vol. 51, pp. 452–457, 2017.

[11] Y. Chen, H. Tian, J. Chen et al., "Highly efficient tandem white organic light-emitting diodes based upon C60/NaT4 organic heterojunction as charge generation layer," *Journal of Materials Chemistry*, vol. 22, no. 17, p. 8492, 2012.

[12] Y. Chen and D. Ma, "Organic semiconductor heterojunctions as charge generation layers and their application in tandem organic light-emitting diodes for high power efficiency," *Journal of Materials Chemistry*, vol. 22, no. 36, pp. 18718–18734, 2012.

[13] L. S. Liao and K. P. Klubek, "Power efficiency improvement in a tandem organic light-emitting diode," *Applied Physics Letters*, vol. 92, no. 22, pp. 1–3, 2008.

[14] T. Cho, C. Lin, and C. Wu, "Microcavity two-unit tandem organic light-emitting devices having a high efficiency," *Applied Physics Letters*, vol. 88, no. 11, pp. 1–3, 2006.

[15] C. Xiang, W. Koo, F. So, H. Sasabe, and J. Kido, "A systematic study on efficiency enhancements in phosphorescent green, red and blue microcavity organic light emitting devices," *Light: Science & Applications*, vol. 2, no. 6, pp. e74–e74, 2013.

[16] M. Kröger, S. Hamwi, J. Meyer et al., "Temperature-independent field-induced charge separation at doped organic/organic interfaces: Experimental modeling of electrical properties," *Physical Review B: Condensed Matter and Materials Physics*, vol. 75, no. 23, 2007.

[17] H. Sun, Q. Guo, D. Yang, Y. Chen, J. Chen, and D. Ma, "High efficiency tandem organic light emitting diode using an organic heterojunction as the charge generation layer: An investigation into the charge generation model and device performance," *ACS Photonics*, vol. 2, no. 2, pp. 271–279, 2015.

22

A Structured Light RGB-D Camera System for Accurate Depth Measurement

Van Luan Tran[1] **and Huei-Yung Lin**[1,2]

[1]*Department of Electrical Engineering, National Chung Cheng University, 168 University Rd., Chiayi 621, Taiwan*
[2]*Advanced Institute of Manufacturing with High-Tech Innovation, National Chung Cheng University, 168 University Rd., Chiayi 621, Taiwan*

Correspondence should be addressed to Van Luan Tran; tranvanluan07118@gmail.com

Guest Editor: Wei Liu

The ability to reliably measure the depth of the object surface is very important in a range of high-value industries. With the development of 3D vision techniques, RGB-D cameras have been widely used to perform the 6D pose estimation of target objects for a robotic manipulator. Many applications require accurate shape measurements of the objects for 3D template matching. In this work, we develop an RGB-D camera based on the structured light technique with gray-code coding. The intrinsic and extrinsic parameters of the camera system are determined by a calibration process. 3D reconstruction of the object surface is based on the ray triangulation principle. We construct an RGB-D sensing system with an industrial camera and a digital light projector. In the experiments, real-world objects are used to test the feasibility of the proposed technique. The evaluation carried out using planar objects has demonstrated the accuracy of our RGB-D depth measurement system.

1. Introduction

In recent years, 3D imaging has received a great value in industrial and consumer applications. Machine vision systems developed with 3D imaging allow faster and more accurate measurement of components at manufacturing whereabouts. Nowadays, RGB-D cameras, such as Microsoft Kinect and Asus Xtion, are very popular due to the ability to provide the depth information directly. However, they have the limitation on accuracy and thus are not suitable for the applications that require accurate shape measurements [1–3]. As a result, the development of real-time RGB-D cameras still receives much attention from researchers and practitioners. The objective is to provide highly accurate RGB-D sensing techniques with more effective implementation approaches in terms of the density of acquired point clouds, time consumption, working environment, noise level, etc.

3D reconstruction based on the structured light technique has been investigated in the past few decades due to its popularity in the manufacturing applications. Structured light systems are suitable solutions for structured light scanning, 3D reconstruction, and 3D sensing with accurate shape measurements [4, 5]. Structured light refers to the process of projecting predesigned known patterns on the scene and capturing the images to calculate the depth for 3D surface reconstruction. It is an important contribution to the development of 3D measurement systems. The patterns projected on the scene can be generated by a projector or other devices [6], and the relationship between the light source and the camera is a crucial factor. The accuracy of 3D reconstruction depends on the correctness of the calibration, which provides the relative pose between the camera and the light source projector.

In recent literature, several works presented the structured light systems for 3D reconstruction and proposed different approaches to deal with the related problems [7–10]. Scharstein *et al.* [11] proposed a method for acquiring high-complexity stereo image pairs with pixel-accurate correspondence information using structured light. Some previous works such as [12–15] described various methods to perform 3D reconstruction and obtained some satisfactory results. However, those techniques require to use precalibrated

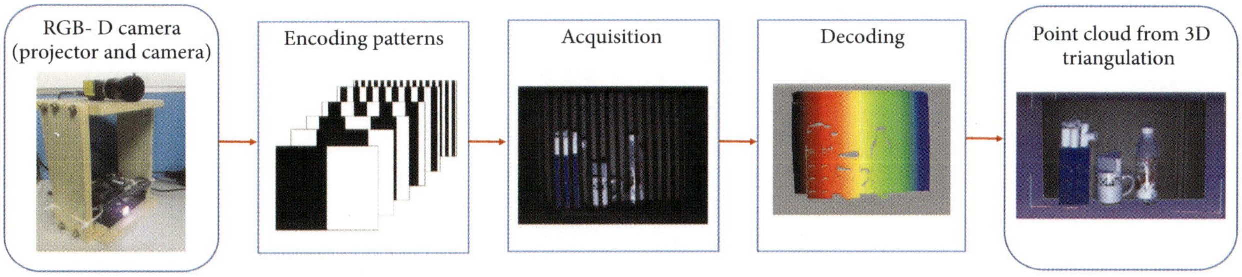

FIGURE 1: The overview of our RGB-D camera system with the structured light technique. The encoding pattern is a gray-code pattern. The acquisition is the images captured by the camera for each pattern in the sequence. The decoding is a coded map.

cameras to find the 3D world coordinates of the projected pattern. Thus, they highly depend on the accuracy of camera calibration and may transfer the error to the projector calibration. In [16], Huang and Tang described a method to perform fast 3D reconstruction using one-shot spatial structured light. Although the method can provide relatively accurate results, the evaluation and analysis were not carried out comprehensively. Some restrictions are also shown in their experiments when performing the tests on complex object surfaces. Cui and Dai [17] proposed a simple and efficient 3D reconstruction algorithm using structured light from 3D computer vision. However, their approach has some limitations on measuring inclined objects, and the 3D information cannot be recovered for the shadow areas.

In this work, we develop an RGB-D camera system based on the structured light technique. A system flowchart is shown in Figure 1. The encoding method is based on the gray-code coding [5], and the 3D reconstruction is achieved by the ray triangulation principle with the estimation of intersection points. The accuracy and density of the obtained point clouds are both high, and therefore it is suitable for the applications such as accurate shape measurements, 3D object recognition, and pose estimation for robotic manipulation.

This article is organized as follows: Section 2 presents a general overview of the structured light system and an accurate calibration method to derive the parameters of the camera-projector system. Section 3 contributes a method to create encoded patterns and decode the captured images from the camera and presents a ray triangulation principle for 3D computation by point intersection. Section 4 provides some experimental results, including the experimental setup, results with several different objects, and the evaluation of the accuracy of the object reconstruction. Finally, Section 5 gives the conclusion.

2. Background

2.1. Structured Light Technique. Currently, the development of structured light systems is in high demand. The structured light technique is based on the principle described in Figure 2. In general, the process of a structured light system can be divided into three basic steps:

(i) **Encoding.** The encoding of the information into a sequence of patterns is performed in the temporal domain. A sequence of structured light patterns depends on the number of required patterns, parameters of the system, and the resolution of the projector and the camera.

(ii) **Acquisition.** The sequence of patterns is projected on the scene by a data projector, and a camera is used to continuously capture the images.

(iii) **Decoding.** The captured pattern-coded images are processed with the recognition of projected patterns to find the corresponding points associated with the projector and the camera.

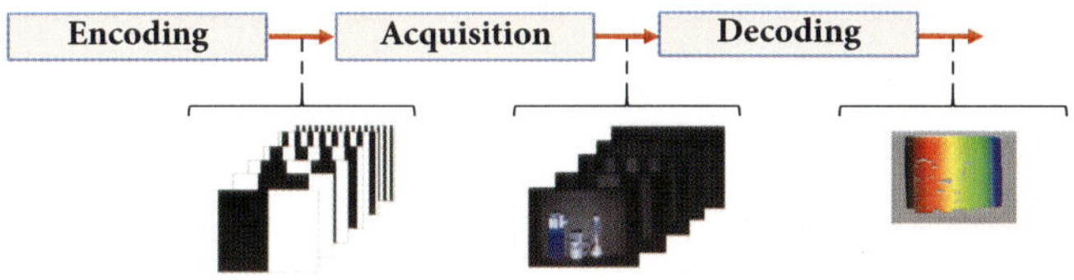

FIGURE 2: The overview of three basic steps in the process of a structured light system [1].

In the implementation, there might be some additional steps depending on the solution of the system designer. It often follows a procedure to create range images, point clouds, or mesh models and possibly integrates several decoded coordinate maps, calibration, and triangulation principle. The calibration is to determine the intrinsic and extrinsic parameters of the camera and the projector, and the reconstruction is usually based on the ray triangulation principle by computing the intersection point.

2.2. Calibration. Calibration is an important issue which greatly affects the accuracy of the results [18]. In the proposed technique, firstly, we find the parameters of the system using the calibration method by Moreno and Taubin [6]. It is a simple and accurate method to calibrate the projector and camera systems. In this method, the projected corner locations are estimated with subpixel precision using local homographies to each corner in the images as illustrated in Figure 3. It includes three main steps as follows:

(i) The camera calibration step to determine the intrinsic parameters of the camera. Camera calibration includes collecting a sequence of images of a planar

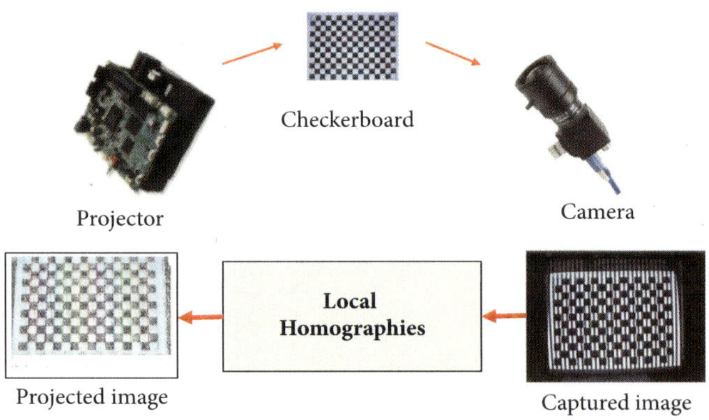

FIGURE 3: An illustration of the structured light system calibration. The captured image was selected in a set of the calibrated images, which are the projector project the patterns on the checkerboard and captured by the camera. The projected image is estimated with subpixel precision using local homographies to each corner in the captured images [1].

checkerboard pattern. The intrinsic camera calibration is derived by estimating the parameters using the perspective camera model [19]. We find the coordinates in the camera image plane for all of the checkerboard corners captured with different pattern orientations. We use OpenCV's findChessboardCorners() function [20] to automatically find the checkerboard corner locations. They are then refined to approach the subpixel accuracy. Finally, OpenCV's calibrateCamera() function is used to derive the calibrated camera parameters.

(ii) The projector calibration step is to determine the intrinsic parameters of the projector. The mathematical model of our projector can be described the same as the camera. But the projector cannot capture the images from its viewpoint to find checkerboard corners. In this situation, we know a relation between the projector and image pixels extracted from the structured light sequences. Thus, we can estimate the checkerboard corner locations in the projector pixel coordinates based on a local homography [6] as an illustration in Figure 3.

(iii) The stereo system calibration step is to derive the extrinsic parameters of the system, which consist of the rotation matrix and the translation vector. We use OpenCV's stereoCalibrate() function with the previously found checkerboard corner coordinates and their projections. The stereo parameters are a rotation matrix R and a translation vector T relating the projector-camera pair.

3. RGB-D Sensing Based Using Structured Light

3.1. Encoding and Decoding Patterns. The gray-code pattern [4] is a sequence of images with black and white stripes created for encoding the scene from the camera viewpoint.

FIGURE 4: The gray-code patterns for the RGB-D camera used by the structure light technique [1].

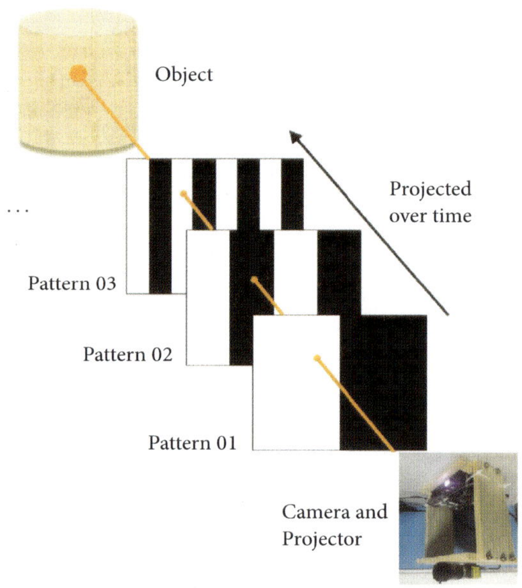

FIGURE 5: The acquisition of the projected patterns on an object. The gray-code pattern is projected by the projector and the scene is captured by the camera [1].

The pattern sequence has two types, one is the horizontal stripe and the other is the vertical stripe, as illustrated in Figure 4. All patterns are projected to a scene or an object as shown in Figure 5. The horizontal patterns consist of 10 pattern images which represent 10-bit values for each pixel. The first pattern is half black and half white, which represents the most significant bit, and the rest patterns are the images that switch between black and white in the columns. After combining all of the 10 horizontal patterns into one image, each column has the same 10 bits with the columns in the same image.

Structured light encoding depends on the resolution of the projector. The information is encoded into a sequence of patterns performed in the temporal domain. Commonly used approaches include gray-code coding and binary-code coding. Gray codes can be calculated by first computing the

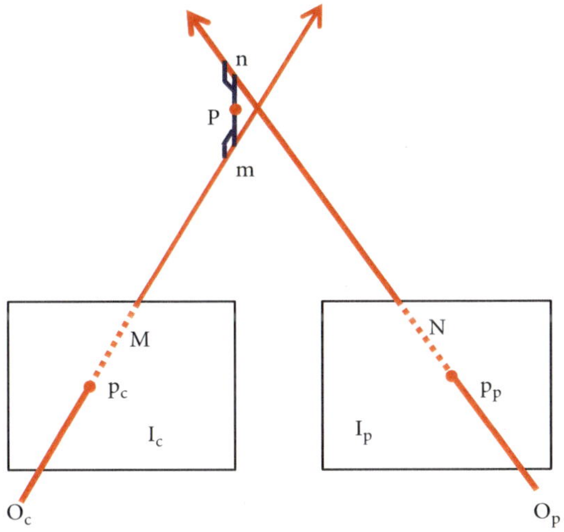

FIGURE 6: The intersection point of two rays from the pixels in the camera and projector coordinates [1].

binary representation of a number and then converting it using the following process: copy the most significant bit as it is, and replace the remaining bits (taking one bit at a time) with the result of an XOR operation of the current bit, with the previous bit of higher significance in the binary form.

For the binary-code coding, only two illumination levels are used and encoded as 0 and 1. The gray-code coding is an alternative to the binary representation, with only one bit change at a time between any two adjacent numbers. If there is an error reading on any changed bit, the value will never be off by more than one unit. In our system, we use a projector with the resolution of 1024 × 768 and decode the pattern with 10 bits ($2^{10} = 1024$ and $2^{10} - \text{offset} = 768$), where the number of vertical patterns is $\lceil \log(1024)/\log(2.0) \rceil = 10$ and the number of horizontal patterns is $\lceil \log(768)/\log(2.0) \rceil = 10$.

The camera captures the images of the projected patterns, and the coding step is to decode each pixel in the captured images into their corresponding decimal number presenting the column and row. It will be used to create a coded map as shown in Figure 1, which presents the corresponding point between the projector and the camera.

3.2. 3D Reconstruction. With a robust projector-camera calibration step, we define the location and orientation of the camera and projector with respect to the world coordinate frame. In the pattern encoding and decoding step, we determine one pixel in image I_c and its corresponding pixel in image I_p. Our reconstruction is based on the ray triangulation principle by the estimation of intersection points [4]. In order to compute the direction vector, two points in a ray are needed. The first point is the camera's center of projection, which is determined based on the extrinsic parameters of the structured light system. The second point is the point corresponding to the pixel from which the ray passes through. One ray passes through O_c and p_c of the left image (I_c), and the other ray passes through O_p and p_p of the right image (I_p), as shown in Figure 6. Here, O is the origin of the system and p is a pixel in the image. The 3D point cloud is obtained from the intersection point P, which is a midpoint of the shortest segment between the rays.

To estimate the intersection point, we consider two rays M and N in 3D space passing through points p_c and p_p with direction vectors $\vec{x}$ and $\vec{y}$, respectively. Let the two closest points on the lines be m and n, as defined in (1) and (2), where g and k are scalar values.

$$m = p_c + g\vec{x} \tag{1}$$

$$n = p_p + k\vec{y} \tag{2}$$

The mn segment connecting rays M and N is perpendicular to the rays, and therefore the dot product of the vectors is equal to 0 as follows:

$$(m - n)\vec{x} = 0 \tag{3}$$

$$(m - n)\vec{y} = 0 \tag{4}$$

With (1) and (2), (3) and (4) are represented by

$$\vec{r} \cdot \vec{x} + g\vec{x} \cdot \vec{x} - k\vec{y} \cdot \vec{x} = 0 \tag{5}$$

$$\vec{r} \cdot \vec{y} + g\vec{y} \cdot \vec{x} - k\vec{y} \cdot \vec{y} = 0 \tag{6}$$

where

$$\vec{r} = p_c - p_p \tag{7}$$

From (5) and (6), the scalar values are calculated by

$$g = \frac{(\vec{r} \cdot \vec{x})(\vec{y} \cdot \vec{y}) - (\vec{y} \cdot \vec{x})(\vec{r} \cdot \vec{y})}{(\vec{y} \cdot \vec{x})(\vec{y} \cdot \vec{x}) - (\vec{y} \cdot \vec{y})(\vec{x} \cdot \vec{x})} \tag{8}$$

$$k = \frac{(\vec{y} \cdot \vec{x})(\vec{r} \cdot \vec{x}) - (\vec{x} \cdot \vec{x})(\vec{r} \cdot \vec{y})}{(\vec{y} \cdot \vec{x})(\vec{y} \cdot \vec{x}) - (\vec{y} \cdot \vec{y})(\vec{x} \cdot \vec{x})} \tag{9}$$

The midpoint of the shortest segment are then estimated by

$$P = \frac{p_c + g\vec{x} + p_p + k\vec{y}}{2} \tag{10}$$

4. Experiments

In the structured light system, the quality of the captured images is important for obtaining a good pattern database to perform the calibration, decoding, and reconstruction. Hence, the resolution of the camera is usually higher than the resolution of the projector. Then, the projection field of view is adjusted inside the field of view of the camera. In the experiments, we use a Flea3 FL3-U3-32S2C from Point Grey Research with the image resolution of 2080 × 1552. The digital light projector is a DLP Light Crafter 4500 projector from Texas Instrument with the resolution of 1024 × 678. Their focus length, resolution, zoom, and direction were selected prior to calibration accordingly to the target of the system. All devices are connected to a host computer. After the system is calibrated, part of the system cannot be moved. We have to

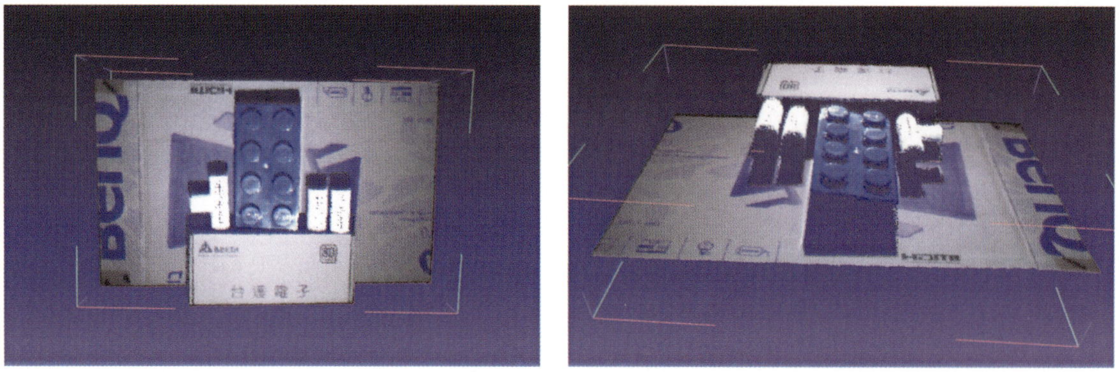

(a) 3D image of the font view (b) 3D image of the upper view

FIGURE 7: 3D reconstruction result with some objects (a lego box, three plastic pipes, and a big box), these objects are low reflecting light.

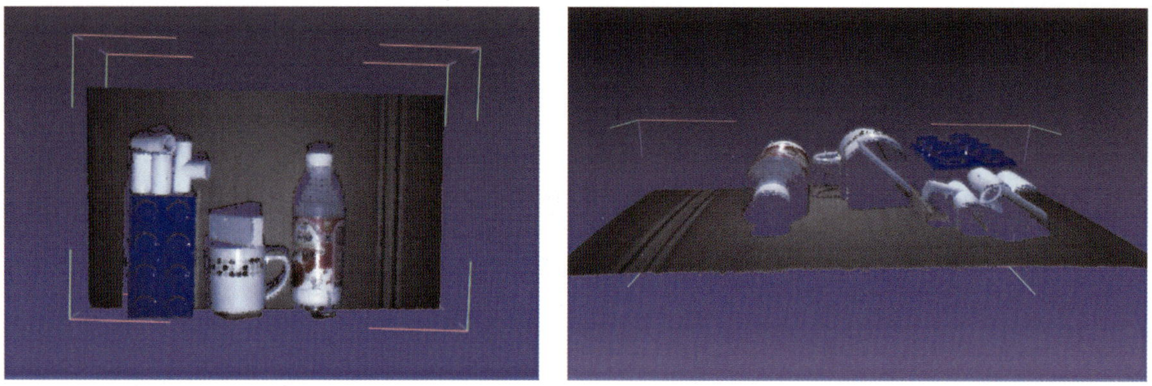

(a) 3D image of the font view (b) 3D image of the upper view

FIGURE 8: 3D reconstruction result with multiple objects (a cup, bottle, small box, lego box, and some plastic pipes), the bottle and cup are with high reflecting light.

keep the distance and orientation between the projector and the cameras intact; otherwise it will be essential to perform a recalibration.

The settings of the camera and projector should be adapted according to the lighting in the scene. Other lighting sources projected directly in the scene should be rejected. If not, the calibrated and reconstructed results will be affected. The system was calibrated with 12 sets of acquired projected checkerboard patterns. An acquisition set includes the images captured by the camera for each pattern in the sequence. After the system is calibrated as exposited in Section 2.2, the calibration result is stored in a .yml file.

For reconstruction, our system includes three main steps. Firstly, our system loads the calibration parameters and projects one acquisition set of patterns on objects. Secondly, decoding the captured pattern-coded images provides a coded map to store the corresponding points between the projector and the camera. Finally, with calibrated parameters and the coded map, we apply the ray triangulation principle to get the 3D point that will be rendered simultaneously with one color image to create an XYZRGB point cloud. In Figures 7 and 8, we present the 3D reconstruction results of several objects. The results successfully measure objects with reflecting light for some of the projected colors. After performing the reconstruction, the 3D information of the reconstructed objects is saved in a .txt file.

The evaluation of the proposed technique is performed by measuring the dimension of a reconstructed checkerboard pattern and its corner points. The checkerboard has the dimension of $399 \times 285\,mm^2$ and each small square has the size of $28.5 \times 28.5\,mm^2$ as demonstrated in Figure 9. After performing the 3D reconstruction for this checkerboard, we can use 3ds Max Design or Meshlab software to examine, as shown in Figure 10. A distance measurement tool is used to measure the dimension of the checkerboard. The accuracy is presented in Table 1, with the errors of the corner points. This table reports that our system can measure the objects with high accuracy. Compared with the algorithms proposed by Moreno *et al.* [6] with Max. error of 0.8546(%). They use a method to estimate the image coordinates of 3D points in the projector image plane and perform the calibration on both projector and camera. With Max. error of 0.1240(%), our proposed method provides the better 3D reconstruction results.

5. Conclusion

In this work, we have developed an RGB-D camera system based on the structured light technique. It contains a camera

TABLE 1: Measuring the accuracy of a reconstructed checkerboard (CB) as shown in Figure 10.

Name	Real CB (mm)	Reconstructed CB (mm)	Error (%)
Width of top checkerboard (AB)	399.000	399.382	0.0957
Width of bottom checkerboard (CD)	399.000	399.495	0.1240
Height of left checkerboard (AC)	285.000	284.746	0.0891
Height of right checkerboard (BD)	285.000	285.298	0.1045
Max. error			0.1240

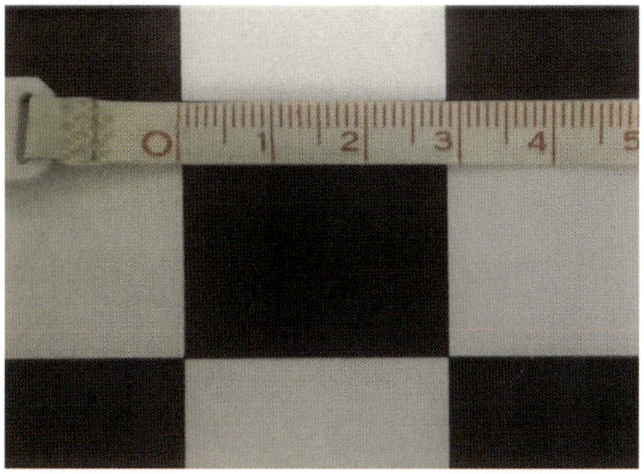

FIGURE 9: The real dimension of each small square has the size of $28.5 \times 28.5\,mm^2$ in the checkerboard reconstructed as shown in Figure 10.

FIGURE 10: The 3D reconstruction of a checkerboard shown with Meshlab software. We used a measuring tool of the Meshlab to measure it.

and a projector to perform accurate shape measurements with high-density point cloud outputs. 3D reconstruction with multiple objects and performance evaluation of the system are carried out in the real-world environment. Our method has high accuracy as presented in the experimental results. In the experiment, we tested the system with different objects to check the surface of reconstruction and accuracy evaluation. The results have demonstrated that the proposed technique is feasible for dense 3D measurement applications.

Disclosure

This publication is an extended version of 2017 International Conference on System Science and Engineering (ICSSE) [1].

Conflicts of Interest

The authors declare that they have no conflicts of interest.

Acknowledgments

The support of this work in part by the Ministry of Science and Technology of Taiwan under Grant MOST 104-2221-E-194-058-MY2 is gratefully acknowledged.

References

[1] V.-L. Tran and H.-Y. Lin, "Accurate RGB-D camera based on structured light techniques," in *Proceedings of the 2017 International Conference on System Science and Engineering, ICSSE 2017*, pp. 235–238, Viet Nam, July 2017.

[2] O. Wasenmüller and D. Stricker, "Comparison of Kinect V1 and V2 Depth Images in Terms of Accuracy and Precision," in *Computer Vision – ACCV 2016 Workshops*, vol. 10117 of *Lecture Notes in Computer Science*, pp. 34–45, Springer International Publishing, Cham, 2017.

[3] J. Smisek, M. Jancosek, and T. Pajdla, "3D with Kinect," in *Proceedings of the 2011 IEEE International Conference on Computer Vision Workshops (ICCV '11)*, pp. 1154–1160, IEEE, Barcelona, Spain, November 2011.

[4] Q. Gu, K. Herakleous, and C. Poullis, "3DUNDERWORLD-SLS: An open-source structured-light scanning system for rapid geometry acquisition," *Computer Vision and Pattern Recognition*, 2016.

[5] D. Moreno, W. Y. Hwang, and G. Taubin, "Rapid Hand Shape Reconstruction with Chebyshev Phase Shifting," in *Proceedings of the 2016 Fourth International Conference on 3D Vision (3DV)*, pp. 157–165, Stanford, CA, USA, October 2016.

[6] D. Moreno and G. Taubin, "Simple, accurate, and robust projector-camera calibration," in *Proceedings of the 2nd Joint 3DIM/3DPVT Conference: 3D Imaging, Modeling, Processing, Visualization and Transmission, 3DIMPVT 2012*, pp. 464–471, Switzerland, October 2012.

[7] C.-Y. Chen, P.-S. Huang, S.-W. Huang, J.-H. Zhang, and B. R. Chang, "Structured light 3D face scanning system," in *Proceedings of the 2nd IEEE International Conference on Consumer Electronics - Taiwan, ICCE-TW 2015*, pp. 344-345, Taiwan, June 2015.

[8] M. Massot-Campos, G. Oliver-Codina, H. Kemal, Y. Petillot, and F. Bonin-Font, "Structured light and stereo vision for underwater 3D reconstruction," in *Proceedings of the MTS/IEEE OCEANS 2015 - Genova*, Italy, May 2015.

[9] M. Piccirilli, G. Doretto, A. Ross, and D. Adjeroh, "A Mobile Structured Light System for 3D Face Acquisition," *IEEE Sensors Journal*, vol. 16, no. 7, pp. 1854-1855, 2016.

[10] D. Lanman, *Grey code patterns for 3D reconstruction by structure light technique*, AIT Computer Vision Wiki, Brown University, 2014.

[11] D. Scharstein and R. Szeliski, "High-accuracy stereo depth maps using structured light," in *Proceedings of the 2003 IEEE Computer Society Conference on Computer Vision and Pattern Recognition*, p. I/202, USA, June 2003.

[12] J. Wang, C. Zhang, W. Zhu, Z. Zhang, Z. Xiong, and P. A. Chou, "3D scene reconstruction by multiple structured-light based commodity depth cameras," in *Proceedings of the 2012 IEEE International Conference on Acoustics, Speech, and Signal Processing, ICASSP 2012*, pp. 5429–5432, Japan, March 2012.

[13] C. Bourgeois-Republique, A. Dipanda, and A. Koch, "A Structured Light System Encoding for an Uncalibrated 3D Reconstruction Based on Evolutionary Algorithms," in *Proceedings of the 2013 International Conference on Signal-Image Technology & Internet-Based Systems (SITIS)*, pp. 124–129, Kyoto, Japan, December 2013.

[14] S. Huang, L. Xie, Z. Wang, Z. Zhang, F. Gao, and X. Jiang, "Accurate projector calibration method by using an optical coaxial camera," *Applied Optics*, vol. 54, no. 4, pp. 789–795, 2015.

[15] K. Herakleous and C. Poullis, "Stripe boundary codes for real-time structured-light range scanning of moving objects," in *Proceedings of the Eighth IEEE International Conference on Computer Vision*, vol. 2, pp. 359–366, Vancouver, Canada.

[16] B. Huang and Y. Tang, "Fast 3D reconstruction using one-shot spatial structured light," in *Proceedings of the 2014 IEEE International Conference on Systems, Man, and Cybernetics, SMC 2014*, pp. 531–536, USA, October 2014.

[17] H. Cui, N. Dai, W. Liao, and X. Cheng, "An accurate reconstruction model using structured light of 3-D computer vision," in *Proceedings of the 7th World Congress on Intelligent Control and Automation, WCICA'08*, pp. 5095–5099, China, June 2008.

[18] M. Fu, Y. Leng, and H. Zhang, "A calibration method for structured light systems based on a virtual camera," in *Proceedings of the 8th International Congress on Image and Signal Processing, CISP 2015*, pp. 57–63, China, October 2015.

[19] S. Zhang and P. S. Huang, "Novel method for structured light system calibration," *Optical Engineering*, vol. 45, no. 8, Article ID 083 601083 6018, 2006.

[20] G. Bradski, "The OpenCV Library," *Dr Dobbs Journal of Software Tools*, 2000.

23

Measuring the Phase of an Optical Field from Two Intensity Measurements: Analysis of a Simple Theoretical Model

Damien P. Kelly

Oryx Optics Ltd, 19 The Elms, Athlone, Co. Westmeath, Ireland

Correspondence should be addressed to Damien P. Kelly; damienpkelly@gmail.com

Guest Editor: Wei Liu

Under the scalar paraxial approximation, an optical wavefield is considered to be complex function dependent on position; i.e., at a given location in space the optical field is a complex value with an intensity and phase. The optical wavefield propagates through space and can be modeled using the Fresnel transform. Lenses, apertures, and other optical elements can be used to control and manipulate the wavefield and to perform different types of signal processing operations. Often these optical systems are described theoretically in terms of linear systems theory leading to a commonly used Fourier optics framework. This is the theoretical framework that we will assume in this manuscript. The problem which we consider is how to recover the phase of an optical wavefield over a plane in space. While today it is relatively straightforward to measure the intensity of the optical wavefield over a plane using CMOS or CCD sensors, recovering the phase information is more complicated. Here we specifically examine a variant of the problem of phase retrieval using two intensity measurements. The intensity of the optical wavefield is recorded in both the image plane and the Fourier plane. To make the analysis simpler, we make a series of important theoretical assumptions and describe how in principle the phase information can be recovered. Then, a deterministic but iterative algorithm is derived and we examine the characteristics and properties of this algorithm. Finally, we examine some of the theoretical assumptions we have made and how valid these assumptions are in practice. We then conclude with a brief discussion of the results.

1. Introduction

Visible light is an electromagnetic field with a wavelength that ranges roughly from ultraviolet (300 nm) to infrared (800 nm). An electromagnetic field is vectorial in nature with three electric and magnetic orthogonal field components. When designing a microwave radio receiver (long wave wavelengths relative to visible light) one must consider how these different magnetic and electric field components interact with each other and with conducting strips of metal and dielectric substrates that are used for impedance matching, mixing, and waveguiding. The resulting equations (that follow from Maxwell's equations) are quite complex and require intensive numerical calculation procedures since the different vectorial components of the EM field interact with each other forming coupled vectorial equations [1].

Fortunately when examining optical problems in the visible regime it is often possible to make the scalar approximation. This means that we can assume that the different vectorial components of the optical wavefield do not "see" each other and hence each vectorial component can be treated independently as a separate scalar problem. This approximation is very useful and quite accurate and can be used to analyze a wide range of important and practical optical problems.

Therefore in this manuscript we assume that a scalar description of light propagation is valid and use the Fresnel transform to relate an optical field in one plane to that in another plane, where the optical planes are separated from each other by an axial distance z. The light sources used are assumed to be of a definite monochromatic temporal frequency and are both temporally and spatially coherent [2, 3]. Optical elements used to shape the form of the illuminating wavefield are assumed to be "thin" and operate on an incident optical field, multiplying it by a function describing the optical element over a plane [4]. The assumptions we have described place us in a Fourier optics framework [4]. Fourier optics is an important branch of optical theory since it allows the development of relatively simple and intuitive models of optical systems. These models provide significant insight into the characteristics of the underlying optical systems and are also reasonably accurate. Furthermore Fourier optics builds important bridges between disciplines in particular allowing optical systems to be interpreted in terms of signal processing operations that are commonly used in electrical engineering, communication theory, and control engineering [4–9]. This type of viewpoint can be extended still further with the development of mixed space/time transforms or mixed space/spatial frequency transforms such as the fractional Fourier transform [10], the Linear Canonical Transform (LCT), and perhaps even more generally the form of Wigner distribution functions [11–17].

We note that in this theoretical framework if we know the intensity and phase of the optical field in a give optical plane, it is then possible to calculate (numerically or under certain circumstances analytically) the field distribution in any other plane using the LCT [18, 19] and hence we can image over a 3D volume, with varying levels of spatial resolution [19].

In a practical sense then it is important to be able to measure experimentally both the intensity and the phase of an optical wavefield in a given plane. Nowadays high quality low noise CCD and CMOS sensors are commercially available and widely used. With these devices it is relatively straightforward to measure the intensity of an optical field in a particular plane. We will now briefly discuss several techniques for indirectly measuring the phase distribution of the field.

1.1. Digital Holography. Digital holography (DH) was first proposed by Gabor himself in [20–22] and the first experimental implementation was done by Goodman et al. in [23]; later pioneering work was done by Yaroslawski, [24], Onural and Scott [25], and Schnars and Jüptner [26]. More recent theoretical and experimental analysis of these systems is given [19, 27–36]. In holography we need both an object wave and a so-called reference wave. Light from the laser source in the optical setup is divided by a beam splitter into a reference and illuminating plane waves. The illuminating plane wave is directed by a mirror so that it illuminates the object we wish to examine. The wavefront interacts with a "thin" object and the resulting object wavefield scatters and propagates a distance z to the CCD/CMOS plane. Meanwhile the reference wave is directed through the optical system where it is recombined with the object wave at a second beam splitter. Thus at the sensor plane we record the interference between the reference and object wavefields thereby recording a hologram. A drawback of this approach is that the additional optics parts are required; a clean stable reference beam is essential; and several undesired "noise" terms, two DC terms and a twin image term, are produced by the recording process. These terms can be removed by capturing several phase shifted intensity images and performing some postprocessing operations. Single shot off-axis holography with a tilted plane reference wave is possible; however the spatial resolution of the resulting real image term (object wavefield) is significantly reduced. The accuracy and limitations of this approach have already been examined in several publications; see, for example, [33, 37].

1.2. Transport of Intensity. These approaches are quite different in nature and require the capture of two separate intensity distributions. Usually these distributions are captured in an image plane and in a slightly defocused plane. Using the so-called transport of intensity equation, see, for example, [38, 39], it is possible to recover an estimate for the phase distribution. There are several interesting aspects of the reconstruction of the phase; firstly the reconstructed phase is not wrapped over 2π and secondly with some reconstruction approaches, namely, an FFT based approach to indefinite integration, see [40], low frequencies can be more difficult to recover.

1.3. Iterative Phase Retrieval. Finally we turn our attention to iterative phase retrieval approaches. Here usually two or more intensity distributions are captured in different LCT domains, or in physical optics terms the intensity is captured in several different optical planes. Often these different optical planes are separated from each other by sections of free-space and combinations of lenses, but also in the image and corresponding Fourier transform planes. The origin of these types of algorithms dates back to the iterative Gerschberg-Saxton (GS) approach first published in the 70s and was further developed Fienup et al. [41, 42]. The initial algorithms assumed that two measurements were made and the intensity distribution was recorded in the image plane and in the Fourier plane. Later algorithms extended these approaches to work with multiple different Fresnel domains [43–47].

The image-Fourier plane (GS) algorithm works as follows:

(1) One starts with an initial guess of the phase distribution in the image plane

(2) We then numerically calculate the Fourier transform of this distribution and compare the calculated intensity in the Fourier plane to the experimentally measured result. The calculated intensity distribution will differ from the experimentally measured distribution because our initial guess at the phase distribution will not be correct

(3) At this stage we replace the incorrect numerically calculated intensity distribution with the correct experimentally measured intensity, while retaining the phase values and numerically propagate the distribution back to the image plane

(4) We do not expect the calculated image intensity to be the same as the measured intensity distribution. Again we replace the numerically calculated distribution with the measured distribution

(5) Steps (2)-(4) are repeated until the algorithm converges, i.e., when there is little difference between the measured and calculated intensity distributions. The similarity between the measured and numerically calculated distributions can be estimated with a root mean square error measure [46].

In [41], Fienup directly compares the behavior of this algorithm to gradient descent optimization approaches. The behavior and properties of these algorithms have been of significant interest in the field. It is generally said that for phase retrieval to work effectively one needs to have a good estimate of the initial phase. When that is the case the algorithms tend to converge to the nearest local minimum which in this case would be one of the global minimums. And hence the algorithm will return the real-physical phase distribution and not a solution that merely satisfies the numerical constraints of the problem but which has no physical meaning. In this manuscript we will explore this question of multiple phase solutions more closely, by examining a simplified version of the phase retrieval problem. A major outstanding issue with these iterative approaches is that algorithm often stagnates in a local minimum yielding a poor estimate for the phase. There is also the possibility that there are several different solutions to this ill-posed problem, at least in the 1D case [48].

In the iterative phase retrieval approach that we have just been discussing the intensity distributions are captured using CCD/CMOS arrays. And so in practice the intensity distribution is only recorded over a finite spatial extent and with a finite number of pixels which limits the spatial resolution and also introduces numerical sampling effects [37, 49]. The numerical procedures that are then used with these 2D arrays of intensity values tend to rely on fast Fourier transform (FFT) based algorithms for their implementation. This requires developing the software and numerical techniques for implementing these algorithms so that they are sufficiently fast and so that other sampling rules and numerical issues can be addressed [49]. This development, while necessary, tends to concentrate attention on the software implementation rather than on other physical and mathematical modeling aspects of the problem.

Therefore in this manuscript we deliberately take a different approach that emphasizes a theoretical analysis using a relatively simple model. Similar to the iterative phase retrieval approach we analyze how the phase may be estimated from two intensity measurements: one made in the image plane and one made in the Fourier plane. We do however make an important simplification. We assume that we can measure the intensity distribution in the Fourier plane over an infinite extent and ignore any sampling or discretization process that occurs due to the finite extent of the pixels in the CCD/CMOS array. Furthermore we assume that the intensity distribution in the image plane consists of a finite set of discrete point sources. With these two simplifications we can derive an analytical solution to the phase retrieval problem.

The manuscript is organized in the following manner: In Section 2, we show, under assumptions, which we have outlined in the text above and which are detailed in Section 2, how the phase values can in principle be determined using an analytical solution. To recover the phase values we must use an iterative procedure which has a finite but large number of steps. In Section 3, we examine an algorithm for implementing the iterative strategy derived in Section 2 and examine how the number of operations depends on the number of contributing point sources in the image plane. We show that there are multiple solutions to the phase retrieval problem. We then examine approaches for reducing the number of possible solutions. In Section 4, we finish with a brief discussion of the presented analysis and highlight some practical difficulties when extending the theoretical model to realistic real-world phase measurements.

2. Analysis of an Idealized Experiment

We begin our analysis by examining the idealized optical system depicted in Figure 1. There is a set of N distinct point sources, each with a known intensity value and hence a known magnitude, a_n, where a_n is the magnitude of point source n. The phase value, ϕ_n, associated with each point source however remains the unknown physical parameter that we wish to measure. So to reiterate, the problem as depicted in Figure 1 is that we are given a set of N point sources with known amplitudes and unknown phase values. We further assume that these points sources are uniformly spaced a distance Δ from each other. We are also given a second measurement, the intensity distribution in the Fourier plane. What we have just described is very similar to the practical iterative phase retrieval experiment described in the Introduction. In a real experiment however these intensity distributions are recorded by CCD/CMOS arrays which have a finite extent, a finite number of pixels, and which sample of the intensity distribution they measure. These are critical factors when considering the performance limitations of these systems; see [33, 36]. They also complicate the theoretical analysis and so now we neglect these factors so that we can develop a simpler theoretical description of the phase retrieval problem.

We imagine that we can measure the intensity in the Fourier domain over an infinite extent and with a sampling step size that is extremely small, so that any sampling effects can be ignored. We also imagine that only a finite number of perfect point sources in the input plane (with known intensity but unknown phase) contribute to the Fourier plane intensity distribution. Hence in this analysis we exclude any physical imposed limiting factors on the ability of our system to make a measurement. This will allow us to significantly simplify the analysis and to concentrate on another and arguable more important effect: the existence of a very high number of alternative solutions to the phase retrieval problem. As we shall see it is possible that many different combinations of phase values can produce identical intensity distributions. Hence this phase retrieval problem is said to be ill-posed.

We refer back to Figure 1, where we have indicated that the contributing point sources are Fourier-transformed by an optical Fourier transform system. This type of optical system is well known; see, for example, [4]. The distribution in the Fourier plane is scaled by both the wavelength, λ, of the light source and the focal length of the lens. It is convenient for our purposes here to set $\lambda f = 1$ so that we can analyze an unscaled Fourier distribution. Here this distribution is incident on an ideal recording material, allowing us to record the continuous intensity distribution. Then we perform a second unscaled inverse Fourier transform on this intensity distribution numerically. We note however that although we perform two successive Fourier operations (a forward and inverse transform), we do not arrive back at the input plane distribution since we perform the second Fourier transform

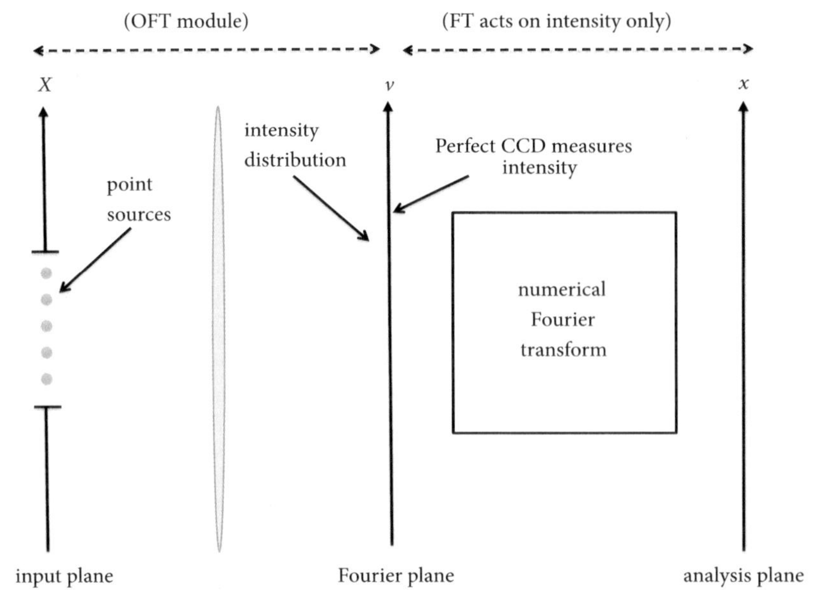

FIGURE 1: Depiction of phase retrieval setup for an idealized phase retrieval experiment, OFT: optical Fourier transform. A set of N point sources with known amplitude and unknown phase are present in the input plane. This distribution is Fourier-transformed by the lens and the intensity of the resulting distribution in the Fourier plane is recorded. We assume that we can measure this intensity distribution over an infinite extent and with extremely high spatial resolution so that we can ignore the numerical effects of sampling due to the CCD/CMOS pixels. We show how this intensity distribution is processed using a second Fourier transform, this time calculated numerically, and we examine its properties in the "analysis plane" (as depicted above). We show how to determine the phase values of the N point sources in the input plane.

operation on the intensity of the Fourier distribution not on its complex amplitude, as in a 4-f imaging system [50].

We thus begin by noting that the Fourier transform of a point source is given by

$$\begin{aligned} &FT\{\delta(x+X_n)\,a_n\exp(j\phi_n)\}(v) \\ &\quad= \exp(-j2\pi X_n v)\,a_n\exp(j\phi_n), \end{aligned} \tag{1}$$

where X_n is the spatial location of point source "n" while the parameters a_n and ϕ_n refer to its phase and amplitude respectively. Hence for a sum of N distinct point sources we can write

$$\begin{aligned} &FT\left\{\sum_{n=1}^{N}\delta(x-X_n)\,a_n\exp(j\phi_n)\right\}(v) \\ &\quad= \sum_{n=1}^{N}\exp(-j2\pi X_n v)\,a_n\exp(j\phi_n). \end{aligned} \tag{2}$$

We are however interested in the intensity distribution of the field in the Fourier plane. We remember that the intensity of a complex number is given by that number times its complex conjugate. With this relationship in mind we now write the conjugate expression for (2)

$$\begin{aligned} &FT\left\{\sum_{n=1}^{N}\delta(x-X_n)\,a_n\exp(j\phi_n)\right\}^{*}(v) \\ &\quad= \sum_{n=1}^{N}\exp(j2\pi X_n v)\,a_n\exp(-j\phi_n), \end{aligned} \tag{3}$$

where "$*$" refers to a complex conjugate operation. We can now write an analytical expression for the intensity distribution in the Fourier plane which when simplified can be written in the following form

$$\begin{aligned} &I_{\mathrm{FT}}(v) \\ &\quad= \sum_{n=1}^{N}a_n^2 \\ &\qquad+ 2\sum_{n=1}^{N-1}\sum_{m=1}^{N-n}a_m a_{m+n}\left\{\cos\left[2\pi(n\Delta)v+\phi_m-\phi_{m+n}\right]\right\}. \end{aligned} \tag{4}$$

From (4) we can see that we have a sum of cosines. For each given value of n there is a specific spatial frequency component $f_s = n\Delta$, and there are $N-1$ terms. It is easier to analyze this double summation by initially concentrating on a single spatial frequency component, i.e., for a specific value of n, and we refer to such a component as $I^n_{\mathrm{FT}}(v)$. Noting the following trigonometric relationship, $\cos(A+B) = \cos(A)\cos(B) - \sin(A)\sin(B)$, we can express $I_{\mathrm{FT}}(v)$ as

$$I_{\mathrm{FT}}(v) = \sum_{n=1}^{N}a_n^2 + \sum_{n=1}^{N-1}I^n_{\mathrm{FT}}(v), \tag{5}$$

where

$$\begin{aligned} I^n_{\mathrm{FT}}(v) &= 2\sum_{m=1}^{N-n}a_m a_{m+n}\cos\left[2\pi n\Delta v+(\phi_m-\phi_{m+n})\right] \\ &= 2\cos(2\pi n\Delta v)\sum_{m=1}^{N-n}a_m a_{m+n}\cos(\phi_m-\phi_{m+n}) \\ &\quad+ 2\sin(2\pi n\Delta v)\sum_{m=1}^{N-n}a_m a_{m+n}\sin(\phi_m-\phi_{m+n}). \end{aligned} \tag{6}$$

We now consider the inverse Fourier transform of $I^n_{\mathrm{FT}}(v)$, which we specifically define as

$$\overline{M}_n(x) = \int_{-\infty}^{\infty} I^n_{\mathrm{FT}}(v) \exp(+j2\pi xv)\, dv, \tag{7}$$

which is now in a spatial domain once again, specifically the "analysis" plane in Figure 1. Noting that

$$\begin{aligned} FT\{\sin(2\pi k_0 x)\}(k) &= \frac{j}{2}\left[\delta(k+k_0) - \delta(k-k_0)\right], \\ FT\{\cos(2\pi k_0 x)\}(k) &= \frac{1}{2}\left[\delta(k+k_0) + \delta(k-k_0)\right], \end{aligned} \tag{8}$$

then

$$\begin{aligned} \overline{M}_n(x) = \delta(x+n\Delta) &\left\{ \sum_{m=1}^{N-n} a_m a_{m+n} \cos(\phi_m - \phi_{m+n}) \right. \\ &\left. - j \sum_{m=1}^{N-n} a_m a_{m+n} \sin(\phi_m - \phi_{m+n}) \right\} + \delta(x-n\Delta) \\ &\cdot \left\{ \sum_{m=1}^{N-n} a_m a_{m+n} \cos(\phi_m - \phi_{m+n}) \right. \\ &\left. + j \sum_{m=1}^{N-n} a_m a_{m+n} \sin(\phi_m - \phi_{m+n}) \right\}. \end{aligned} \tag{9}$$

We see from (9) that the complex numbers (arising from the summation of sines and cosines) multiplying the $\delta(x - n\Delta)$ and $\delta(x + n\Delta)$ components are complex conjugates of each other. Hence we see that

$$Re\left\{\overline{M}_n(n\Delta)\right\} = \sum_{m=1}^{N-n} a_m a_{m+n} \cos(\phi_m - \phi_{m+n}), \tag{10}$$

and that

$$Im\left\{\overline{M}_n(n\Delta)\right\} = -\sum_{m=1}^{N-n} a_m a_{m+n} \sin(\phi_m - \phi_{m+n}). \tag{11}$$

We also remember that the iterator n above spans the range $1 \le n \le N - 1$. Examining (10) and (11) we recognize that the only unknowns are the values for the phase parameters, ϕ_n. Both the initial amplitudes a_n and the LHS of each equation are given. The LHS values are found by calculating the Fourier transform of the intensity distribution that we measure in the Fourier plane, see the "analysis plane" in Figure 1, and determining the value of the resulting complex function at the locations $x = n\Delta$.

We also note that the larger the value of n, the lower the number of terms from (10) and (11) that contribute to both $Im\{\overline{M}_n(n\Delta)\}$ and $Re\{\overline{M}_n(n\Delta)\}$. When we set about trying to find ϕ_n from (10) and (11), we start by first setting $n = N - 1$, in which case the equations reduce to the following:

$$Re\left\{\overline{M}_{N-1}\left[(N-1)\Delta\right]\right\} = a_1 a_N \cos(\phi_1 - \phi_N), \tag{12}$$

and

$$Im\left\{\overline{M}_{N-1}\left[(N-1)\Delta\right]\right\} = -a_1 a_N \sin(\phi_1 - \phi_N). \tag{13}$$

If we arbitrarily set $\phi_1 = 0$, then we can calculate ϕ_N from the following

$$\phi_N = \arctan\left[\frac{\sin(\phi_N)}{\cos(\phi_N)}\right], \tag{14}$$

which means that we now have values for both ϕ_1 and ϕ_N.

Now setting $n = N - 2$, (10) and (11) will reduce to the following:

$$\begin{aligned} &Re\left\{\overline{M}_{N-2}\left[(N-2)\Delta\right]\right\} \\ &\quad = a_1 a_{N-1} \cos(\phi_1 - \phi_{N-1}) + a_2 a_N \cos(\phi_2 - \phi_N), \end{aligned} \tag{15}$$

and

$$\begin{aligned} &Im\left\{\overline{M}_{N-2}\left[(N-2)\Delta\right]\right\} \\ &\quad = -a_1 a_{N-1} \sin(\phi_1 - \phi_{N-1}) - a_2 a_N \sin(\phi_2 - \phi_N). \end{aligned} \tag{16}$$

It is instructive to consider the next step in the procedure before we generalize the results. Setting $n = N - 3$, and substituting into (10) and (11), we arrive at

$$\begin{aligned} &Re\left\{\overline{M}_{N-3}\left[(N-3)\Delta\right]\right\} - a_2 a_{N-1} \cos(\phi_2 - \phi_{N-1}) \\ &\quad = a_1 a_{N-2} \cos(\phi_1 - \phi_{N-2}) + a_3 a_N \cos(\phi_3 - \phi_N), \end{aligned} \tag{17}$$

and

$$\begin{aligned} &Im\left\{\overline{M}_{N-2}\left[(N-1)\Delta\right]\right\} + a_2 a_{N-1} \sin(\phi_2 - \phi_{N-1}) \\ &\quad = -a_1 a_{N-2} \sin(\phi_1 - \phi_{N-2}) - a_3 a_N \sin(\phi_3 - \phi_N). \end{aligned} \tag{18}$$

Again we see that we can calculate the LHS of (17) and (18) from an experimental measurement and by using values for ϕ_2 and ϕ_{N-1} that were estimated in the previous iteration step. We summarize our theoretical results so far:

(i) We set $\phi_1 = 0$ and calculate ϕ_N using (14)

(ii) Once ϕ_1 and ϕ_N are determined, we need in turn to determine ϕ_2 and ϕ_{N-1} from (12) and (13)

(iii) If we can find values for ϕ_2 and ϕ_{N-1}, then we need to repeat the process to determine ϕ_3 and ϕ_{N-2}

(iv) We repeat this iterative process $N/2 - 1$ times to find all the unknown phase values

To proceed further we must therefore find a solution to the following coupled set of equations

$$\Omega_R = a\cos(\alpha) + b\cos(\beta - \Theta), \tag{19}$$

$$\Omega_I = a\sin(\alpha) - b\sin(\beta - \Theta), \tag{20}$$

where α and β are the unknown phase values for a particular step in the iterative procedure. Using a software package like Mathematica it is possible to find an analytical solution for these equations; however the solution is not unique: there are 8 possible coupled values for α and β.

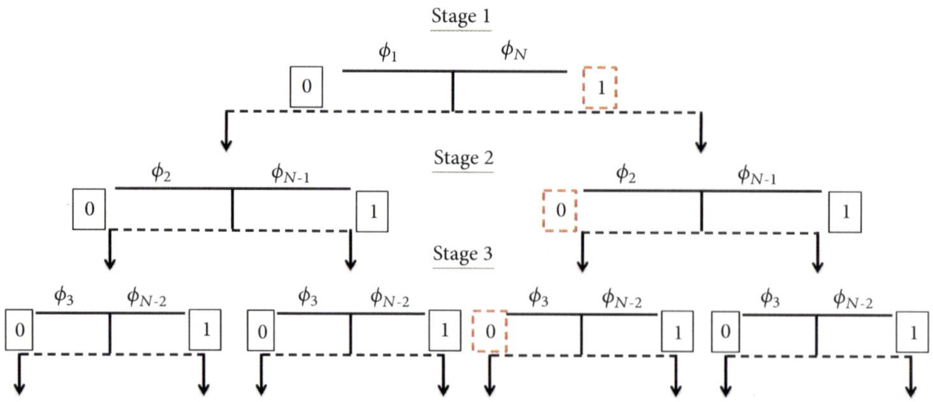

FIGURE 2: Decision tree illustrating the choices made by the algorithm as it steps through the unknown phase values. At each stage of the algorithm, a choice between two paths, Path 0 or Path 1, is made. A particular path through the solution space is then uniquely identified with a bit sequence. For example, if the dashed red boxes are chosen at each stage of the algorithm being chosen, this corresponds to the path [1 0 0 .....].

3. Algorithm Analysis

In this section we will examine some of the implications of the theoretical analysis we have just derived. First we will examine how we may reduce the number of possible solutions at each level of the iteration, reducing them from eight to two. To find a phase value for each contributing point source the algorithm must move through $N/2 - 1$ steps. At each step a decision on which solution to take must be made; we examine how this can be considered as a set of different possible paths through the solution space. We examine the character of the solutions produced by the algorithm. Finally we examine how to prune (reduce) the solution space by ruling out "unphysical" solutions and outline an algorithm for implementing this procedure.

3.1. From Eight Solutions to Two. We note that it is possible to rewrite the coupled equations, (19) and (20), in the following form

$$\Omega_R = as_1 + bs_2, \tag{21}$$

$$\Omega_I = a\sqrt{1-s_1^2} - b\sqrt{1-s_2^2}, \tag{22}$$

where $\cos(\alpha) \longrightarrow s_1$, $\cos(\beta - \Theta) \longrightarrow s_2$ and where we use the trigonometric relation $\cos^2(A) + \sin^2(A) = 1$ and hence we can write the $\sin(\alpha)$ in terms of the $\cos(\alpha)$, i.e., $\sin(\alpha) = \sqrt{1 - [\cos(\alpha)]^2}$. Using (21) to find an expression for s_2 and subbing into (22) we can express s_1 as a quartic equation with four possible solutions [51]. For each value of s_1 there are two possible angles for α that satisfy $s_1 = \cos(\alpha)$ and hence eight possible solutions for α and similarly eight solutions for β.

This is where the difficulties with our phase retrieval approach begin in earnest. Although it is possible to find an analytical solution to the coupled equations (19) and (20), the solution has eight different possible solutions for the unknown pair: (α, β). In reality however only one of the solution pairs corresponds to the actual correct physical result. If we randomly choose from these solutions we will have a probability of 1/8 that we have chosen the physically correct answer.

Having decided on a particular choice of (ϕ_2, ϕ_{N-2}), we can then set $n = N - 3$ and repeat the process. Hence at each step we have a 1/8 chance of guessing the physically correct solution, and for N point sources we will need to make approximately $N/2 - 1$ guesses at each step of the algorithm. Hence the chance that we correctly guess the correct phase for each point source is $(1/8)^{N/2-1}$, which would seem to be a vanishingly small probability. In practical experimental phase retrieval systems we may wish to find N phase values where $N \approx 10^6$.

Fortunately, we have found from numerical simulations that although there are eight solutions possible in principal, including solutions with complex values, in practice when we substitute real physical parameters into our system of equations several of these solutions yield the same answer. We have found that at each step in the algorithm only two solutions are produced. This is the case for all simulations we have run. We do not pursue this question further here, concluding that at each step in an iterative algorithm we are able to find two solutions that are physically valid and satisfy (19) and (20).

3.2. Solution Paths. From the previous section we have found that it is possible at each stage in the iterative procedure to reduce the number of solutions to two possible answers. At each stage in the algorithm then a "binary" choice has to be made about which solution to choose. If we have N contributing point sources then we must move through $N/2-1$ stages. In Figure 2, we depict how to track our progress through such a decision tree, in this example making the following choices at each stage of the algorithm, $[1, 0, 0, \ldots]$; see the red dashed boxes in Figure 2. In general then there are $2^{(N/2-1)}$ possible paths in total and each possible path is identified as a unique binary number. For example, if $N = 8$, then there are 2^3 or 8 possible solutions and if $N = 16$, there are 2^7 or 128 solutions while if $N = 32$ the number of possible solutions jumps to 32,768.

3.2.1. Results of Numerical Simulations. We have run numerous numerical simulations of the equations derived in Section 2 with some interesting results that we now summarize.

(i) The correct phase solution is always found among the other different solutions

(ii) The overwhelming majority of solutions have complex numbers for the phase and hence can be excluded as real physical solutions to the phase retrieval problem

(iii) There are often several other real valued solutions in addition to the actual real solution

(iv) We have run several numerical simulations with $N = 32$. In one run we found that there are 32,768 solutions of which 32,764 are complex and the 2 remaining solutions are real valued, one of which is the correct phase solution. On a second run we found that only 32,762 solutions were complex while the 6 remaining were real valued, one of which was the correct solution.

We conclude that if we had unlimited computing resources and time it would in principle be possible to calculate all possible solutions even when N is very large, for example, $N \approx 10^6$. In practice however this calculation would result in a solution space of 2^{499999}, which is not feasible with standard computing resources. When N is relatively small and all the possible solutions can be calculated we have found that the vast majority are complex valued, and only a small number of solutions are real valued.

3.3. Pruning the Solution Space. From the preliminary analysis we have reported in the previous section, we find that the overwhelming majority of the solutions yield phase values that are complex and hence "unphysical" in nature and can be discarded. This opens up the opportunity of pruning the solution space. We will now outline a strategy that can be employed to greatly reduce the search space. This solution depends on unique binary number that identifies each particular path through the search space and we note that it will be a $2^{(N/2-1)}$-bit number. As we move through the solution space we will need a variable, which we refer to as *location* to keep track of where we are in the search tree. We will also need an iterator variable k to refer to the "stage" of the algorithm; see Figure 2.

From Figure 2 we see that at the first stage, i.e., when $k = 1$, we identify values for ϕ_1 and ϕ_N and when $k = 2$, we identify possible candidates for ϕ_2 and ϕ_{N-1}, which we denote by S_0 and S_1. We initially set *location* to be all zeros and of length $2^{(N/2-1)}$. For example if $N = 8$ then initially *location* would be set to *location* $= 000$. We also need to address the individual bits in *location* specifically. We do this using the following notation *location*$[n]$, where n refers to n^{th} individual bit. We illustrate this with the following example: Consider the case when *location* $= 10010$ and then *location*$[1] = 1$, *location*$[2] = 0$, and *location*$[4] = 1$.

The algorithm then runs as follows:

(i) For a given N, set $k = 1$ and *location* $= 000\ldots$, where the length of *location* is $2^{(N/2-1)}$,

(ii) WHILE $k < N/2$,

(1) Using Eq. (20) and Eq. (21), find S_0 and S_1
(2) IF *location*$[k] = 0$, then $(\phi_k, \phi_{N-k+1}) = S_0$, ELSE $(\phi_k, \phi_{N-k+1}) = S_1$
(3) IF (ϕ_k, ϕ_{N-k+1}) are real valued then $k = k + 1$,
(4) ELSE:
IF *location*$[k] = 0$ then set *location*$[k] = 1$,
ELSE $k = k - 1$, and *location*$[k] = 1$

(iii) Return *location*

This algorithm will run through the search space checking the solutions. If a complex solution is found it will choose the alternate solution. For a given stage in the algorithm if both S_0 and S_1 are complex the algorithm will go back up the search tree, updating *location* and choosing a different path through the search space. This algorithm will return only the first real valued solution it finds but can be modified to return all solutions that are physically significant.

When employing the "pruning" approach to search the solutions space we must still search a very large space; however the depth of the search per path will be greatly reduced; see, for example, [52]. This will significantly reduce the number of numerical operations performed. Whether the phase retrieval problem can be solved depends on the rate of branching (or the depth of the search) before a given path can be excluded from further search. From a preliminary analysis based on simulations results we find that the depth when one must search a path before an incorrect complex value is returned as the phase estimate can be modeled using statistical methods. We have found that the average path "length" approximately follows a negative exponential distribution similar to speckle intensity statistics [37, 53].

4. Conclusion

In this manuscript we have examined the problem of measuring the phase of an optical wavefield. In the Introduction, we briefly reviewed several approaches for making this measurement before concentrating on iterative phase retrieval. In these phase retrieval problems an attempt is made to estimate the phase of wavefield using usually two or more intensity distributions captured in different Linear Canonical Transform planes. An iterative FFT based algorithm is employed that tunes/modifies the phase values iteratively so that the difference between a numerically calculated intensity distribution and an experimentally captured intensity distribution is minimized. Fienup [41] compares this algorithmic approach to a gradient descent algorithm. As is often the case with such optimization algorithms, they can stagnate at local minima, producing phase estimates that are quite far from the correct physical solution. This is particularly the case when the initial guess at the phase distribution is quite far from this correct physical solution [46]. This is an important issue and different approaches for improving these algorithms are an active research topic. There is however understandably an emphasis on the numerical aspects of these algorithms.

Here we have deliberately chosen to examine a simplified version of the phase retrieval problem, assuming that we can make perfect measurements of the intensity distribution in an image and its corresponding Fourier plane. We assume that the distribution in the image plane consists of N point sources with unknown phase values. Importantly we assume that we can measure the intensity distribution in the Fourier plane over an infinite extent and with an infinitely fine resolution; in short we assume that we can perfectly measure the continuous intensity distribution. We have found the following under these idealized conditions:

(1) It is possible in principal to find solutions to the phase retrieval problem that satisfy the physical constraints of the problem. The algorithm for finding these solutions works iteratively but deterministically and will find among the many solutions returned the physically correct solution

(2) The number of solutions is enormous; if there are N contributing point sources there are $2^{(N/2-1)}$ possible solutions; however the overwhelming majority of these solutions are complex valued and hence "unphysical" and can be excluded as potential solution candidates

(3) We can modify the algorithm so that "unphysical" solutions are excluded which we refer to as pruning. Using a pruning approach we can significantly reduce the space of possible solutions. The performance of the pruning approach can be approximately estimated using statistical techniques. A critical factor determining whether the algorithm can be computed in a realistic time is the "branching" rate or depth of search before a given solution can be excluded

It is important to also highlight the significant practical shortcomings of this theoretical analysis.

4.1. Physical Model. We model the field whose phase values we wish to recover as a set of point sources (with known amplitude) in an image plane which are separated uniformly from each other by a fixed distance, Δ. Of course an optical wavefield is continuous in nature and so this representation is only an approximation of the real physical field. Nevertheless in standard iterative phase retrieval problems, the intensity value measured at each pixel is considered to be a point source; hence the description of the field that we use is in keeping with the standard approach in the literature.

There is however a difference in the modeling in this paper for the intensity distribution recorded in the Fourier plane. In this paper we effectively assume that we can measure the continuous Fourier plane intensity; i.e., it is as if we can record and measure the intensity over an infinite plane in the Fourier domain and with an infinitely fine spatial resolution. Making these assumptions means that we can develop convenient analytical equations relating the unknown phase values directly to experimental measurements. These analytical equations are used then to develop the iterative algorithm so that the phase values can be estimated. This contrasts with real experimental systems in that the intensity in the Fourier plane is measured only at a finite number of locations, i.e., at each pixel, and is therefore sampled. And secondly the Fourier intensity distribution is only measured over the finite extent of the CCD/CMOS array. This summarizes the theoretical differences between the analysis presented in this manuscript and the standard theoretical treatment of the iterative phase retrieval problem.

We note however that it is possible to replicate experimentally the theoretical conditions we have assumed in the manuscript. We can achieve this by making several individual measurements of the Fourier plane intensity distribution where we move the CCD/CMOS array between measurements. Then using standard digital signal processing operations we synthetically increase both the spatial resolution and the spatial extent of the measurement by stitching together the single intensity measurements.

4.2. Effects of Noise. A significant shortcoming of this theoretical approach is the effect that even small amounts of measurement noise would have on the algorithm. In every experiment there will be measurement error, for example, electronic noise in the readout of a CCD or CMOS device. If a measurement is made on a variable that has a low signal level, the relative error tends to be larger [36]. In the situation described here, we find that intensity measurements associated with higher spatial frequencies have a lower number of contributing point sources. For example, the highest spatial frequency component of the measurement has contributions from only two point sources. In general (but not always) these measurements will contain the most noise. These measurements however are used to directly estimate ϕ_1 and ϕ_N, which in turn feed iteratively back into the algorithm and are repeatedly used to estimate the remaining phase values.

So we can see that to turn this theoretical approach into a practical solution would require careful experimental measurements and the development of additional signal processing algorithms to overcome unavoidable noise in these measurements. Nevertheless we hope that by approaching the phase retrieval problem from a different direction we stimulate some new interesting ideas and insights and also shed some light on the character of the solution space for the phase distributions that are recovered from intensity measurements.

4.3. Extension to 2D. In this manuscript we deliberately focused on a simple 1D theoretical model (we only consider the x and z spatial coordinates) so that we could examine different aspects of the phase retrieval problem. We choose to set ϕ_1 to zero and then iteratively solve for the other phase values. Extension to 2D, where we consider both x and y as well as the z spatial coordinates, is not necessarily straightforward and will change depending on the geometry of the contributing point sources. For example, if the contributing point sources are arranged in a square lattice, then the highest spatial frequency components arise from the two point sources that are spatially separated from each other by the largest amount, i.e., the vertices of the square. The square has four vertices; we can assign a phase value of zero to only one of those vertices, which acts as an effective reference value for the others. Sequential illumination with lines of 1D light could also be considered whereupon the 1D analysis considered here could be more easily extended.

Conflicts of Interest

The authors declare that they have no conflicts of interest.

Acknowledgments

The author gratefully acknowledges the EU commission (H2020-TWINN-2015 HOLO, project Nr. 687328), for supporting his attendance at Summer School "Digital Holography and Related Techniques" organized in Stuttgart, 21-23 June 2017. Part of the work described in this paper results from discussions that the author had during this meeting.

References

[1] J. D. Jackson, *Classical Electrodynamics*, Wiley, New York, NY, USA, 1962.

[2] P. Meystre, *Physics Today*, vol. 61, no. 12, pp. 59-60, 2008.

[3] D. P. Kelly, A simple optical system for interpreting coherence theory , ArXiv e-prints 1609.05734,.

[4] J. Goodman, *Introduction to Fourier Optics*, New York, McGraw-Hill, 2nd edition, 1966.

[5] E. ONeill, *Introduction to Statistical Optics , Dover books on physics*, Dover Publications, 2004.

[6] C. E. Shannon, "A mathematical theory of communication," *Bell Labs Technical Journal*, vol. 27, pp. 379–423, 1948.

[7] C. E. Shannon, "Communication in the presence of noise," *Proceedings of the IRE*, 1998.

[8] D. Slepian, "On Bandwidth," *Proceedings of the IEEE*, vol. 64, no. 3, pp. 292–300, 1976.

[9] A. Stern and B. Javidi, "Shannon number and information capacity of three-dimensional integral imaging," *Journal of the Optical Society of America A: Optics, Image Science & Vision*, vol. 21, no. 9, pp. 1602–1612, 2004.

[10] C. Shen, J. Tan, C. Wei, and Z. Liu, "Coherent diffraction imaging by moving a lens," *Optics Express*, vol. 24, no. 15, pp. 16520–16529, 2016.

[11] A. W. Lohmann, "Image rotation, wigner rotation, and the fractional fourier transform," *Journal of the Optical Society of America A: Optics and Image Science, and Vision*, vol. 10, no. 10, pp. 2181–2186, 1993.

[12] Z. Zalevsky, D. Mendlovic, and A. W. Lohmann, "Understanding superresolution in Wigner space," *Journal of the Optical Society of America A: Optics, Image Science & Vision*, vol. 17, no. 12, pp. 2422–2430, 2000.

[13] J. T. Sheridan, B. Hennelly, and D. Kelly, "Motion detection, the Wigner distribution function, and the optical fractional Fourier transform," *Optics Expresss*, vol. 28, no. 11, pp. 884–886, 2003.

[14] A. Stern and B. Javidi, "Sampling in the light of Wigner distribution," *Journal of the Optical Society of America A: Optics, Image Science & Vision*, vol. 21, no. 3, pp. 360–366, 2004.

[15] D. P. Kelly, B. M. Hennelly, and J. T. Sheridan, "Magnitude and direction of motion with speckle correlation and the optical fractional Fourier transform," *Applied Optics*, vol. 44, no. 14, pp. 2720–2727, 2005.

[16] B. M. Hennelly and J. T. Sheridan, "Generalizing, optimizing, and inventing numerical algorithms for the fractional Fourier, FREsnel, and linear canonical transforms," *Journal of the Optical Society of America A: Optics, Image Science & Vision*, vol. 22, no. 5, pp. 917–927, 2005.

[17] A. Stern, "Uncertainty principles in linear canonical transform domains and some of their implications in optics," *Journal of the Optical Society of America A: Optics, Image Science & Vision*, vol. 25, no. 3, pp. 647–652, 2008.

[18] H. T. Yura and S. G. Hanson, "Optical beam wave propagation through complex optical systems," *Journal of the Optical Society of America A: Optics, Image Science & Vision*, vol. 4, no. 10, p. 1931, 1987.

[19] D. P. Kelly, J. J. Healy, B. M. Hennelly, and J. T. Sheridan, "Quantifying the 2.5D imaging performance of digital holographic systems," *Journal of the European Optical Society: Rapid Publications*, vol. 6, 2011.

[20] D. Gabor, "Microscopy by Reconstructed Wave-Fronts," *Proceedings of the Royal Society A Mathematical, Physical and Engineering Sciences*, vol. 197, no. 1051, pp. 454–487, 1949.

[21] D. Gabor, "Microscopy by reconstructed wave fronts: II," *Proceedings of the Physical Society, Section B*, vol. 64, no. 6, pp. 449–469, 1951.

[22] D. Gabor, "Light and information," in *Progress in Optics, Vol. I*, pp. 109–153, North-Holland, Amsterdam; Interscience, New York, 1961.

[23] J. W. Goodman and R. W. Lawrence, "Digital image formation from electronically detected holograms," *Applied Physics Letters*, vol. 11, no. 3, pp. 77–79, 1967.

[24] L. P. Yaroslavskii and N. S. Merzlyakov, *Methods of Digital Holography*, Springer US, Boston, MA, 1980.

[25] L. Onural and P. D. Scott, "Digital decoding of in-line holograms," *Optical Engineering*, vol. 26, no. 11, pp. 1124–1132, 1987.

[26] U. Schnars and W. Jüptner, "Direct recording of holograms by a ccd target and numerical reconstruction," *Applied Optics*, vol. 33, no. 2, pp. 179–181, 1994.

[27] G. Pedrini, P. Fröning, H. J. Tiziani, and M. E. Gusev, "Pulsed digital holography for high-speed contouring that uses a two-wavelength method," *Applied Optics*, vol. 38, no. 16, pp. 3460–3467, 1999.

[28] W. Osten, T. Baumbach, and W. Jüptner, "Comparative digital holography," *Optics Expresss*, vol. 27, no. 20, pp. 1764–1766, 2002.

[29] P. Picart, J. Leval, D. Mounier, and S. Gougeon, "Some opportunities for vibration analysis with time averaging in digital Fresnel holography," *Applied Optics*, vol. 44, no. 3, pp. 337–343, 2005.

[30] G. Pedrini, I. Alexeenko, W. Osten, and U. Schnars, "On-line surveillance of a dynamic process by a moving system based on pulsed digital holographic interferometry," *Applied Optics*, vol. 45, no. 5, pp. 935–943, 2006.

[31] D. Claus, P. Bryanston-Cross, and B. Timmerman, "Digital Double Exposure Holography by Means of Polarization Optics," in *Proceedings of the Digital Holography and Three-Dimensional Imaging*, p. PMA7, Vancouver, Canada, 2007.

[32] P. Picart and J. Leval, "General theoretical formulation of image formation in digital Fresnel holography," *Journal of the Optical Society of America A: Optics and Image Science, and Vision*, vol. 25, no. 7, pp. 1744–1761, 2008.

[33] D. P. Kelly, B. M. Hennelly, N. Pandey, T. J. Naughton, and W. T. Rhodes, "Resolution limits in practical digital holographic systems," *Optical Engineering*, vol. 48, no. 9, 2009.

[34] D. P. Kelly, D. S. Monaghan, N. Pandey, T. Kozacki et al., "Digital Holographic Capture and Optoelectronic Reconstruction for 3D Displays," *International Journal of Digital Multimedia Broadcasting*, vol. 2010, Article ID 759323, 14 pages, 2010.

[35] D. Claus, D. Iliescu, J. Watson, and J. Rodenburg, "Comparison of different digital holographic setup configurations," in *Proceedings of the Digital Holography and Three-Dimensional Imaging, DH 2012*, p. DM4C.3, USA, May 2012.

[36] Y. Wu, J. P. Ryle, S. Liu, D. P. Kelly, and A. Stern, "Experimental evaluation of inline free-space holography systems," *Applied Optics*, vol. 54, no. 13, pp. 3991–4000, 2015.

[37] D. P. Kelly and D. Claus, "Filtering role of the sensor pixel in Fourier and Fresnel digital holography," *Applied Optics*, vol. 52, no. 1, pp. A336–A345, 2013.

[38] M. R. Teague, "Deterministic phase retrieval: a Green's function solution," *Journal of the Optical Society of America*, vol. 73, no. 11, p. 1434, 1983.

[39] C. J. R. Sheppard, "Three-dimensional phase imaging with the intensity transport equation," *Applied Optics*, vol. 41, no. 28, pp. 5951–5955, 2002.

[40] D. P. Kelly, L. Megel, T. Meinecke et al., "A theoretical comparison of Fresnel based digital holography and phase retrieval from the transport of intensity equation," in *Proceedings of the SPIE Optical Engineering + Applications*, p. 88330G, San Diego, California, United States.

[41] J. R. Fienup, "Phase retrieval algorithms: a comparison," *Applied Optics*, vol. 21, no. 15, pp. 2758–2769, 1982.

[42] J. R. Fienup, "Reconstruction of a complex-valued object from the modulus of its Fourier transform using a support constraint," *Journal of the Optical Society of America A: Optics, Image Science & Vision*, vol. 4, no. 1, p. 118, 1987.

[43] P. F. Almoro and S. G. Hanson, "Random phase plate for wavefront sensing via phase retrieval and a volume speckle field," *Applied Optics*, vol. 47, no. 16, pp. 2979–2987, 2008.

[44] P. F. Almoro, P. N. Gundu, and S. G. Hanson, "Numerical correction of aberrations via phase retrieval with speckle illumination," *Optics Expresss*, vol. 34, no. 4, pp. 521–523, 2009.

[45] P. F. Almoro, G. Pedrini, P. N. Gundu, W. Osten, and S. G. Hanson, "Phase microscopy of technical and biological samples through random phase modulation with a diffuser," *Optics Expresss*, vol. 35, no. 7, pp. 1028–1030, 2010.

[46] L. Megel, D. P. Kelly, T. Meinecke, and S. Sinzinger, "terative phase retrieval and the important role played by initial conditions," in *Fringe 2013*, W. Osten, Ed., pp. 123–128, Springer Berlin Heidelberg, Berlin, Heidelberg, 2014.

[47] Z. Liu, C. Guo, J. Tan, Q. Wu, L. Pan, and S. Liu, "Iterative phase-amplitude retrieval with multiple intensity images at output plane of gyrator transforms," *Journal of Optics*, vol. 17, no. 2, p. 025701, 2015.

[48] R. Barakat and G. Newsam, "Necessary conditions for a unique solution to two-dimensional phase recovery," *Journal of Mathematical Physics*, vol. 25, no. 11, pp. 3190–3193, 1984.

[49] D. P. Kelly, "Numerical calculation of the Fresnel transform," *Journal of the Optical Society of America A: Optics, Image Science & Vision*, vol. 31, no. 4, p. 755, 2014.

[50] D. P. Kelly, J. T. Sheridan, and W. T. Rhodes, "Fundamental diffraction limitations in a paraxial 4-f imaging system with coherent and incoherent illumination," *Journal of the Optical Society of America A: Optics and Image Science, and Vision*, vol. 24, no. 7, pp. 1911–1919, 2007.

[51] P. Strobach, "The fast quartic solver," *Journal of Computational and Applied Mathematics*, vol. 234, no. 10, pp. 3007–3024, 2010.

[52] P. H. Winston, *Artificial Intelligence*, Addison-Wesley Publishing Co., Reading, Mass.-London-Amsterdam, 1977.

[53] J. Dainty, "I The Statistics of Speckle Patterns," vol. 14 of *Progress in Optics*, pp. 1–46, Elsevier, 1977.

On the Creation of Solitons in Amplifying Optical Fibers

Christoph Mahnke, Alexander Hause, and Fedor Mitschke

Institut für Physik, Universität Rostock, 18059 Rostock, Germany

Correspondence should be addressed to Fedor Mitschke; fedor.mitschke@uni-rostock.de

Academic Editor: Murugan Senthil Mani Rajan

We treat the creation of solitons in amplifying fibers. Strictly speaking, solitons are objects in an integrable setting while in real-world systems loss and gain break integrability. That case usually has been treated in the perturbation limit of low loss or gain. In a recent approach fiber-optic solitons were described beyond that limit, so that it became possible to specify how and where solitons are eventually destroyed. Here we treat the opposite case: in the presence of gain, new solitons can arise from an initially weak pulse. We find conditions for that to happen for both localized and distributed gain, with no restriction to small gain. By tracing the energy budget we show that even when another soliton is already present and copropagates, a newly created soliton takes its energy from radiation only. Our results may find applications in amplified transmission lines or in fiber lasers.

1. Introduction

Solitons are fascinating objects. They arise from a variety of nonlinear wave equations; here we will concentrate on the Nonlinear Schrödinger Equation (NLSE) and fiber-optic solitons as these represent the only type of solitons that has already seen commercial application [1]. Fiber-optic solitons are light pulses which balance the fiber's dispersion with its nonlinearity such as to stabilize their shape; this makes them eminently suitable as signalling light pulses in optical data transmission. For any other type of soliton a similar interplay of effects produces a similar self-stabilization.

For the NLSE, Zakharov and Shabat found the soliton solution in their ground-breaking paper [2] (called ZS hereafter). This was followed by an equally important paper by Satsuma and Yajima [3] (hereafter, SY) where the pertaining initial-value problem was solved. Both together established the basics of solitons in fibers as they were suggested in [4]; experimentation commenced a few years later [5].

When it comes to real-world settings rather than the idealized context of the integrable NLSE, one has to deal with the impact of power loss on solitons. This issue was treated with perturbation methods by several authors [6–11]. However, such approach requires that the loss be weak and can cover neither strong attenuation coefficients nor long distances with weak coefficients. Moreover, it entirely misses the eventual decay of the soliton. For a long time, investigations of lossy fibers beyond the weak-loss limit were confined to numerical simulations.

We could recently demonstrate [12] that SY can be used to cover lossy fiber by interpreting continuous loss as a sequence of infinitely many infinitesimal localized losses, each of which can be treated by SY. It became clear, among other things, what the mechanism for the eventual death of a soliton is. While that paper concentrated on loss, the total accumulated loss factor (called $\Gamma < 1$) can easily be used to describe gain, by letting $\Gamma > 1$.

Can solitons be amplified without creating radiation in the process? As shown in [13], that is possible in a very special set of conditions: the soliton needs to have a particular chirp, the fiber parameters must vary along the distance to a certain specification (tapering), and there are constraints on the gain mechanism. Here we consider conventional (unchirped) solitons in conventional (nontapered) fibers, without assumptions about the gain mechanism. Then, radiation-free amplification is possible only in the adiabatic limit, a case of little use in practical terms.

In the general case of more-than-vanishing gain, the question naturally and unavoidably arises: if one starts with a weak (nonsolitonic) pulse and boosts its power, can a soliton be created? Or, alternatively, if one starts with a soliton and amplifies its power, can a second soliton arise?

The answer is straightforward only if one has localized (stepwise) gain acting on an unchirped pulse as this special case can be treated with SY. For chirped pulses and distributed gain the answer is not obvious at all. We point out that clarification of this issue may be relevant to several contexts. In today's fiber links, optical gain is often provided by Raman amplification which is distributed over a long fiber length. Can a soliton, after suffering from some severe perturbation, be restored to obtain solitonic properties again? Or if one considers a fiber laser, if the gain fiber segment is fed with a weak pulse, can a soliton arise? (In the case of a laser, resonator boundary conditions apply; our treatment covers processes in the gain fiber only.)

In the present paper we address the problem of soliton creation from gain, both localized and distributed. Where closed solutions do not exist, we resort to fits of observed behavior, to arrive at statements with some predictive power.

2. Basic Definitions

The Nonlinear Schrödinger Equation (NLSE) in normalized form is [2, 14, 15]

$$i\frac{\partial}{\partial\xi}u + \frac{1}{2}\frac{\partial^2}{\partial\tau^2}u + |u|^2 u = 0, \tag{1}$$

where ξ is distance, τ is time, and $u = u(\xi,\tau)$ is the amplitude envelope in a comoving frame of reference. Equation (1) describes the propagation of light in optical fibers in the absence of loss and gain, at anomalous dispersion. Its fundamental soliton solution has the form

$$u(\xi,\tau) = \eta\,\text{sech}(\eta\tau)\exp\left(i\frac{\eta^2}{2}\xi\right), \tag{2}$$

when it rests in the center of the frame of reference. η is the scaling parameter that sets both peak amplitude and inverse duration. That η appears in both reflects the soliton's hallmark balance between dispersive and nonlinear effects, a fact that has also been expressed as an "area theorem" [16] that the product of peak amplitude and width—or, alternatively, peak power times width squared—takes a constant value [14, 15, 17]. Frequently the same is expressed using characteristic length scales for dispersion (L_D) and nonlinearity (L_{NL}) [14, 15]; in our dimensionless units they both take the form $L_D = L_{\text{NL}} = \eta^{-2}$. η^2 appears in the phase term indicating a nonlinear phase rotating at a rate proportional to peak power. The soliton energy is $E_{\text{sol}} = 2\eta$.

We will investigate the evolution of a pulse that is initially not a soliton. In its fullest generality the problem would be intractable, and so we make one simplifying assumption: we consider pulses that have sech amplitude envelope and are initially unchirped. To describe nonsolitonic pulses, we break the link between peak amplitude and inverse duration by introducing an amplitude scaling parameter M. As pulse parameters may evolve during propagation, we abbreviate M_ξ for $M(\xi)$. Hence, the initial condition (the launch pulse at $\xi = 0$) is written as

$$u(0,\tau) = M_0\eta\,\text{sech}(\eta\tau) \stackrel{!}{=} N\eta\,\text{sech}(\eta\tau). \tag{3}$$

On the RHS we use the soliton order N which compares the relative strength of dispersive and nonlinear effects [14, 15]. The integer number closest to N is the soliton count [3]. $M_\xi = N$ if and only if the pulse is chirp-free. As we always consider chirp-free initial conditions, we do not need to make the distinction at the launch point, but during propagation, M_ξ may and will differ from N. For $M_0 = 1$ the soliton is recovered; in the limit $M_0 \to 0$ the propagation is linear, that is, purely dispersive.

In Section 3.2 we will consider pulses launched with $0 < N < 1/2$, motivated by the well-known fact obtained from Inverse Scattering theory [2] that no soliton is formed if $N < 1/2$. Subsequently, in Section 4.2 we will consider pulses launched with $1/2 < N < 3/2$, that is, from the regime, where there is exactly one soliton, to determine whether a second soliton can arise from gain.

If the gain is localized, it follows from SY that the gain enhances the power at least up to the threshold $N_{\text{thr}} = 1/2$ for the first soliton to appear, $N_{\text{thr}} = 3/2$ for the second, and so on. The minimum required gain $\Gamma = E_{\text{out}}/E_{\text{in}}$ for that is $\Gamma = (N_{\text{thr}}/N)^2$. If the gain is not localized but distributed, more of it is required to reach threshold because there are two opposing trends: the pulse energy grows exponentially with distance ξ as $E_p = E_p(0)\exp(\gamma\xi)$ where we introduce a dimensionless gain coefficient $\gamma > 0$. A characteristic dimensionless gain length is $L_\gamma = \gamma^{-1}$. The gain can be represented in (1) by replacing the RHS with $(i\gamma/2)u$. Energy growth brings the pulse closer to the threshold of soliton generation. At the same time, the pulse undergoes dispersive broadening which renders it temporally broader, lets the peak power droop, and—most importantly—creates a chirp. Power boost helps chirp hinder soliton creation. Therefore we must first discuss the formation of chirp in Section 3.1.

3. The Situation without Gain or Loss

3.1. A Note on Chirp. We start with a comment on the linear, purely dispersive limit. The sech shape typical for solitons is mathematically less convenient than a Gaussian: an initially Gaussian pulse, upon dispersion, remains Gaussian everywhere, except for rescaling, but underneath the Gaussian envelope a linear chirp (quadratic phase profile) develops. According to [14] but written in dimensionless units, its evolution is given by

$$u_D(\xi,\tau) = \frac{A}{\sqrt{1 - i\eta^2\xi}}\exp\left[-\frac{(\eta\tau)^2}{2(1 - i\eta^2\xi)}\right] \tag{4}$$

with arbitrary amplitude A. At center ($\tau = 0$),

$$|u_D(\xi,0)| = \frac{A}{\sqrt[4]{1 + \eta^4\xi^2}},$$
$$\varphi_D(\xi,0) = -\frac{1}{2}\arctan\left(\eta^2\xi\right). \tag{5}$$

This is to be compared to an initially sech-shaped pulse which undergoes some shape variations. Pulse wings exhibit some wiggles around $\xi_0 = \pi/2$ (after one soliton period), but asymptotically the shape is not much different from a chirped sech again at least in its central part. Unfortunately, no closed analytic expression is available to write this evolution. We therefore make the simplifying assumption for the linear regime that the pulse envelope never strays far from the sech shape; indeed assume that the shape is constant except for scaling. This forces us to take into account the different spreading rates of the two shapes by introducing a scaling factor κ for the amplitude and phase evolution in the form

$$|u_D(\xi,0)| = A\frac{1}{\sqrt[4]{1+\kappa\eta^4\xi^2}},$$
$$\varphi_D(\xi,0) = -\frac{1}{2}\arctan\left(\kappa\eta^2\xi\right). \tag{6}$$

By varying κ for the best match of both evolutions we find $\kappa = 0.42$ [12].

We now proceed by considering a pulse amplitude that is not vanishingly small. Then there is a nonlinear contribution to the pulse evolution. It was observed early on in numerical work [11] that in the anomalous-dispersion regime considered here this contribution always serves to reduce the broadening. Here, however, we need to evaluate the chirp.

The effect of chirp on the soliton content of a pulse has been studied before. In 1979 Hmurcik and Kaup [16] started from the observation that the McCall-Hahn area theorem in self-induced transparency only applies to unchirped pulses and found that chirped pulses need to be stronger in order to generate the same effect. Desem and Chu [18] 1986 specifically talked about fiber-optic solitons, considered ZS soliton eigenvalues, and found how for (in current usage of terminology) different soliton orders N the power going into the soliton is diminished when chirp is introduced, to the point that a soliton vanishes at some particular value of the chirp, beyond which all power goes into radiation (the soliton order is closely related to the area theorem). Also in 1986, Blow and Wood [19] pursued the relation between the solitonic energy eigenvalue and the pulse's bandwidth. In 1987, Mamistov and Sklyarov [20] determined threshold values of phase modulation beyond which the soliton count in a pulse would be reduced. Kaup et al. [21] 1994 introduced a simplified pulse shape (rectangular amplitude profile, piecewise-linear antisymmetric phase profile) which allowed getting further insight. They found that, for different soliton orders, at the point where the soliton vanishes the phase slope was always close to a characteristic value.

In all cases the results indicate that comparing an unchirped with a chirped pulse of the same energy, a lesser fraction of energy is available for soliton formation in the chirped case. This will become apparent in detail below.

3.2. Evolution of an $N < 1/2$ sech Pulse. Chirp is detrimental to soliton formation; energy gain is conducive to it. With energy growing exponentially with distance and the chirp growing at a much lesser rate (this will be detailed in Figure 3), there will always be a crossover point where the advantage from gain outweighs the disadvantage from chirp. The smaller the gain factor, the longer the distance over which the chirp can develop. Our discussion will thus concentrate on the run-up distance to soliton formation, that is, the position in the amplifying fiber ξ_B where a soliton is created.

The propagation of a "weak" pulse in this framework is fully described by just three parameters: amplitude scaling factor M_ξ, soliton scaling parameter $\eta(\xi) = \eta_\xi$, and chirp coefficient $C(\xi) = C_\xi$. The latter is defined as

$$C = -\left.\frac{\partial^2\varphi(\tau)}{\partial\tau^2}\right|_{\tau=0}, \tag{7}$$

where $\varphi(\tau)$ is the optical phase. A chirped sech pulse at position ξ is then written as

$$u_\xi(\tau) = M_\xi\eta_\xi\,\text{sech}\left(\eta_\xi\tau\right)\exp\left(-i\frac{C_\xi}{2}\tau^2\right). \tag{8}$$

We now follow the evolution of such pulse. It will broaden with distance due to dispersion, but depending on N the nonlinear Kerr effect will counteract and slow down the broadening [11]. Due to energy conservation, the peak amplitude $\eta_\xi M_\xi$ will droop accordingly.

In the linear limit, by way of our approximation, the amplitude evolution is well described [12] by

$$|u(\xi,0)| = \frac{\eta N}{\sqrt[4]{1+\kappa\left(\eta^2\xi\right)^2}} \tag{9}$$

with a small amplitude factor in the numerator. In the denominator, the relevant variable is $\eta^2\xi$ because the relevant length scales L_{NL} and L_D both scale as η^{-2}. To find a useful expression for the evolution for amplitudes larger than in the linear limit we numerically checked how $|u|$ scales with N and find that

$$|u(\xi,0)| = \frac{\eta N}{\sqrt[4]{1+\kappa\left(\eta^2\xi\right)^{(2-N^2)}}} \tag{10}$$

makes a very good fit. Data in Figure 1 test this assertion for $\eta = 1$ and several values of N; we have also checked for different η values (not shown). A comparison is shown between plots of (10) (solid curves) and a full numerical simulation which provides a quite accurate reference (dashed curves). The agreement is convincing, confirming that our approximations are quite reasonable.

Energy conservation ensures that

$$2\eta N^2 = 2\eta_\xi M_\xi^2, \tag{11}$$

and the peak amplitude evolves as

$$|u(\xi,0)| = \eta_\xi M_\xi. \tag{12}$$

Combining (10), (11), and (12) we have

$$M_\xi = N\sqrt[4]{1+\kappa\left(\eta^2\xi\right)^{(2-N^2)}}, \tag{13}$$

$$\eta_\xi = \left(\frac{N}{M_\xi}\right)^2\eta. \tag{14}$$

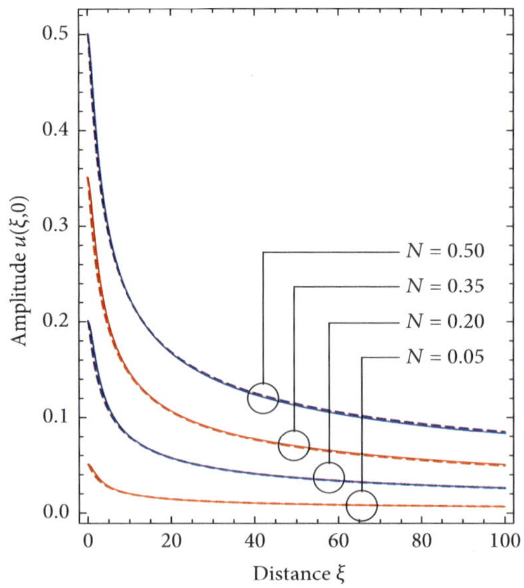

FIGURE 1: Amplitude evolution of weak sech pulses ($N = 0.05, 0.2, 0.35$, and 0.5, as indicated) with $\eta = 1$. Dashed curves: numerical data; solid curves: predictions from (10).

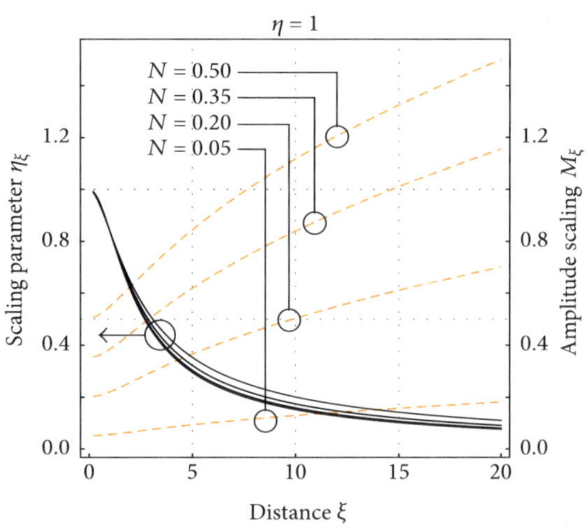

FIGURE 2: Evolution of amplitude scaling factor M_ξ (orange, dashed curves) and scaling parameter η_ξ (black solid curves) with distance, for $N = 0.05, 0.2, 0.35$ and 0.5, as indicated.

In Figure 2, (13) and (14) are plotted for $\eta = 1$ and for $N = 0.05, \ldots, 0.5$. In all cases η_ξ falls with distance (reflecting the broadening), while M_ξ grows. One might suspect that with growing M_ξ a soliton might be created once threshold is reached at $M_\xi = 1/2$. However, that is evidently not the case because the number of solitons, here zero, is guaranteed to be preserved. The discussion is only complete with consideration of the chirp, and it is the development of a chirp that counteracts soliton creation.

We concentrate on the central part of the pulse (between the 80% amplitude points) where the phase is very nearly parabolic. From numerical propagation simulation data we obtain C_ξ by least-squares fit. We let the chirp coefficient $C(0, N, \eta) = 0$ at launch for all N, η. As the pulse propagates, C rises linearly, peaks, and then rolls down as $C(\xi) \propto 1/\xi$ when for $\xi \to \infty$ the pulse duration diverges. By systematic variation of both η and N we determined an expression capturing dependencies on these parameters. The expected inversion symmetry $C(\xi) = -C(-\xi)$ is also built in. This results in

$$C(\xi, N, \eta) = \frac{a\eta^2 \left(1 - N^2/4\right)}{\eta^2 |\xi| + b(1 + N/2) + 1/\left(\eta^2 |\xi|\right)} \operatorname{sgn}(\xi) \quad (15)$$

with fit parameters $a = 1.07$ and $b = 1.40$. Figure 3 shows plots of (15) (solid curves), in comparison to numerical data (dotted). It is clear that (15) is a good approximation over the range of interest, that is, $0 \leq N \leq 0.5$ and $0.5 \leq \eta \leq 1.5$. One observation to be made is that, during its dispersive propagation, a weak sech pulse with $\eta = 1$ will never exceed $C \approx 0.4$ which is quite moderate. The other is that chirp values are highest for $N \to 0$: nonlinearity not only slows broadening as asserted in [11], but also reduces chirp.

3.3. Soliton Content of Chirped sech-Shaped Pulses. We are now positioned to determine the soliton content of the pulse under consideration. It is well known [2, 3, 15] that an unchirped sech pulse with $1/2 < N < 3/2$ has pulse energy $E_p = 2\eta N^2$ and contains exactly one soliton with energy $E_{sol} = 2\eta(2N - 1)$; the balance is radiation with energy $E_{rad} = 2\eta(N - 1)^2$. We employed the Direct Scattering Transform (DST) technique [22–24] to obtain solitonic eigenvalues from pulses as in (8). Results for various chirp values are shown in Figure 4 as data points. We are not concerned here with the real part of the eigenvalue as all velocities are zero; the imaginary part μ represents the energy of the pertaining solitons, $E_{sol} = 4\mu$.

This figure is similar to Figure 1 in [18] (with $\phi = C/2$) and to Figure 1 in [25] (with $\phi_0 = C/2$). It shows that the power contained in a chirped pulse can only partially be converted into a soliton: the more the chirp, the lower the soliton energy. For any N there is a "critical" chirp C_{crit} beyond which soliton formation is not possible.

At $C = 0$ all traces start at the well-known value

$$\mu(C = 0, N, \eta) = \left(N - \frac{1}{2}\right)\eta = \mu_0 \quad (16)$$

and then roll down. Different from [25] we observe that this occurs first quadratically, but beyond the diagonal auxiliary line, linearly. In the quadratic regime, an ansatz

$$\mu(C, \mu_0, \eta) = -\frac{C^2}{8\eta^2\mu_0} + \mu_0 \quad (17)$$

fits well, and the same is true for

$$\mu(C, \mu_0, \eta) = -\frac{C}{2.5\eta} + 1.335\,\mu_0 \quad (18)$$

in the linear regime; the crossover occurs at

$$\mu_\chi = \frac{C}{2.5\,\eta}. \quad (19)$$

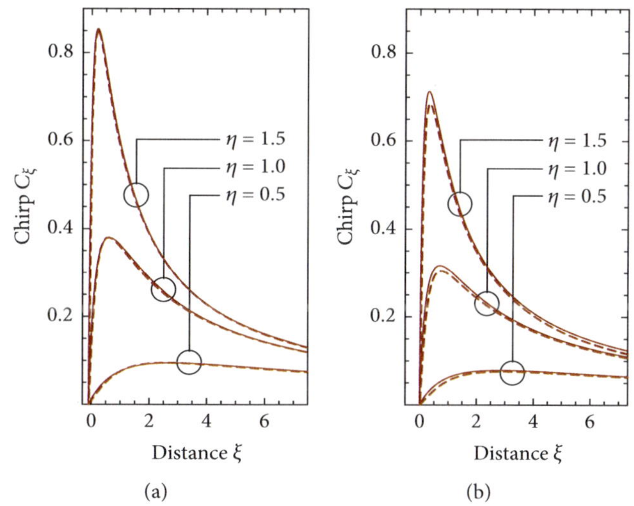

FIGURE 3: Chirp evolution for $\eta = 0.5$ (bottom trace) to $\eta = 1.5$ (top trace) in increments of 0.5. Solid curves represent the approximation (15); dashed curves are numerical data. (a) $N = 0.001$ approximates linear propagation. (b) $N = 0.5$ represents the most extreme case considered here.

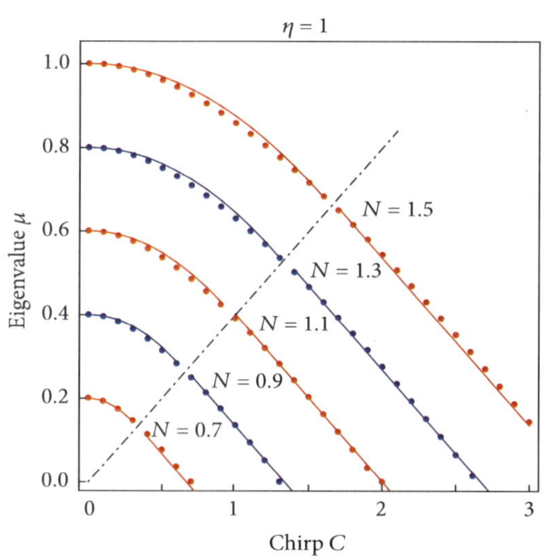

FIGURE 4: Soliton energy eigenvalues when the chirp is varied. Data shown as dots for $\eta = 1$ and $N = 0.7$ (lowest trace); 0.9; 1.1; 1.3; and 1.5 (top trace). Lines: fits with (17) and (18).

Using these fits, the critical chirp value (at $\mu = 0$) is

$$C_{\text{crit}} = 3.34\,\eta^2 \left(N - \frac{1}{2}\right). \tag{20}$$

This result coincides with (20) in [25] (their numerical prefactor was 3.28 which agrees within numerical precision); we remark that our two-part function (17) and (18) fits the data, including those in [25], significantly better than the fit given there.

We convinced ourselves that our fit functions hold well for all combinations of $0.5 \le N \le 1.5$ and $0.1 \le \eta \le 10$. As an example we specify for the fundamental soliton with standard values $\eta = 1$, $N = 1$ (the reference soliton): it disappears at $C_{\text{crit}} = 1.67$, and at the crossover point its chirp amounts to $C_\chi = C_{\text{crit}}/2$.

4. Amplified Propagation

We will consider gain in its simplest form, that is, as multiplication of amplitude with a certain gain factor, but without complications like gain saturation, frequency-dependent (selective) gain, and so on. In a purely linear system, it is immaterial whether gain acts at the beginning or at the end of the propagation or is distributed over the distance: both dispersion and gain are linear processes, and they commute. As soon as there are nonlinear phase shifts (or other consequences of nonlinearity), it makes a great difference whether gain is localized or distributed (for the analogous case of loss this was demonstrated in [12]).

4.1. Starting from Pure Radiation. For weak pulses, that is, pulses with powers below the threshold of soliton generation or just at the threshold at most ($0 < N < 1/2$), we may approximate that gain acts only on the amplitude scaling factor M_ξ, without immediate impact on pulse width, chirp, or other shape characteristics. We start by discussing limiting cases.

High Gain ($\gamma \to \infty$). If amplification is applied more or less stepwise before dispersive spreading has set on, the problem is a straightforward application of [3]: all that is required to generate a soliton is to boost M_ξ at least to the threshold value of $M_\xi = N = 1/2$, that is, $\Gamma = 1/(2N)$.

$$M_\xi = N \exp\left(\frac{\gamma}{2}\xi\right) \longrightarrow \xi = \ln\left(\frac{M_\xi^2}{N^2}\right) L_\gamma. \tag{21}$$

For example, starting from an unchirped sech pulse with $N = 0.4$, a first soliton arises at $\xi_B = 0.4463 L_\gamma$. (The threshold

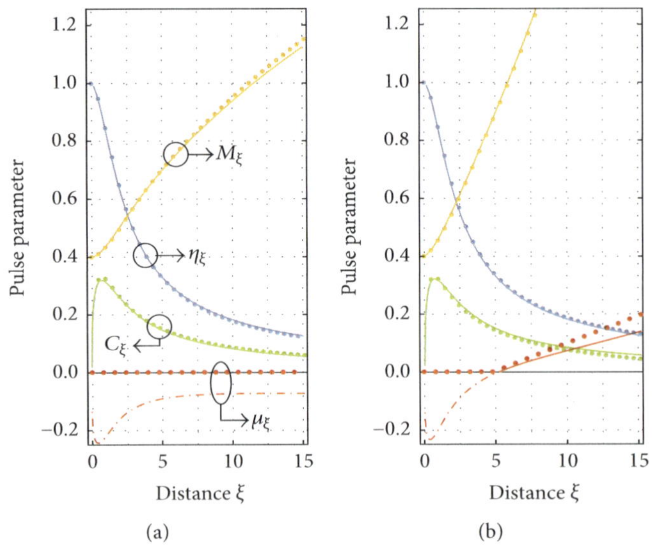

FIGURE 5: Impact of gain: evolution of amplitude scaling parameter M_ξ (orange), scaling parameter η_ξ (blue), chirp C_ξ (green), and soliton content μ_ξ (red) as calculated here (solid/dash-dotted lines) in comparison to numerical simulation (dots). (a) For the integrable NLSE system with $\gamma = 0$ and (b) with gain ($\gamma = 0.1$), soliton appears at $\xi_B \approx 5$. Initial parameters: $N = 0.4$, $\eta = 1$, and $C = 0$.

for a second soliton would be at $M_\xi = 3/2$ and in the same example would be reached at $\xi_B = 2.644L_\gamma$.) Incidentally, a pure soliton is obtained when the gain lifts M_ξ to 1. If amplification occurs only after some spreading and chirp has developed, the quoted distance is a lower bound.

Adiabatic Gain ($\gamma \to 0$). The signature of the adiabatic case is that the gain length is very much longer than the characteristic dispersion length, so that $L_D/L_\gamma \to 0$. In this limit, no soliton is generated; all of the energy remains in the radiation. We conclude that a soliton count of zero is maintained during adiabatic gain: no soliton will be generated, ever.

Finite Gain. We introduce a gain term into (13),

$$M_\xi = N\sqrt[4]{1+\kappa\left(\eta^2\xi\right)^{(2-N^2)}}\exp\left(\frac{\gamma}{2}\xi\right), \tag{22}$$

with gain parameter $\gamma > 0$; it is also possible to let $\gamma < 0$ in which case γ is Beer's coefficient of absorption. The chirp coefficient C_ξ and scaling parameter η_ξ at position ξ can be calculated using the initial values of N and η from (15) and

$$\eta_\xi = \frac{\eta}{\sqrt{1+\kappa\left(\eta^2\xi\right)^{(2-N^2)}}}. \tag{23}$$

The eigenvalue component representing the solitonic energy μ_ξ can be obtained from the calculated local values given by (17) and (18).

$$\mu_\xi = -\frac{C_\xi^2}{8\eta_\xi^3\left(M_\xi - 1/2\right)} + \left(M_\xi - \frac{1}{2}\right)\eta_\xi \quad \text{if } C_\xi < 1.67\,\eta_\xi^2\left(M_\xi - \frac{1}{2}\right),$$

$$= -\frac{C_\xi}{2.5\eta_\xi} + 1.335\left(M_\xi - \frac{1}{2}\right)\eta_\xi \quad \text{if } C_\xi > 1.67\,\eta_\xi^2\left(M_\xi - \frac{1}{2}\right). \tag{24}$$

In Figure 5(a) the evolution as predicted by this approximation is shown for the integrable NLSE system with $\gamma = 0$ and at $N = 0.4$. Parameters M_ξ (solid orange line), η_ξ (solid blue line), C_ξ (solid green line), and μ_ξ (dash-dotted red line) are shown. In comparison, corresponding simulation results are shown as dots with matching color. Note that formally the "soliton content" μ_ξ comes out negative when there is no soliton present, while numerics always correctly yields a zero in such case. Theoretical predictions and simulations agree well.

Figure 5(b) shows the same situation as in Figure 5(a) but with a gain of $\gamma = 0.1$. The soliton content μ_ξ is shown as solid red curve for meaningful positive values predicting the existence of a soliton. After a certain distance $\xi_B \approx 5$ the soliton content becomes positive, indicative of the birth of a soliton. This demonstrates that with our formalism the location of the birth of a soliton can be predicted. The slight difference between the predicted positive soliton content and numerical data for ξ well above the threshold arises from the increasing deviation of the numerical pulse from a pure hyperbolic secant shape.

The position of the soliton's birth, ξ_B, is obtained by letting $\mu_\xi = 0$ in (24) and inserting (15), (22), and (23); we do not spell out the resulting rather unwieldy expression here. ξ_B depends on γ and N: large values of either render it short. Predictions of $\xi_B(N, \gamma)$ are shown in Figure 6 as solid curves; numerical results from DST are also shown (dots). It is obvious that the general trends are represented well, but there is a tendency that our equations find ξ_B systematically high.

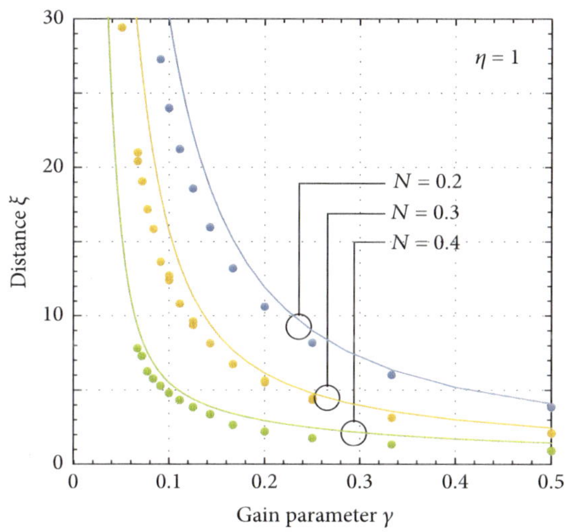

FIGURE 6: The distance until a first soliton appears (ξ_B) depends on the gain factor γ. Shown for $N = 0.2$ (top trace) to $N = 0.4$ (bottom trace) in increments. Solid, predictions; dots, numerical results.

This is particularly true for long distances, surely because then pulse shape differences can play out more. Nevertheless, the agreement with numerics is still good enough for predictions to be useful.

4.2. Starting from One Soliton Plus Radiation. After discussing how gain can bring a weak ($N < 1/2$, "subsolitonic") pulse up to the solitonic threshold, we now consider initial conditions $1/2 < N < 3/2$ so that we start from a pulse which contains both one soliton and some nonsolitonic radiation. The energy E_{rad} of the latter is given by the difference of the pulse energy E_p and the soliton energy E_{sol} [3, 12]:

$$E_{\text{rad}}(0) = 2\eta N^2 - 2\eta(2N-1) = 2\eta(N-1)^2. \tag{25}$$

During amplified propagation,

$$E_{\text{rad}}(\xi) = 2\eta_\xi M_\xi^2 - 4\mu_\xi. \tag{26}$$

A new (second) soliton will always be created from the radiative part once its energy reaches a threshold, as discussed in the previous section. Note that the radiation has two contributions: (i) from the initial condition as in (25) [3, 12], except when $N = 1$; (ii) from continuous generation during amplified propagation, except in the adiabatic limit [12]. This is a consequence of the continuous perturbation of the soliton's characteristic balance between nonlinearity and dispersion.

The main difference to the case of loss as discussed in [12] is that during its propagation the situation becomes more and more diabatic in the lossy fiber, but approaches adiabaticity in the amplifying fiber. This is because in the case of gain the scaling parameter η_ξ grows so that the ratio L_D/L_γ decreases. As shown in [12], the rate of generation of radiative energy is proportional to that ratio. As a result, the radiative fraction of the pulse energy saturates within a few dispersion lengths. For the $N = 1$ soliton we find that it approaches

$$\frac{E_{\text{rad}}(\xi)}{E_p(\xi)} \approx \left(\frac{L_D(0)}{L_\gamma}\right)^2 = \left(\frac{\gamma}{\eta^2}\right)^2. \tag{27}$$

The situation is compounded by the fact that both contributions to radiation and also the soliton are subject to gain all the time. We will again discuss limiting cases first.

High Gain ($\gamma > \eta^2$). Then $L_\gamma < L_D$; in the limit $\gamma \to \infty$ we approach the case of stepwise gain as in SY. No significant change of η_ξ and C_ξ can occur over L_γ; only M_ξ is raised. Then, the second soliton arises once $M_\xi = 3/2$. Using (21) we can predict for the above example of an initially unchirped $N = 0.6$ sech pulse that for $\gamma \to \infty$ the second soliton appears at $\xi_B = 1.833\, L_\gamma$.

Adiabatic Gain ($\gamma \to 0$). In the limit, the soliton does not shed energy to radiation. Thus, both soliton and radiation energy grow exponentially, without redistribution among them. Also, their temporal overlap is minimal because $L_D \ll L_\gamma$, and radiation is rapidly broadened. In consequence, even after a vast propagation distance no further solitons will arise (i.e., $\xi_B \to \infty$).

Small Gain ($\gamma < \eta^2$). The equilibrium between nonlinearity and dispersion is approximately maintained because the soliton can rearrange its width and peak power at a rate that nearly keeps up with the change. Therefore, if there is any sizable amount of radiative energy from the initial condition, it will dominate.

In [12] we derived an expression for the amplitude evolution of a sech pulse in the presence of initial radiation. Here an extended version in dimensionless form is given including the impact of distributed gain depending on γ.

$$|u(\xi,0)| = \eta \exp(\gamma\xi)\left[(2N-1) + \frac{1-N}{\sqrt[4]{1+\kappa(\eta^2\xi)^2}} \times \cos\left\{\frac{(2N-1)^2\eta^2}{2}\exp\left(\gamma\xi^{1+\epsilon}\right)\xi + \frac{1}{2}\arctan\left(\kappa\eta^2\xi\right)\right\}\right]. \tag{28}$$

The first term on the RHS is the exponentially growing amplitude, under the assumption that the sech shape is basically preserved, and taking the pulse narrowing into account. Inside the square bracket, the first parenthesis pertains to the initial soliton. The prefactor of the cosine term models the dispersive broadening. The argument of the cosine contains one term for the nonlinear phase rotation and one for the Guoy shift [12]; both together define the beating between soliton and radiation. Incidentally, we find empirically that the nonlinear phase, which follows the peak amplitude, is modelled slightly better with a correction $\epsilon = 0.03$

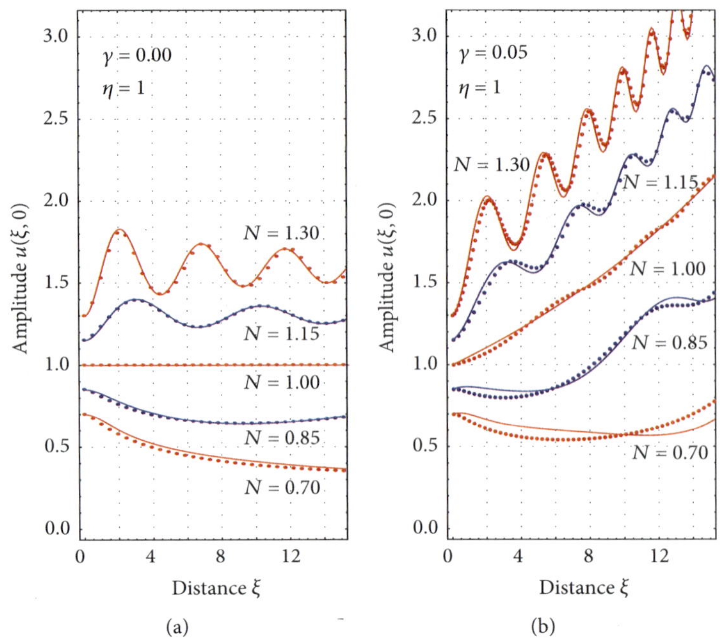

FIGURE 7: (a) Evolution of pulse amplitude from model (solid curve) and numerics (dots) for initial values of $N = 0.7, \ldots, 1.3$. (b) Same propagation with included gain ($\gamma = 0.05$, $\epsilon = 0.03$).

as indicated, apparently because the sech shape is not strictly preserved.

We put (28) to a numerical test. We considered the propagation of an initially $1/2 < N < 3/2$ sech pulse, first without gain by setting $\gamma = 0$. This is shown in Figure 7(a) for values of $N = 0.7, \ldots, 1.3$. Again solid curves (dots) correspond to theoretical prediction (numerical data), respectively. Pulse parameters oscillation due to beating of the solitonic and dispersive part of the pulse. Incidentally, this figure closely resembles Figure 2 in SY [3], but they only had numerical data. Here the agreement between numerical data and (28) is very good. The oscillation frequency is constant because in the absence of gain the nonlinear phase rotates at a constant rate.

In contrast, Figure 7(b) shows similar data but with weak gain ($\gamma = 0.05$). As the amplitude builds up (on average), the oscillation frequency also increases. Again numerics and theory agree well.

From the amplitude the pulse parameters M_ξ and η_ξ can be calculated using

$$M_\xi = \frac{\eta N^2 \exp(\gamma\xi)}{|u(\xi, 0)|},$$
$$\eta_\xi = \left(\frac{N}{M_\xi}\right)^2 \exp(\gamma\xi)\,\eta. \quad (29)$$

In Figure 8 the parameter evolution for an initially unchirped $N = 1.2$ sech pulse is shown for $\gamma = 0$ (constant energy; (a)) and $\gamma = 0.05$ (gain; (b)).

M_ξ and η_ξ agree well with numerics. The chirp parameter C is here only obtained from numerics. It is obvious that the chirp oscillates around zero with moderate values resulting in only a slight change of the soliton content. Thus for the calculation of the soliton content the chirp was neglected resulting in the simple expression

$$\mu_\xi = \eta_\xi \left(M_\xi - \frac{1}{2}\right). \quad (30)$$

The soliton eigenvalue is shown in red in Figure 8: predictions of (30) (solid) and numerical results (dotted) are compared and show good agreement. The gain causes the scaling factor η_ξ to rise, but not the amplitude scaling factor M_ξ which undulates around unity and represents a fundamental soliton.

4.3. Generation of a Second Soliton. So far we discussed the evolution of the first soliton; now we turn to a discussion of the second soliton. We traced the energy partition between solitons and radiation by evaluating the nonlinear spectrum. For Figure 9 an unchirped sech pulse ($\eta = 1$, $N = 1.3$) was launched into a fiber with $\gamma = 0.1$ and propagated numerically up to $\xi = 10.0$. Undulations indicate beating between constituents, and there may be some amount of energy swapping. That notwithstanding so much is clear: both soliton 1 and radiation grow from their initial value up to the onset of soliton 2 at $\xi_B = 8$ (arrow); beyond that point the radiative energy drops. The second soliton is clearly created from radiation energy, without depletion of the first soliton.

As an example of an actual pulse shape we show the field envelope at $\xi = 9.32$ (Figure 9(b)). The pulse shape consists of a narrow, tall peak atop a broad, low-amplitude pedestal. One can fit a sech pulse with $M_\xi = 0.965$ and $\eta_\xi = 4.43$ with flat phase to the narrow spike, and $M_\xi = 1.22$ and $\eta_\xi = 0.185$ with parabolic phase ($C = 0.076$) to the pedestal. Using (18) and (30) this suggests eigenvalues of $\mu_1 = 2.06$ (narrow peak) and $\mu_2 = 0.014$ (pedestal). DST analysis yields $\mu_1 = 2.024$ and

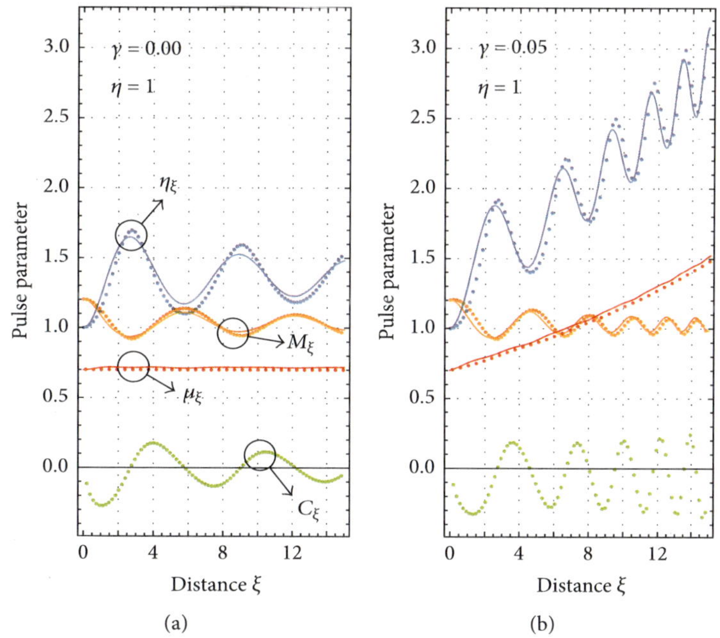

FIGURE 8: Parameter evolution and soliton content (red) of an initial unchirped $N = 1.2$ sech pulse in the case without gain (a) and with gain (b). Again solid curves/dots correspond to the model/numerics, respectively.

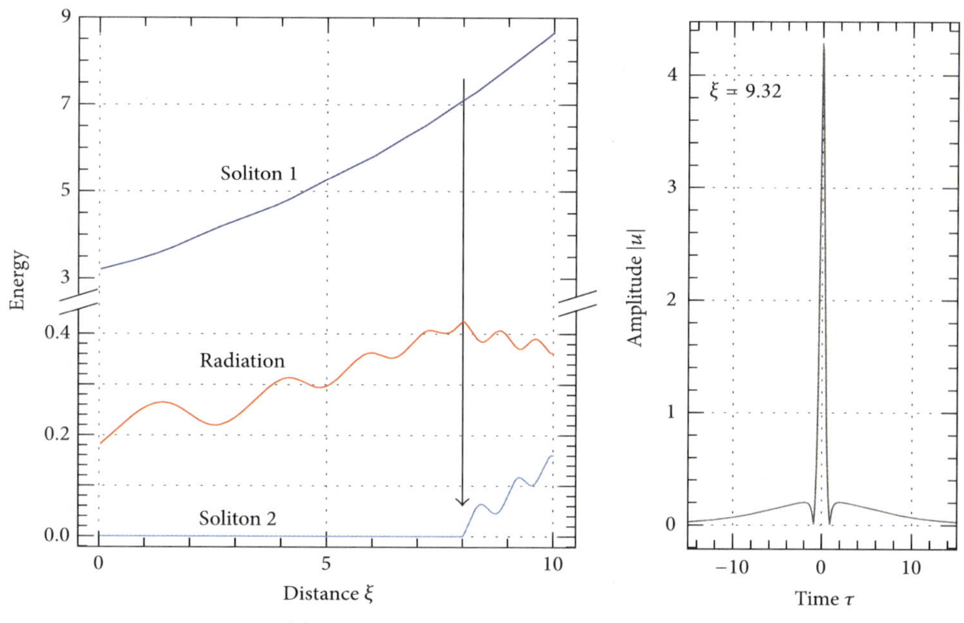

FIGURE 9: (a) Evolution of the energies of both solitons and radiation at $\gamma = 0.1$, calculated from Direct Scattering Transform. Initial conditions were $N = 1.3$ and $\eta = 1.0$. Note the break in the scale. The point of creation of the second soliton is highlighted by an arrow. (b) Example of a pulse envelope taken at $\xi = 9.32$ where the pulse consists of two solitons and radiation.

$\mu_2 = 0.029$, respectively. In view of approximations made, the agreement is reassuring. One is led to conclude that the high peak almost exclusively represents the higher-energy soliton, while the second soliton and radiation together form the pedestal. Such identification is possible and meaningful because there is a large difference in solitonic energies (almost two orders of magnitude); the resulting different widths lead to minimal temporal overlap.

In Figure 10 we assess the whole range of γ values numerically for four values of N. Arrows on the right mark the predicted emergence length obtained from (21) for high values of γ.

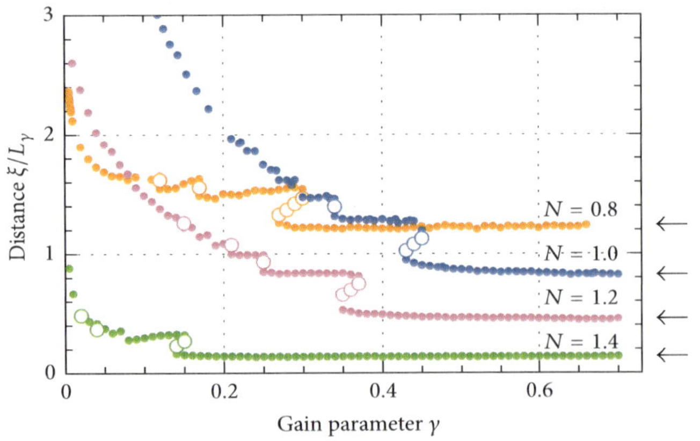

FIGURE 10: Position of creation of the second soliton (dots) as a function of gain parameter γ, for several initial soliton orders N. Open circles: disappearance of the second soliton; see text. Distance is normalized to $L_\gamma = 1/\gamma$. Arrows mark distances as predicted by (21).

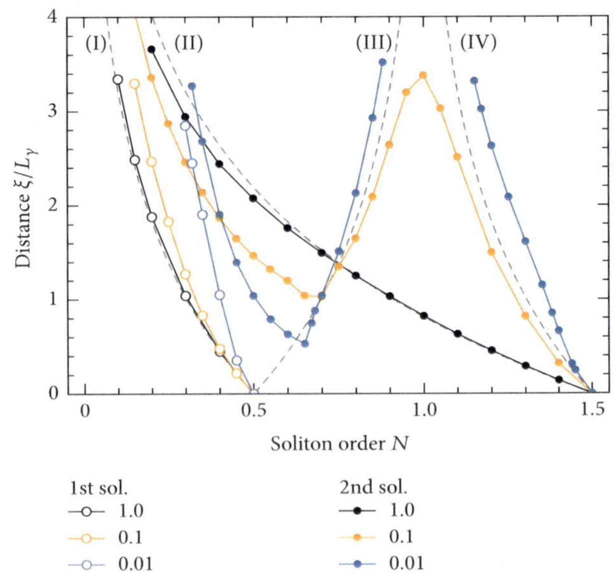

FIGURE 11: Propagation distance until a new soliton is generated. Distance is normalized to L_γ; therefore, values also represent the accumulated gain up to that point. Parameter: γ as indicated. For reference, auxiliary lines (I–IV) indicate certain limiting cases; see text.

Solitons arise from the radiative energy once there is "enough of it." This implies that, beginning at the point of soliton creation, radiative energy is depleted; usually the depletion rate is initially faster than the restoration rate by gain. This is in fact visible in all calculations. Indeed, the partition of energy between radiation and soliton has some oscillation with the beat frequency as seen in Figure 9. We note that the new soliton arises near a radiation maximum in the oscillation cycle. Indeed, this can go so far that the subsequent downswing of the oscillation, combined with the depletion, can cause such a rapid sagging of radiative energy that it falls below the threshold again. In such case the just-created soliton disappears again and returns during the next cycle of oscillations. Accordingly the energy of the emerged second soliton does not rise monotonically but in an oscillating manner. Such on-and-off soliton creation appears in Figure 10 as a folding over of the curves; note the open circles for the reverse branches. Arrows mark distances as predicted by (21).

The normalized position of first and second soliton generation is shown in an overview in Figure 11. Note that position $\xi/L_\gamma = \xi\gamma$ so that it also represents the total accumulated amplification $\ln(\Gamma)$. We inserted auxiliary lines to indicate the limiting cases in which the radiation from the initial condition is amplified in a single step. Line (I) applies to the creation of the first soliton (see (21) with $M_\xi = 1/2$); it diverges at $N \to 0$ for lack of power. Similarly, (II) is for the creation of two solitons at once at $M_\xi = 3/2$; this curve also diverges at $N = 0$ and continues to $N = 3/2$. Data are shown for several values of γ. They show the onset of the first soliton to be above (I) by a difference that grows with decreasing γ. This is expected because lower gain necessitates a longer distance to accumulate the same amplification, but then dispersion will develop more chirps. The onset of two solitons near (II) occurs below the curve because solitons generated from chirped pulse take lower energy values; thus, the first soliton does not fully deplete the radiative energy.

Lines (III) and (IV) appear in the realm where the initial condition contains one soliton plus some radiation, that is, for $1/2 < N < 3/2$, with the exception of $N = 1$, where the initial condition does not contain radiation. Note that (IV) is a shifted version of (I), and (III) is a mirrored version thereof. These curves refer to the situation where a second soliton is formed from the radiative part. This assumes relatively independent evolution of soliton and radiation, justified for small gain ($\gamma \ll 1$) because dispersion broadens the radiative part rapidly so that the mutual overlap is minimal. Therefore, data are closer to (III) and (IV) when the gain is low, while for high gain, data tend towards (II) as expected. There is an interesting transition between both limiting cases, where near $N = 1$ where the initial condition has no radiation; the radiation generated during propagation becomes dominant.

5. Conclusion

It is well known from SY [3] that if one considers an unchirped sech pulse of certain width and scales up its energy, above some threshold it contains a soliton, and above certain higher thresholds more solitons appear. The thresholds are given by the amount of radiative energy.

We were curious to see how this translates into the case of an optical fiber with gain. That situation is more complex because (i) chirp develops during propagation, and (ii) radiation can come from two different sources: the initial condition and the generation during amplified propagation. These circumstances render it less than obvious whether gain leads to the creation of solitons.

By writing an expression for the chirp and by tracking the energy budget, we determined the distance for a new soliton to appear. We verified that the soliton is formed from radiative energy; if another soliton is already present, its energy is not depleted. As long as the fiber length exceeds the creation distance, a soliton will be created under almost all circumstances, the sole exception being the limiting case of adiabatic amplification where the creation distance diverges. As prior investigations treated the problem within the adiabatic approximation, predictions as given here have not been possible before.

In any realistic setting, solitonic data transmission will incorporate both loss and gain in the fiber. In the design of such systems it is mandatory to understand the impact of those factors well, and not just in the somewhat unrealistic adiabatic limit. Our results may also be of interest for investigations of amplification in a fiber laser when the formation of solitonic pulses is desired.

Conflicts of Interest

The authors declare that there are no conflicts of interest regarding the publication of this paper.

References

[1] W. Forysiak, "Dispersion managed solitons and real-world link installation," in *Proceedings of the 2003 IEEE LEOS Annual Meeting Conference Proceedings*, p. 842, usa, October 2003.

[2] V. E. Zakharov and A. B. Shabat, "Exact Theory of Two-Dimensional Self-Focusing and One-Dimensional Self-Modulation of Waves in Nonlinear Media," *Soviet Physics—JETP*, vol. 34, pp. 62–69, 1972.

[3] J. Satsuma and N. Yajima, "Initial value problems of one-dimensional self-modulation of nonlinear waves in dispersive media," *Progress of Theoretical and Experimental Physics*, pp. 284–306, 1974.

[4] A. Hasegawa and F. Tappert, "Transmission of stationary nonlinear optical pulses in dispersive dielectric fibers. I. Anomalous dispersion," *Applied Physics Letters*, vol. 23, no. 3, pp. 142–144, 1973.

[5] L. F. Mollenauer, R. H. Stolen, and J. P. Gordon, "Experimental observation of picosecond pulse narrowing and solitons in optical fibers," *Physical Review Letters*, vol. 45, no. 13, pp. 1095–1098, 1980.

[6] A. Hasegawa and Y. Kodama, "Signal Transmission by Optical Solitons in Monomode Fiber," *Proceedings of the IEEE*, vol. 69, no. 9, pp. 1145–1150, 1981.

[7] D. J. Kaup, "A perturbation expansion for the Zakharov-Shabat inverse scattering transform," *SIAM Journal on Applied Mathematics*, vol. 31, no. 1, pp. 121–133, 1976.

[8] V. I. Karpman and E. M. Maslov, "Perturbation theory for solitons," *Soviet Physics—JETP*, vol. 73, no. 2, pp. 281–291, 1977.

[9] Y. Kodama and A. Hasegawa, "Amplification and reshaping of optical solitons in glass fiber-II," *Optics Expresss*, vol. 7, no. 7, pp. 339–341, 1982.

[10] K. J. Blow and N. J. Doran, "High bit rate communication systems using non-linear effects," *Optics Communications*, vol. 42, no. 6, pp. 403–406, 1982.

[11] K. J. Blow and N. J. Doran, "The asymptotic dispersion of soliton pulses in lossy fibres," *Optics Communications*, vol. 52, no. 5, pp. 367–370, 1985.

[12] F. Mitschke, A. Hause, and C. Mahnke, "Solitons in fibers with loss beyond small perturbation," *Physical Review A: Atomic, Molecular and Optical Physics*, vol. 96, no. 1, Article ID 013826, 2017.

[13] G. M. Melchor, M. A. Granados, and G. H. Corro, "On the problem of ideal amplification of optical solitons," *Quantum Electronics*, vol. 32, no. 11, pp. 1020–1028, 2002.

[14] G. P. Agrawal, *Nonlinear Fiber Optics*, Academic Press, Waltham, Mass, USA, 5th edition, 2013.

[15] F. Mitschke, *Fiber Optics: Physics and Technology*, Springer, Heidelberg, Germany, 2nd edition, 2016.

[16] L. V. Hmurcik and D. J. Kaup, "Solitons created by chirped initial profiles in coherent pulse propagation," *Journal of the Optical Society of America*, vol. 69, no. 4, pp. 597–604, 1979.

[17] L. F. Mollenauer and J. P. Gordon, *Solitons in Optical Fibers: Fundamentals and Applications*, Academic Press, Burlington, Mass, USA, 2006.

[18] C. Desem and P. L. Chu, "Effect of chirping on solution propagation in single-mode optical fibers," *Optics Expresss*, vol. 11, no. 4, pp. 248–250, 1986.

[19] K. J. Blow and D. Wood, "The evolution of solitons from non-transform limited pulses," *Optics Communications*, vol. 58, no. 5, pp. 349–354, 1986.

[20] A. I. Mamistov and Yu. M. Sklyarov, "Influence of regular phase modulation on formation of soliton pulses," *Soviet Journal of Quantum Electronics*, vol. 17, pp. 500–504, 1987.

[21] D. J. Kaup, J. El-Reedy, and B. A. Malomed, "Effect of a chirp on soliton production," *Physical Review E: Statistical, Nonlinear, and Soft Matter Physics*, vol. 50, no. 2, pp. 1635–1637, 1994.

[22] G. Boffetta and A. R. Osborne, "Computation of the direct scattering transform for the nonlinear Schroedinger equation," *Journal of Computational Physics*, vol. 102, no. 2, pp. 252–264, 1992.

[23] M. I. Yousefi and F. R. Kschischang, "Information transmission using the nonlinear Fourier transform, Part II: Numerical methods," *Institute of Electrical and Electronics Engineers Transactions on Information Theory*, vol. 60, no. 7, pp. 4329–4345, 2014.

[24] F. Mitschke, C. Mahnke, and A. Hause, "Soliton content of fiber-optic light pulses," *Applied Sciences (Switzerland)*, vol. 7, no. 6, article no. 635, 2017.

[25] D. Anderson, M. Lisak, and T. Reichel, "Asymptotic propagation properties of pulses in a soliton-based optical-fiber communication system," *Journal of the Optical Society of America B: Optical Physics*, vol. 5, no. 2, pp. 207–210, 1988.

Permissions

All chapters in this book were first published in IJO, by Hindawi Publishing Corporation; hereby published with permission under the Creative Commons Attribution License or equivalent. Every chapter published in this book has been scrutinized by our experts. Their significance has been extensively debated. The topics covered herein carry significant findings which will fuel the growth of the discipline. They may even be implemented as practical applications or may be referred to as a beginning point for another development.

The contributors of this book come from diverse backgrounds, making this book a truly international effort. This book will bring forth new frontiers with its revolutionizing research information and detailed analysis of the nascent developments around the world.

We would like to thank all the contributing authors for lending their expertise to make the book truly unique. They have played a crucial role in the development of this book. Without their invaluable contributions this book wouldn't have been possible. They have made vital efforts to compile up to date information on the varied aspects of this subject to make this book a valuable addition to the collection of many professionals and students.

This book was conceptualized with the vision of imparting up-to-date information and advanced data in this field. To ensure the same, a matchless editorial board was set up. Every individual on the board went through rigorous rounds of assessment to prove their worth. After which they invested a large part of their time researching and compiling the most relevant data for our readers.

The editorial board has been involved in producing this book since its inception. They have spent rigorous hours researching and exploring the diverse topics which have resulted in the successful publishing of this book. They have passed on their knowledge of decades through this book. To expedite this challenging task, the publisher supported the team at every step. A small team of assistant editors was also appointed to further simplify the editing procedure and attain best results for the readers.

Apart from the editorial board, the designing team has also invested a significant amount of their time in understanding the subject and creating the most relevant covers. They scrutinized every image to scout for the most suitable representation of the subject and create an appropriate cover for the book.

The publishing team has been an ardent support to the editorial, designing and production team. Their endless efforts to recruit the best for this project, has resulted in the accomplishment of this book. They are a veteran in the field of academics and their pool of knowledge is as vast as their experience in printing. Their expertise and guidance has proved useful at every step. Their uncompromising quality standards have made this book an exceptional effort. Their encouragement from time to time has been an inspiration for everyone.

The publisher and the editorial board hope that this book will prove to be a valuable piece of knowledge for researchers, students, practitioners and scholars across the globe.

List of Contributors

Hukum Singh, A. K. Yadav, Sunanda Vashisth and Kehar Singh
Department of Applied Sciences, The Northcap University, Sector 23-A, Gurgaon 122 017, India

Yan Ren, Zhipeng Qin, Guoqiang Xie, Zhen Qiao, Jingui Ma, Peng Yuan and Liejia Qian
Key Laboratory for Laser Plasmas (Ministry of Education), Collaborative Innovation Center of IFSA (CICIFSA), School of Physics and Astronomy, Shanghai Jiao Tong University, Shanghai 200240, China

Shikai Wang, Chunlei Yu and Lili Hu
Key Laboratory of High Power Laser Materials, Shanghai Institute of Optics and Fine Mechanics, Chinese Academy of Sciences, Shanghai 201800, China

Rajesh Yadav and Gurjit Kaur
Department of ECE, School of ICT, Gautam Buddha University, Greater Noida, India

Jun Huang
College of Modern Science and Technology, China Jiliang University, Hangzhou, Zhejiang 310018, China

Bowen Wang and Kai Ni
Institute of Optoelectronic Technology, China Jiliang University, Hangzhou, Zhejiang 310018, China

Doai Le Van, Khoa Dinh Xuan and Bang Nguyen Huy
Vinh University, Le Duan 182, Vinh City, Vietnam

Phuong Le Thi Minh
Vinh University, Le Duan 182, Vinh City, Vietnam
Saigon University, 273 An Duong Vuong, Ho Chi Minh City, Vietnam

Moses Ndunda
University of Nairobi Kenya, Chiromo Campus, OSA Student Member and Optics Transmission Engineer in Liquid Telecom Kenya, Nairobi, Kenya

Alix Dehayem-Kamadjeu
Department of Physics and Researcher in Applied Optics Research Group in University of Nairobi, Chiromo Campus, Nairobi, Kenya

David Waswa
Fibre Optics and Laser Research Group in Department of Physics, University of Eldoret, Eldoret, Kenya

Dhoumendra Mandal
Department of Physics, Saldiha College, Saldiha, Bankura,West Bengal 722173, India

Sumana Mandal and Mrinal Kanti Mandal
Department of Physics, NIT Durgapur, West Bengal 713209, India

Sisir Kumar Garai
Department of Physics, MUCWomen's College, Burdwan, West Bengal 713104, India

Amer Kotb
The Guo China-US Photonics Laboratory, Changchun Institute of Optics, Fine Mechanics, and Physics, Chinese Academy of Sciences, Changchun 130033, China
Department of Physics, Faculty of Science, University of Fayoum, Fayoum 63514, Egypt

Yasser Mohamed
Department of Physics, Faculty of Science, University of Fayoum, Fayoum 63514, Egypt

Beibei Zhao, Weiwei Ma, Hao Yu, Xiaobo Qian, Jingya Wang and Anhua Wu
Synthetic Single Crystal Research Center, Key Laboratory of Transparent and Opto-Functional Inorganic Materials, Shanghai Institute of Ceramics, Chinese Academy of Sciences, Shanghai, China

Fanming Zeng
School of Materials Science and Engineering, Changchun University of Science and Technology, Changchun 130022, China

Lingchen Kong and Guoqiang Xie
Key Laboratory for Laser Plasmas (Ministry of Education), Collaborative Innovation Center of IFSA (CICIFSA), Department of Physics and Astronomy, Shanghai Jiao Tong University, Shanghai 200240, China

Liangbi Su
Synthetic Single Crystal Research Center, Key Laboratory of Transparent and Opto-Functional Inorganic Materials, Shanghai Institute of Ceramics, Chinese Academy of Sciences, Shanghai, China
State Key Laboratory of High Performance Ceramics and Superfine Microstructure, Shanghai Institute of Ceramics, Chinese Academy of Sciences, Shanghai, China

Jingxin Ding
Synthetic Single Crystal Research Center, Key Laboratory of Transparent and Opto-Functional Inorganic Materials, Shanghai Institute of Ceramics, Chinese Academy of Sciences, Shanghai, China
School of Materials Science and Engineering, Changchun University of Science and Technology, Changchun 130022, China

Zhipeng Qin, Guoqiang Xie, Jingui Ma, Peng Yuan and Liejia Qian
Key Laboratory for Laser Plasmas (Ministry of Education), Collaborative Innovation Center of IFSA (CICIFSA), School of Physics and Astronomy, Shanghai Jiao Tong University, Shanghai 200240, China

Jian Zhang
Key Laboratory of Transparent and Opto-Functional Advanced Inorganic Materials, Shanghai Institute of Ceramics, Chinese Academy of Sciences, Shanghai 201899, China

Huizhen Yang, Zhen Zhang and Jian Wu
School of Electronic Engineering, Huaihai Institute of Technology, Lianyungang 222005, China
School of Information and Electrical, China University of Mining and Technology, Xuzhou 221008, China

Xianghao Meng and Zhaohua Wang
Beijing National Laboratory for Condensed Matter Physics, Institute of Physics, Chinese Academy of Sciences, Beijing 100190, China

Ninghua Zhang and Jiangfeng Zhu
School of Physics and Optoelectronic Engineering, Xidian University, Xi'an 710071, China

Zhiyi Wei and Wenlong Tian
School of Physics and Optoelectronic Engineering, Xidian University, Xi'an 710071, China
Beijing National Laboratory for Condensed Matter Physics, Institute of Physics, Chinese Academy of Sciences, Beijing 100190, China

Farid Atry
Department of Biomedical Engineering, University of Wisconsin–Madison, Madison, WI, USA

Israel Jacob De La Rosa and Ramin Pashaie
Electrical Engineering Department, University of Wisconsin–Milwaukee, Milwaukee, WI, USA

Kevin R. Rarick
Department of Pediatrics, Medical College of Wisconsin, Milwaukee, WI, USA

Lin Hu, Yunxiu Shui, Yaohui Dai, Haiyu Wu, Gang Zhu and Yan Yang
College of Mechanical Engineering, Chongqing University of Technology, Chongqing 400054, China

Anine Crous, Elvin Chizenga, Natasha Hodgkinson and Heidi Abrahamse
Laser Research Centre, Faculty of Health Sciences, University of Johannesburg, Doornfontein, Johannesburg 2028, Room 5308, John Orr Building, South Africa

Deepak Kedia and Monika Nehra
Department of Electronics and Communication Engineering, Guru Jambheshwar University of Science & Technology, Hisar, Haryana 125001, India

Meiqin Wang, Qianwen Chen, Dong Wang and Sui Wei
Key Laboratory of Intelligent Computing and Signal Processing, Ministry of Education, Anhui University, Hefei, Anhui Province 230039, China

Cheng Zhang
Key Laboratory of Intelligent Computing and Signal Processing, Ministry of Education, Anhui University, Hefei, Anhui Province 230039, China
Key Laboratory of Modern Imaging and Displaying Technology, Anhui University, Hefei, Anhui Province 230039, China

Long Jin and Xingqiang Zhang
Department of Basic Science, Hubei University of Automotive Technology, Shiyan 442002, China

Monika Nehra and Deepak Kedia
Department of Electronics and Communication Engineering, Guru Jambheshwar University of Science & Technology, Hisar, Haryana 125001, India

Israel J. De La Rosa and Ramin Pashaie
Electrical and Computer Engineering Department, University of Wisconsin-Milwaukee, Milwaukee, Wisconsin, 53201, USA

Mehdi Azimipour
Vision Science and Advanced Retinal Imaging Laboratory (VSRI), UC Davis Eye Centre, Sacramento, California, 95817, USA

Patrick K. Cullen and Fred J. Helmstetter
Department of Psychology, University of Wisconsin-Milwaukee, Milwaukee, Wisconsin 53201, USA

Akanksha Jetly and Rajesh Mehra
Electronics and Communication Department, NITTTR, Chandigarh, India

Van Luan Tran
Department of Electrical Engineering, National Chung Cheng University, 168 University Rd., Chiayi 621, Taiwan

Huei-Yung Lin
Department of Electrical Engineering, National Chung Cheng University, 168 University Rd., Chiayi 621, Taiwan
Advanced Institute of Manufacturing with High-Tech Innovation, National Chung Cheng University, 168 University Rd., Chiayi 621, Taiwan

Damien P. Kelly
Oryx Optics Ltd, 19 The Elms, Athlone, Co. Westmeath, Ireland

Christoph Mahnke, Alexander Hause and Fedor Mitschke
Institut für Physik, Universität Rostock, 18059 Rostock, Germany

Index